Java SE 程序设计

李瑞生 何珍祥 主 编

张 静 副主编

中国铁道出版社
CHINA RAILWAY PUBLISHING HOUSE

内 容 简 介

本书从实用的角度出发,全面介绍 Java SE 程序设计知识,通过大量案例引导读者系统掌握 Java SE 程序设计技能。全书共有 15 章内容:第 1 章和第 2 章,主要介绍 Java 的发展、特点、Java 程序运行机制与虚拟机、Java SE 程序开发环境、Java 语法等内容;第 3~5 章,主要介绍 Java 面向对象编程基础、Java 继承与多态、Java 抽象类与接口等内容;第 6~8 章,主要介绍 Java 数组与枚举、 Java 常用工具类、异常处理机制等内容;第 9~14 章,主要介绍 Java 组件与事件编程、Java 数据库编程、Java 输入输出流、Java 多线程编程、Java 网络编程、Java 泛型与集合类等 Java SE 高级程序设计内容;第 15 章主要介绍 Java Applet 编程基础知识。

本书编者在总结长期 Java 程序设计教学经验的基础上,不断凝练内容和案例,内容安排上环环相扣,案例丰富实用,所有例题代码均在 JDK+Eclipse 环境下调试运行成功,同时给出了详尽的注释。

本书适合作为 Java 软件开发从业人员的入门培训教材和参考书,也可作为普通高等院校计算机类、信息类本科专业 Java SE 程序设计教材。

图书在版编目(CIP)数据

Java SE 程序设计 / 李瑞生,何珍祥主编. —北京:中国铁道出版社,2016.12(2017.8重印)
 ISBN 978-7-113-22015-0

Ⅰ. ①J… Ⅱ. ①李… ②何… Ⅲ. ①JAVA 语言-程序设计 Ⅳ. ①TP312.8

中国版本图书馆 CIP 数据核字(2016)第 188332 号

书　　名:	Java SE 程序设计
作　　者:	李瑞生　何珍祥　主编
策　　划:	潘晨曦　孙晨光　　　　　读者热线:(010)63550836
责任编辑:	秦绪好　鲍　闻
封面设计:	刘　颖
封面制作:	白　雪
责任校对:	汤淑梅
责任印制:	郭向伟

出版发行:中国铁道出版社(100054,北京市西城区右安门西街 8 号)
网　　址:http://www.51eds.com
印　　刷:虎彩印艺股份有限公司
版　　次:2016 年 12 月第 1 版　　　2017 年 8 月第 2 次印刷
开　　本:787mm×1092mm　1/16　印张:22.5　字数:558 千
书　　号:ISBN 978-7-113-22015-0
定　　价:52.00 元

版权所有　侵权必究

凡购买铁道版图书,如有印制质量问题,请与本社教材图书营销部联系调换。电话:(010)63550836
打击盗版举报电话:(010)51873659

前 言

Java 不仅是一种程序设计语言，而且是一种软件开发的标准、规范、平台和技术的集合。Java 包括 Java SE、Java EE、Java ME 三种平台，各个平台下都包含相关的 API、服务组件和基础服务等，允许开发和部署在桌面、服务器、嵌入式环境和实时环境中使用的 Java 应用程序；支持 Web 应用程序及面向服务体系结构的企业级应用；支持移动应用开发和用于家电、智能设备、手持设备、通信设备等嵌入式应用程序的开发等。Java 以其纯面向对象、跨平台、高性能、分布性和可移植性等特点，已成为目前网络编程、手机应用开发等领域的主流编程技术。随着网络向着云计算、物联网的方向发展，Java 技术具有更加广阔的应用市场和应用前景。

近年来，社会对 Java 技术人才的需求持续增长，因此，Java 程序设计系列课程已成为各个普通高等院校计算机类、信息类专业学生的程序设计主干课程之一，Java 程序设计能力成为计算机类、信息类专业软件技术人才的基本能力要求。

Java SE 程序设计是学习 Java 编程技术的起点和基础。编者在总结多年教学经验的基础上，按照学生的认知规律和学习特点精心组织了本书的内容，并通过大量的实例由浅入深地讲解了 Java SE 程序设计的相关知识和编程方法。全书共分为 15 章：第 1 章 Java 简介与 Java SE 程序开发环境；第 2 章 Java 语法基础；第 3 章 Java 面向对象程序设计基础；第 4 章 Java 继承与多态；第 5 章 Java 抽象类与接口；第 6 章 Java 数组与枚举；第 7 章 Java 常用工具类；第 8 章 异常处理机制；第 9 章 Java 组件与事件编程；第 10 章 Java 数据库编程；第 11 章 Java 输入、输出流；第 12 章 Java 多线程编程；第 13 章 Java 网络编程；第 14 章 Java 泛型与集合类；第 15 章 Java Applet 编程简介。各章节的例题都在 JDK+Eclipse 环境下调试成功。各章节的安排和内容取舍注重可读性和实用性，理论知识点以实用、够用为原则，减少冗余繁复；每章节的例题和习题都经过精心考虑，做到既能帮助理解编程理论，又具有实战性、启发性。

本书章节层次合理清晰、内容充实、实例丰富、易学易用。本书既可作为普通高等院校计算机类、信息类本科专业 Java SE 程序设计的教材，也可作为 Java 软件开发从业人员的入门培训教材。

本书由甘肃政法学院李瑞生、何珍祥任主编，张静任副主编，由李瑞生对全书进行统稿，何珍祥负责本书的校对工作。本书中第 1、3、4、5、9、10、12、13、14 章由李瑞生编写，第 2 章由何珍祥编写，第 6、7、8、11、15 章由张静编写。

在本书的编写过程中,得到了甘肃政法学院信息工程学院院长安德智教授、陆军副教授、武光利副教授的指导和帮助。同时,得到了中国铁道出版社编辑部老师们的大力支持和帮助。在此,对以上各位老师的指导和帮助表示真挚的感谢。

由于编者水平有限,缺点和欠妥之处难免,恳请读者指正。

编者

2016 年 9 月

目 录

第 1 章 Java 简介与 Java SE 程序开发环境 1
- 1.1 Java 的发展 1
- 1.2 Java 的 3 种平台 2
- 1.3 Java 的特点 2
- 1.4 Java 程序的运行机制与 Java 虚拟机 3
  - 1.4.1 Java 程序的运行机制 3
  - 1.4.2 Java 虚拟机 3
- 1.5 Java 程序分类 4
- 1.6 JDK 的安装与配置 4
  - 1.6.1 JDK 的安装 4
  - 1.6.2 JDK 环境变量的配置 5
- 1.7 JDK 环境下开发 Java 程序 6
  - 1.7.1 Java Application 的开发 6
  - 1.7.2 Java Applet 的开发 10
- 1.8 Eclipse 环境下开发 Java 程序 12
  - 1.8.1 在 Eclipse 环境下开发 Java Application 12
  - 1.8.2 在 Eclipse 环境下开发 Java Applet 14
- 1.9 其他几个问题 15
- 小结 16
- 习题 1 16

第 2 章 Java 语法基础 18
- 2.1 Java Application 程序的基本输入、输出方法 18
  - 2.1.1 Java 向显示器（控制台）输出数据 18
  - 2.1.2 Java 从键盘读入数据 19
  - 2.1.3 用 Scanner 灵活读入数据 20
- 2.2 Java 注释 20
- 2.3 标识符与关键字 21
  - 2.3.1 标识符 21
  - 2.3.2 关键字 22
- 2.4 Java 数据类型 22
  - 2.4.1 基本数据类型 22
  - 2.4.2 常量与变量 23
  - 2.4.3 数据类型转换 26
  - 2.4.4 引用数据类型 26
- 2.5 运算符与表达式 26
  - 2.5.1 算术运算符及其表达式 27
  - 2.5.2 关系运算符 28
  - 2.5.3 逻辑运算符 28
  - 2.5.4 位运算符 28
  - 2.5.5 赋值运算符及其表达式 29
  - 2.5.6 复合赋值运算符及其表达式 29
  - 2.5.7 条件运算符及其表达式 30
  - 2.5.8 instanceof 运算符及其表达式 30
- 2.6 Java 语句 30
- 2.7 Java 程序流程控制 31
  - 2.7.1 顺序结构 31
  - 2.7.2 选择结构 31
  - 2.7.3 循环结构 36
  - 2.7.4 流程跳转语句 37
- 小结 39
- 习题 2 39

第 3 章 Java 面向对象程序设计基础 ... 41
- 3.1 面向对象概述 41
  - 3.1.1 类与对象的概念 41
  - 3.1.2 面向对象的主要特性 42
- 3.2 Java 类与对象 42

3.2.1 Java 类的定义 42
3.2.2 构造方法与对象的创建 43
3.2.3 对象声明及创建的内存模型 45
3.2.4 对象间的赋值 46
3.3 方法重载 47
3.4 this 关键字 48
3.5 static 与静态成员 50
   3.5.1 静态变量 50
   3.5.2 静态方法 51
   3.5.3 静态代码块 52
3.6 参数传递 54
3.7 类的关联与依赖关系 56
3.8 package 与 import 关键字 57
   3.8.1 Java 中的包 57
   3.8.2 用 package 关键字自定义包 58
   3.8.3 用 import 关键字导入包 58
3.9 JavaBean 60
3.10 jar 命令的用法 62
   3.10.1 将应用程序打包为 jar 文件 63
   3.10.2 生成 jar 文件扩展类库 64
   3.10.3 Eclipse 环境下的文件打包 65
小结 66
习题 3 66

第 4 章 Java 继承与多态 69
4.1 Java 继承 69
4.2 权限修饰符 70
4.3 子类继承性 72
4.4 变量覆盖与方法重写 74
4.5 super 的用法 76
4.6 对象的上下转型 80
4.7 instanceof 关键字 82
4.8 多态 83
4.9 final 修饰符 84
小结 85

习题 4 85

第 5 章 Java 抽象类与接口 87
5.1 抽象方法与抽象类 87
5.2 Java 接口 89
5.3 Java 接口回调 91
5.4 内部类 92
5.5 匿名类与匿名对象 95
5.6 面向抽象（接口）编程 99
5.7 接口的一个应用——工厂模式 101
5.8 Java 内置注解简介 103
小结 106
习题 5 106

第 6 章 Java 数组与枚举 108
6.1 一维数组 108
6.2 数组间赋值 110
6.3 Arrays 类中处理数组的系统方法 111
6.4 Java 二维数组的定义及使用 112
6.5 Java 对象数组 115
6.6 Java 枚举类型 117
   6.6.1 用 enum 定义枚举类型 117
   6.6.2 枚举类型的构造方法 118
   6.6.3 在 switch 结构中使用枚举类型 118
小结 119
习题 6 119

第 7 章 Java 常用工具类 121
7.1 Object 类及其常用方法 121
7.2 基本数据类型包装类 125
7.3 String 类 126
7.4 StringBuffer 类 131
7.5 正则表达式 132
   7.5.1 正则表达式简介 132
   7.5.2 Pattern 与 Macther 类 133
7.6 字符串解析方法 135
7.7 日期时间类 138
7.8 Math 类 139

7.9	BigInteger 类	141
7.10	Random 类	141
7.11	其他常用类	142
7.12	Class 类与 Java 的反射机制简介	144
小结		150
习题 7		150

第 8 章 异常处理机制 ... 152

8.1	Java 异常处理概述	152
8.2	Java 异常类	152
8.3	Java 异常处理语法	153
8.4	强制检查异常和非强制检查异常	156
8.5	用户自定义异常	159
小结		161
习题 8		161

第 9 章 Java 组件与事件编程 ... 162

9.1	Java AWT 与 Swing 简介	162
9.2	容器和组件	162
9.3	Java 布局管理器	164
	9.3.1 FlowLayout	165
	9.3.2 BorderLayout	166
	9.3.3 GridLayout	166
	9.3.4 CardLayout	167
	9.3.5 GridBagLayout	169
	9.3.6 BoxLayout	171
	9.3.7 空布局	172
9.4	Java Swing 常用的中间容器	173
9.5	Java 事件编程机制	175
9.6	事件监听器对象的几种实现	180
	9.6.1 窗体类自身实现相应事件监听器接口的方式	180
	9.6.2 自定义外部类实现相应事件监听接口的方式	181
	9.6.3 自定义外部类继承相应的事件适配器类的方式	182
	9.6.4 匿名类实现事件监听器	183
9.7	Swing 的常用组件及其事件编程	184

	9.7.1 JButton	184
	9.7.2 JLabel	184
	9.7.3 JTextField	185
	9.7.4 JTextArea	185
	9.7.5 JRadioButton	185
	9.7.6 JCheckBox	185
	9.7.7 JComboBox	186
	9.7.8 JList	186
	9.7.9 JMenuBar、JMenu 与 JMenuItem	189
	9.7.10 JPopupMenu	190
	9.7.11 JToolBar	190
	9.7.12 JTable	193
	9.7.13 JTabbedPane	196
	9.7.14 JTree	198
	9.7.15 JDialog、JOptionPane 与 JFileChooser	200
	9.7.16 JSlider	203
	9.7.17 JprogressBar	205
	9.7.18 Timer	205
	9.7.19 键盘事件示例	207
9.8	其他几个应用	209
	9.8.1 更换窗体标题栏图标	209
	9.8.2 让窗体在屏幕上居中显示	209
	9.8.3 将窗体显示为任务栏图标	210
9.9	字体与颜色	211
	9.9.1 Font 类	211
	9.9.2 Color 类	211
9.10	GUI 图形绘制	214
	9.10.1 Graphics 类	214
	9.10.2 Canvas 类	215
小结		218
习题 9		218

第 10 章 Java 数据库编程 ... 220

10.1	JDBC 简介	220
10.2	JDBC API	220

10.3	MySQL 简介	223
10.4	数据库基本操作的 SQL 语法	224
10.5	MySQL 的使用	224
10.6	JDBC 数据库基本操作	226
10.7	运用 JavaBean 进行数据库操作	233
10.8	数据库的批处理与事务操作	237
10.9	JDBC 操作 Access 数据库	247
小结		249
习题 10		249

第 11 章 Java 输入、输出流 250

11.1	Java 输入、输出流概述	250
11.2	File 类的应用	251
11.3	输入、输出流类	252
11.4	文件字节输入、输出流类	254
11.5	文件字符输入、输出流类	256
11.6	字节数组输入、输出流类	257
11.7	过滤流类	257
11.8	随机访问文件	259
11.9	Serializable 接口与对象序列化	260
11.10	标准输入、输出流	262
11.11	文件对话框	263
11.12	用 Desktop 类打开文件	266
小结		266
习题 11		267

第 12 章 Java 多线程编程 268

12.1	程序、进程与线程	268
12.2	Java 多线程机制	269
12.3	Java 多线程实现的方法	269
12.4	线程的生命周期与状态转换	271
12.5	Java 多线程调度机制	272
12.6	Thread 类	273
12.7	线程的让步	276
12.8	线程的联合	277
12.9	多线程的互斥与同步	278
12.9.1	线程的互斥	278
12.9.2	互斥线程的协调	280

12.10	守护线程	283
12.11	线程之间的通信流类	284
小结		286
习题 12		286

第 13 章 Java 网络编程 288

13.1	Java 网络编程概述	288
13.2	InetAddress 类的应用	289
13.3	URL 类的应用	290
13.3.1	URL 简介	290
13.3.2	URL 类的常用方法	290
13.4	URLConnection 类	292
13.5	TCP、UDP、端口与套字符	294
13.6	基于 TCP 的 Socket 网络编程	294
13.6.1	ServerSocket 类	295
13.6.2	Socket 类	295
13.6.3	Socket 编程的一般流程	296
13.7	基于 UDP 的网络编程	305
13.7.1	DatagramPacket 类	305
13.7.2	DatagramSocket 类	306
13.7.3	基于 UDP 网络编程的一般流程	306
13.8	基于组播的网络编程	312
小结		315
习题 13		315

第 14 章 Java 泛型与集合类 316

14.1	泛型	316
14.1.1	泛型概述	316
14.1.2	泛型类	317
14.1.3	泛型构造方法	318
14.1.4	泛型方法	320
14.1.5	泛型通配符	320
14.1.6	泛型接口	322
14.1.7	子类泛型	323
14.1.8	引入泛型的好处	323
14.2	集合类与接口	324
14.2.1	Collection 接口	324
14.2.2	Iterator 接口	324
14.2.3	List 接口	325

14.2.4　ArrayList 类 325
14.2.5　LinkedList 类 327
14.2.6　Collections 类 329
14.2.7　Set 接口 331
14.2.8　HashSet 类 331
14.2.9　TreeSet 类 332
14.2.10　EnumSet 类 334
14.2.11　Map 接口 334
14.2.12　HashMap 类 335
14.2.13　SortedMap 接口与
　　　　 TreeMap 类 336
14.2.14　EnumMap 类 337
小结 338
习题 14 338

第 15 章　Java Applet 编程简介 340

15.1　Java Applet 简介 340
15.2　Apple 程序的编写方法 340
15.3　Applet 类的主要方法 342
15.4　<applet>标记的属性及 Applet
　　　参数传递 344
15.5　Applet 的组件与事件处理 345
小结 348
习题 15 348

参考文献 349

第 1 章　Java 简介与 Java SE 程序开发环境

【本章内容提要】
- Java 的发展；
- Java 的 3 种平台；
- Java 的特点；
- Java 程序的运行机制与 Java 虚拟机；
- Java 程序分类；
- JDK 的安装与配置；
- JDK 环境下开发 Java 程序；
- Eclipse 环境下开发 Java 程序。
- 其他几个问题。

1.1　Java 的发展

1990 年 12 月，Sun 公司就由 James Gosling 等人发起一个叫做 Green 的开发小组，主要目标是为家用消费类电子产品开发一个分布式代码系统，起初的名字叫做 Oak，但是有一家公司已经用了这个名字，于是就以一个盛产咖啡的小岛名 Java 命名了。1996 年，Sun 公司发布了 JDK（Java Development Kit）1.0，其中包括了 JRE（Java Runtime Environment）及 JDK 两部分。1997 年推出的 JDK1.1，其中增加了 JIT（即时编译）功能，有效提高了 Java 程序运行效率。1998 年底，Sun 公司发布了 JDK1.2，这是 Java 发展史上的一个里程碑。1999 年 6 月，Sun 公司将 Java 分为三个版本 J2SE（标准版）、J2EE（企业版）及 J2ME（微型版）。从 2000 年到 2004 年，JDK1.3、JDK1.4、J2EE1.3、J2SE1.4 等陆续被发布。2004 年 9 月，J2SE1.5 被发布，成为 Java 语言发展史上的又一个里程碑，为了表示该版本的重要性，J2SE1.5 更名为 Java SE 5.0。2005 年 6 月，JavaOne 大会召开，Sun 公司公开 Java SE 6。此时，将 Java 的各版本更名，以取消其中的数字"2"：J2EE 更名为 Java EE，J2SE 更名为 Java SE，J2ME 更名为 Java ME。2006 年 12 月，JDK 6.0 正式版发布。2009 年，Sun 公司被 Oracle 公司收购，2011 年，Oracle 发布了 Java 7.0 的正式版，2014 年又发布了 Java 8.0 的正式版。

目前，Java 的开发与应用涉及很多方面，包括桌面级的开发、网络开发和嵌入式开发等。在动态网站和企业级开发中，Java 作为一种主流编程语言占到了很大份额；在嵌入式方面的发展更是迅速，现在流行的手机游戏，几乎都是应用 Java 语言开发的。

1.2　Java 的 3 种平台

Java 开发平台有 Java SE（Java　Platform Standard Edition，Java 平台标准版）、Java EE（Java Platform Enterprise Edition，Java 平台企业版）及 Java ME（Java　Platform Micro Edition，Java 平台微型版）3 种。

Java SE：它允许开发和部署在桌面、服务器、嵌入式环境和实时环境中使用的 Java 应用程序。Java SE 包含了构成 Java 语言的核心类库和基础开发环境，并为 Java EE 和 Java ME 提供开发基础支持。

Java EE：企业版本帮助开发和部署可移植、健壮、可伸缩且安全的服务器端 Java 应用程序。Java EE 是在 Java SE 的基础上构建的，它提供 Web 服务、组件模型、管理和通信 API，可以用来构建 Web 应用程序以及面向服务体系结构的企业级应用。

Java ME：为在移动设备和嵌入式设备（如手机、PDA、电视机顶盒等）上运行的应用程序提供一个健壮且灵活的环境。Java ME 包括灵活的用户界面、健壮的安全模型、许多内置的网络协议以及对可以动态下载的联网和离线应用程序的丰富支持。

1.3　Java 的特点

Java 具有入门简单、面向对象、跨平台、安全性、并发机制等一系列特点。

- 跨平台性：依赖于 Java 独有的虚拟机机制，使得 Java 源程序经编译后形成的字节码不依赖于任何具体的平台，只要具有 JRE（Java Runtime Environment），就可以经解释后执行。因此，Java 程序具备"一次编译，到处运行"的跨平台特性。
- 简单性：Java 语言语法借鉴了 C、C++的语法。但没有显式地运用指针、运算符重载、多重继承等难以理解的概念，并通过实现垃圾自动收集机制简化了程序设计者的内存管理工作。
- 面向对象：Java 语言是纯面向对象的编程语言，它实现了面向对象编程思想中的抽象、封装、继承及多态等基本特性。同时，它本身提供了强大丰富的系统类与接口包。
- 解释执行：Java 的设计者设计 Java 的主要目的，就是希望可以做到，"编写一次，到处运行"，Java 源程序被编译成字节码，这种节码在任何 Java 运行环境中由 Java 虚拟机解释执行，这种方式保证了 Java 的与平台无关性和可移植性。解释执行与即时编译技术的完美结合，提供了相当高的运算性能。
- 安全性：Java 安全性包括语言级安全性、编译时安全性、运行时安全性、可执行代码安全性。语言级安全性指 Java 的数据结构是完整的对象，这些封装过的数据类型具有安全性；编译时要进行 Java 语言和语义的检查，保证每个变量对应一个相应的值及编译后生成 Java 类；当 Java 源程序经过编译后形成的字节码被加载到内存后，Java 解释器中的代码校验器对字节码的合法性进行检查，若发现有破坏 Java 安全性的操作，则给出错误信息并拒绝执行该程序；Java 类在网络上使用时，对它的权限进行了设置，保证了被访问用户的安全性。
- 健壮性：Java 致力于检查程序在编译和运行时的错误，其类型检查可以检查出许多开发早期出现的错误；Java 支持自动内存管理，减少了内存出错的可能性；Java 采用对象数组并能检测数组边界，避免了覆盖数据的可能；基于异常（Exception）处理机制，在编译时，能揭示出可能出现但未被处理的异常，帮助程序员正确地进行选择，以防止系统的崩溃。

- 多线程：Java 支持用户编写包含多条执行线索的程序，每条执行线索对应一个线程，每个线程线程具有独立的任务，通过 Java 虚拟机的调度并发地完成多个任务。
- 动态特性：Java 的动态特性表现在两个方面：其一是在 Java 语言中，可以简单、直观地查询运行时的信息；其二是可以将新代码加入到一个正在运行的程序中。
- 分布性：Java 拥有广泛的能轻易地处理 TCP/IP 协议的运行库，这使得在 Java 中比在 C 或 C++中更容易建立网络连接。Java 应用程序可以借助 URL 通过网络开启和存取对象，就如同存取一个本地文件系统一样简单。

1.4 Java 程序的运行机制与 Java 虚拟机

1.4.1 Java 程序的运行机制

各种计算机程序在运行机制上，大体可被分为编译型和解释型两种。而 Java 程序需要经过编译及解释执行两个阶段。

一个 Java 程序要经过编辑、编译、解释执行等几个阶段。首先，运用 Java 语言编写的程序以"主类名.java"的形式保存，被称为 Java 源程序；其次，Java 源程序被 JDK 提供的编译命令 javac 编译后，形成的"主类名.class"文件，该文件被称为字节码文件；最后，Java 字节码文件被 Java 虚拟机 JVM（Java Virtual Machine）解释为可在不同平台下执行的指令并执行，如图 1.1 所示。

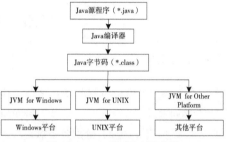

图 1.1 Java 程序运行过程

Java 字节码文件是一种和任何具体机器环境及操作系统环境无关的中间代码，是 Java 源文件由 Java 编译器编译后生成的目标代码文件。编程人员和计算机都无法直接读懂字节码文件，它必须由 JVM 来解释执行，因此说 Java 是一种在编译基础上进行解释执行的语言。

1.4.2 Java 虚拟机

Java 程序是如何做到"一次编译，到处运行"呢？这正是通过 JVM 来实现的。JVM 可以被理解成一个用软件模拟的、以字节码为机器指令的虚拟计算机,它屏蔽了 Java 程序底层运行平台（操作系统和硬件）的差别，在执行字节码时，把字节码解释成适合于不同具体平台上的机器指令并将机器指令交给本地的操作系统来运行。

在运行 Java 程序时，首先会启动 JVM，然后由它来负责解释执行 Java 的字节码，并且 Java 字节码也只能运行于 JVM 之上。只要在不同的计算机上安装了针对特定具体平台的 JVM，Java 程序就可以运行，而不用考虑当前具体的硬件平台及操作系统环境，也不用考虑字节码文件是在何种平台上生成的。JVM 把这种不同软硬件平台的具体差别隐藏起来，从而实现了真正的二进制代码级的跨平台移植，Java 的跨平台特性正是通过在 JVM 中运行 Java 程序实现的。

Java 程序通过 JVM 可以实现跨平台特性，但 JVM 是不跨平台的。也就是说，不同操作系统之上的 JVM 是不同的，如 Windows 平台之上的 JVM 与 UNIX 平台上的 JVM 是不同的。

1.5 Java 程序分类

针对 Java SE、Java EE 以及 Java ME 等不同平台，Java 程序可以被分成如下几类。

- Java 应用程序（Java Application）：是必须包括一个 main 方法的独立完整的桌面应用程序，main 方法作为程序的入口。
- Java 小应用程序（Java Applet）：将其编译形成的字节码嵌入在网页 (.html) 文件中并被客户端浏览器加载执行的一种 Java 程序。
- Java Servlet：在 Java EE 平台下开发的，运行在 Web 服务器端的 Java 程序，为了能够支持 Servlet 的运行，Web 服务器端必须安装包含 Java 虚拟机的服务器软件。
- 手机上运行的 Java ME 应用程序 MIDlet（Mobile Information Devices let），即移动信息设备小程序。

Java Application 与 Java Applet 是 Java SE 平台下两种基本的程序类型。但是目前，Java Applet 的应用较少，所以，本书中绝大多数程序是 Java Application 程序。

1.6 JDK 的安装与配置

1.6.1 JDK 的安装

本书以 Windows 8 操作系统环境，JDK1.7（JDK7）为例，在 JDK1.7 的官方下载地址 http://www.oracle.com/technetwork/java/javase/downloads/jdk7-downloads-1880260.html 下载 JDK1.7。根据操作系统是 32 位或 64 位的，分别要下载对应的 jdk-7u79-windows-i586.exe（32 位）或者 jdk-7u79-windows-x64.exe（64 位）开发包（见图 1.2）。

Windows x86	138.31 MB	jdk-7u79-windows-i586.exe
Windows x64	140.06 MB	jdk-7u79-windows-x64.exe

图 1.2 JDK1.7 开发包

下面以 jdk-7u79-windows-x64 的安装与配置为例，介绍 JDK1.7 的安装与配置方法。

双击安装程序，按照安装向导（见图 1.3）提示一步步完成即可。安装过程中可以更改安装路径（见图 1.4），然后单击“下一步”按钮继续完成安装，当弹出图 1.5 所示的向导页面时，单击其中的“后续步骤”按钮后可浏览和下载 JDK1.7 的开发文档等资源（见图 1.6），同时，可进一步了解 JDK1.7 的其他特性等。

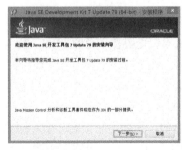

图 1.3 JDK1.7 安装向导

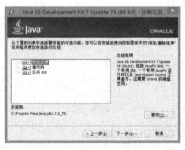

图 1.4 JDK1.7 安装路径选择

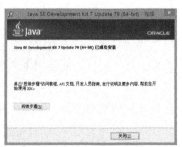

图 1.5 JDK1.7 安装完成

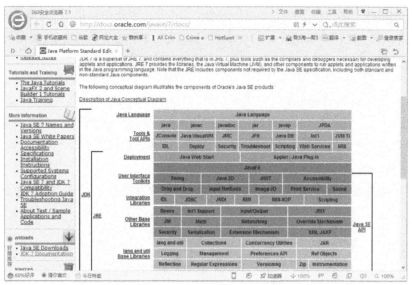

图 1.6　JDK 文档页面

JDK1.7 安装完成后，其安装路径下会包含图 1.7 所示的 jdk1.7.0_79 以及 jre7 两个文件夹。其中，JRE7（Java SE Runtime Environment）提供了运行 Java 程序不可或缺的环境，jdk1.7.0_79 中产生以下几个主要子目录，它们各自的主要作用如表 1.1 所示。

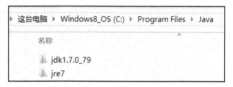

图 1.7　JDK 安装路径下的文件夹

表 1.1　jdk1.7.0_79 中主要子目录及其各自的主要作用

名称	作用	名称	作用
bin	实用工具程序及 JDK 命令集合	jre	Java 运行环境文件
include	系统 C 语言的头文件等	lib	Java 类库文件

安装好 JDK 后，是否就意味着可以用 JDK 的命令来编译、执行 Java 程序了呢？为了验证这一点，打开 Windows 命令提示符窗口，在其中键入 JDK 命令 javac 并回车。系统会提示"'javac'不是内部或外部命令，也不是可执行程序或批处理文件"，如图 1.8 所示。这就说明，系统还不认识 JDK 命令。因此，还需要配置环境变量。

图 1.8　在命令提示符下键入 JDK 命令

1.6.2　JDK 环境变量的配置

JDK 安装完成后，需要配置环境变量。为什么要配置环境变量？这是因为我们需要使用

"C:\Program Files\Java\jdk1.7.0_79\bin"路径下的javac、java等命令来编译、解释执行Java程序。但操作系统并不知道这些JDK命令存放在哪里,这就需要通过配置环境变量,告诉操作系统这些命令的存放路径所在。

配置环境变量的步骤如下:
- 在桌面"计算机"图标上右击,选择"属性"命令,在弹出的窗口中单击"高级系统设置"按钮,在"系统属性"对话框的"高级"选项卡中单击"环境变量"按钮,选择系统中原有的Path环境变量并对其进行编辑,即在原有的Path环境变量值的前面添加";C:\Program Files\Java\jdk1.7.0_79\bin;",如图1.9所示。

注意:不要覆盖系统中原有的Path变量值,否则会导致有些系统命令无法执行。

- 在图1.10所示的对话框中新建系统环境变量并将其命名为classpath,其值为";C:\Program Files\Java\jre7\lib\rt.jar; C:\Program Files\Java\jdk1.7.0_79\lib\tools.jar;"。

对Path的设置,是为了告诉操作系统JDK平台提供的Java编译器(javac.exe)、Java解释器(java.exe)等命令在Java安装目录的bin文件夹中。对classpath的设置,是为了使编译程序能够找到用户定义的类和系统类所在的包。

图1.9 设置Path

图1.10 设置classpath

当配置好环境变量后,重新打开Windows命令提示符窗口,在其中键入javac命令并回车(如图1.11所示),发现javac的命令参数选项逐一列出,就表明JDK环境变量配置好了。

图1.11 配置环境变量后测试JDK命令

1.7 JDK环境下开发Java程序

1.7.1 Java Application的开发

Java源程序是以.java为扩展名,可以用各种Java集成开发环境中的源代码编辑器来编写,也可以用其他文本编辑工具如Windows中的记事本、JCreator等编辑Java源程序。

Java Application程序的开发可以分为以下三个步骤:

（1）编辑源程序

利用记事本等编辑工具完成源程序编辑，形成"主类名.java"的 Java 源文件。

（2）编译源程序

使用 JDK 的 javac.exe 命令将源文件编译生成字节码文件"主类名.class"。

（3）运行程序

对于 Java Application，用 JDK 的 java 命令对字节码文件进行解释执行。

【例题 1_1】第一个 Java Application 程序。

（1）打开记事本，在其中编辑如下程序：

```java
public class MyFirstApplication {
    public static void main(String[] args) {
        System.out.println("这是我的第一个 Java Application 程序！");
        System.out.println("This is my first Java Application.");
    }
}
```

（2）将以上程序保存为 MyFirstApplication.java，保存时一定要在"另存为"对话框的"保存类型"列表框中选择"所有文件"选项，如图 1.12 所示，否则会默认保存为.txt 文件。假定保存路径选择 E:\Java Pro。

图 1.12 保存 Java 源程序

（3）打开 Windows 命令提示符窗口，将路径转到当前保存源程序文件的 E:\Java Pro 下。在当前命令提示符下键入 E:并回车，将根目录转到 E:\目录下。

（4）用 cd 命令进入 Java Pro 子目录下（见图 1.13），继而键入 javac MyFristApplication.java 并回车对源程序进行编译，编译过程中若有语法错误，则会报错。

（5）若没有语法错误，编译后在 Java Pro 子目录下会生成对应的字节码文件 MyFirstApplication.class；再输入 java MyFristApplication 并回车来解释执行程序，得到正确的输出结果（见图 1.14）。

图 1.13 用 javac 命令编译 Java 源程序　　　图 1.14 用 java 命令执行 Java 程序

程序说明：

- 类是组成 Java 源程序的基本单位，一个 Java 源程序是由若干个类组成的。本例只有 MyFirstApplication 一个类。
- "public class MyFirstApplication"被称为 Java 类的声明部分，其中，public 及 class 都是 Java 的关键字，public 表明该类的访问权限为公有，class 是定义类的关键字 MyFirstApplication 是类名标识符。

- 第一个大括号和最后一个大括号以及它们之间的内容叫作类体。该类体中只包含一个 public static void main(String[] args) 方法。一个 Java Application 程序必须有且仅有一个类含有 main 方法，包含有 main 方法的类被称为应用程序的主类。在一个 Java 应用程序中 main 方法必须被声明为 public static void 修饰的。String args[]声明一个字符串类型的数组 args[ ] 用来存储来自于命令行的参数。
- public static void main(String[] args) 方法是所有的 Java Application 程序执行的入口，当执行 Java Application 时，整个程序将从这个 main 方法的方法体的第一个语句开始执行。
- System.out.println（字符串）是 Java 向屏幕输出的基本方法。
- 如果源程序文件中包含有多个类，那么只能有一个类前面可以加 public 修饰（一般是在主类前加 public）。源程序文件以主类名.java 保存。

如果在一个源程序中有多个类定义，则在编译时将为每个类生成一个.class 文件。为了说明这一点，再举一个例子。

【例题 1_2】 定义一个学生类，在 Example1_2 类中创建学生类对象并调用其方法输出学生信息。

```java
public class Example1_2 {
    public static void main(String[] args) {
        Student zs=new Student();           //给 Student 类创建对象 zs
        zs.stuno="20150001";                //给 zs 对象的成员变量 stuno 赋值
        zs.stuname=" 张三";                 //给 zs 对象的成员变量 stuname 赋值
        zs.output();                        //zs 对象调用 output()方法
    }
}
class Student{                              //定义 Student 类
    String stuno;
    String stuname;                         //成员变量定义
    public void output()                    //成员方法定义
    {
        System.out.println("学号: "+stuno+" 姓名"+stuname);
    }
}
```

程序说明：
- 对于本例中对象创建、成员变量、成员方法的定义等问题，在本章制作简单了解即可，不必深究。
- 程序包含两个类，Example1_2 以及 Student 类，其中 Example1_2 是主类，所以，保存该源程序文件时，应以 Example1_2.Java 的形式进行保存。
- 在命令提示符窗口 E：\Java Pro 下键入 javac Example1_2.java 并回车，若没有语法错误，会在当前源程序目录下生成两个字节码文件 Example1_2.class 以及 Student.class。

【例题 1_3】 先读入字符串 "I am a student,I study hard."，将其以

```
****************************
I am a student,I study hard.
****************************
```

的形式显示在控制台。

```java
import java.util.Scanner;
public class Example1_3 {
    public static void main(String[] args) {
        System.out.println("请输入字符串: ");
```

```
        Scanner in=new Scanner(System.in);
        String s=in.nextLine();                    //读入一行
        System.out.println("***************************");
        System.out.println(s);
        System.out.println("***************************");
    }
}
```
程序运行结果如图 1.15 所示。

图 1.15　Java Application 的基本输入/输出

程序说明：
- 本程序主要显示了 Java Application 基本输入的方法。
- 该程序中的 System.in 代表标准输入流，可以从来自于标准输入设备键盘的数据。Scanner 类是 java.util 系统类包中的类，在创建其对象时包装了 System.in，从而可以使用 Scanner 类的 nextLine()方法按行读取键盘的输入。
- 要使用 Java 系统类包中的类时，要用 import 关键字将对应包导入进来，如 import java.util.Scanner;，且该语句应放在本源程序所有类定义之前。

【例题 1_4】　编写程序实现通过命令行读入两个整数，求其和并将结果输出。
```
public class Example1_4 {
    public static void main(String[] args) {
        int sum=0;
        int n1=Integer.parseInt(args[0]);
        int n2=Integer.parseInt(args[1]);
        sum=n1+n2;
        System.out.println("sum="+sum);
    }
}
```
程序运行结果如图 1.16 所示。

图 1.16　获取命令行参数并对其求和

程序说明：
- public static void main(String[] args)中的 String[] args 是一个 String 类型的一维数组，用于接收命令行参数。将这些参数对应存储在 args[0]，args[1]，…，args[n]中。
- 由于 args[0]，args[1]，…，args[n]中的数据都是 String 类型，不能参与算数运算，因此用 Integer.parseInt(args[0])等方法将对应的 String 类型数据转化为基本数据类型 int 型数据并进行算数运算。这些方法在后续章节中会陆续展开介绍，此处只做了解。
- 在执行该程序时，应在主类名后给出命令行参数 200,300，主类名与每个命令行参数之间有且仅有一个空格隔开。

【例题 1_5】 通过输入对话框接收数据并将其输出。

```
public class Example1_5 {
    public static void main(String[] args) {
        String str=javax.swing.JOptionPane.showInputDialog("请输入一个字符串: ");
        System.out.println("输入对话框中输入的字符串是: ");
        System.out.println(str);
    }
}
```

程序运行结果如图 1.17（a）、（b）、（c）所示。

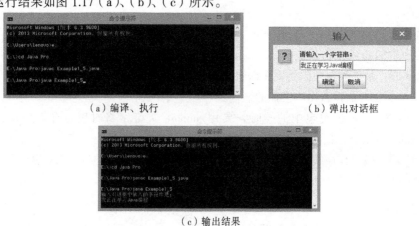

（a）编译、执行　　　　　　　　　（b）弹出对话框

（c）输出结果

图 1.17　Example1_5 的运行结果

javax.swing 包中的 JOptionPane 调用其静态方法 showInputDialog()可以产生一个输入对话框，在其中输入字符串，单击"确定"按钮后，该方法返回该字符串。

例题 1_3、例题 1_4、例题 1_5 分别演示了 Java Application 程序接收数据的几种常用方法，要灵活应用。

1.7.2　Java Applet 的开发

Java Applet 是另一类非常重要的 Java 程序。开发一个 Java Applet 程序需经过：编写源文件、编译源文件生成字节码、通过浏览器加载运行字节码三个步骤。

1. 编辑 Applet 源程序

Java Applet 程序的编写和编译与 Java Application 相类似，两者的区别主要在于其执行方式的不同。Java Application 是从其中的 main()方法开始运行的，Java Applet 没有 main()方法，不能独立运行,需被浏览器加载运行。

2. 编译 Applet 源程序

Java Applet 程序的编译方式与 Application 完全一样,即使用 javac 命令来编译 Applet 源程序,形成字节码文件。

3. 执行 Applet 源程序

Applet 的执行方式与 Application 完全不同,Applet 由浏览器或 JDK 命令 apppletviewer.exe 来加载执行。

Applet 程序的字节码文件必须嵌入 HTML 文件中才能够被浏览运行,因此必须编写相应的 HTML 文件,一般格式如下:
```
<html>
    <applet code=Applet源文件名.class width=800 height=600> </applet>
</html>
```

【例题 1_6】编写 Applet 程序,在 Applet 中以不同颜色绘制输出 "I am a student,I study hard.","我一边喝着咖啡,一边学习 Java"两个字符串。

首先,在记事本中编辑 Example1_6,并将其以 Example1_6.java 进行保存。程序如下:
```
import java.applet.Applet;
import java.awt.Color;
import java.awt.Graphics;
public class Example1_6 extends Applet//自定义的Applet类必须继承系统Applet类
 {
    public void paint(Graphics g) {
       g.setColor(Color.ORANGE);         //设置画笔颜色
       g.drawString("I am a student,I study hard.", 100, 40);    //绘制字符串
       g.setColor(Color.BLUE);
       g.drawString("我一边喝着咖啡,一边学习Java", 100, 100);
    }
}
```
其次,在记事本中编辑.html 格式文件并以 Example1_6.html 为名保存在与 Example1_6.java 相同的目录下。Example1_6.html 的写法如下:
```
<html>
    <applet code=Example1_6.class width=800 height=600>
    </applet>
</html>
```
最后,在命令提示符下用 javac 命令编译 Example1_6.java,若编译成功,则继续在命令提示符下键入 appletviewer Example1_6.html 并回车 [见图 1.18(a)],得到结果 [见图 1.18(b)]。也可以在浏览器浏览 Example1_6.html,但由于有些浏览器设置禁用 Applet,需要对浏览器进行相应设置后方可浏览。

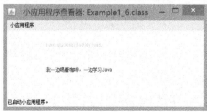

(a)编译、执行 Applet 程序　　　　　　　　(b)Applet 绘制字符串

图 1.18　Applet 运行结果

程序说明：
- 一个 Applet 类必须用 extends 关键字继承包含在 java.applet 系统类包中的 Applet 类。
- 需要重写 paint 方法，其方法参数 Graphics 是 Java 中用于绘图的类，g 是 Graphics 类的对象，g 可以形象地被理解为"一支画笔"。
- 画笔对象 g 调用 setColor 方法对其设置颜色，setColor 方法的参数是 Java 颜色类 Color 的常量，表示常用颜色。
- 画笔对象 g 调用 drawString 方法在 Applet 中绘制输出字符串，如 g.drawString("I am a student,I study hard.", 100, 40)中第一个参数是要输出的字符串内容，第 2、3 个整型参数表示绘制的字符串相对于 Applet 窗体左上角为坐标原点(0,0)点的 x、y 坐标位置。

1.8　Eclipse 环境下开发 Java 程序

Eclipse 是一个开放源代码的、基于 Java 的、可扩展的集成开发环境。就其本身而言，它只是一个框架和一组服务，用于通过插件组件构建开发环境。

在 http://www.eclipse.org/downloads/index.php 下载 eclipse-java-mars-R-win32-x86_64。将其解压到某目录下。双击运行 eclipse.exe 即可启动 Eclipse，第一次启动 Eclipse 时，会弹出如图 1.19 所示的对话框，让用户选择 Java 项目存放的工作空间。当选好工作空间后，单击"OK"

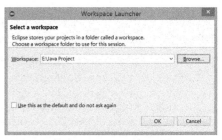

图 1.19　选择工作空间

按钮，进入如图 1.20 所示的 Eclipse 开发环境的主窗口，关闭欢迎页标签，进入图 1.21 所示的 Eclipse 主窗口。

图 1.20　Eclipse 欢迎页　　　　　　　　　图 1.21　Eclipse 主窗口

1.8.1　在 Eclipse 环境下开发 Java Application

在如图 1.21 所示的 Eclipse 主窗口中单击"File"→"New"→"Java Project"命令，弹出图 1.22 所示的对话框，键入项目名称 Ch1，单击"Finish"按钮。在如图 1.23 所示的 Ch1 项目的 src 上右击，选择"New"→"Package"命令，弹出图 1.24 所示的对话框，键入新建包名 ch1.example1，单击"Finish"按钮，形成图 1.25 所示的项目结构。

图 1.22　新建项目

图 1.23　项目结构

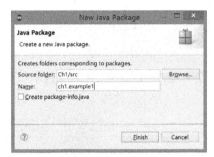
图 1.24　新建包

在包 ch1.example1 上右击，选择"New"→"Class"命令，新建类 MyFirstApplication，如图 1.26 所示，同时勾选"public static void main(String[] args)"复选框，表示该类包含主方法，单击"Fininsh"按钮。在 Eclipse 编辑区编辑完整的源程序并在 MyFirstApplication.java 上右击，选择"Run As"→"Java Application"命令或用 Run 菜单中的 Run 命令或工具栏中的 Run 按钮执行程序，在控制台输出结果，如图 1.27 所示。

图 1.25　项目结构

图 1.26　新建类

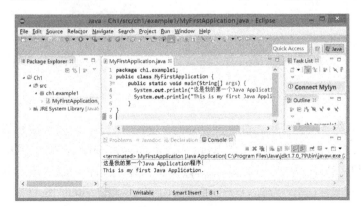
图 1.27　编辑并执行程序

说明：Eclipse 环境下，可以直接在项目下新建一个类，而无须先新建包。但好的习惯是在新建一个类之前，先新建包，再在包中新建类，这样，可以避免同一个项目中多个同名类名的冲突，并可以方便地、分门别类地管理多个类。有关包的用法，将在第 3 章中详细讨论。

1.8.2　在 Eclipse 环境下开发 Java Applet

在 Ch1 项目中新建包 ch1.example2，在该包上右击，选择"New"→"Class"命令，新建一个类 MyFirstApplet，编辑源程序（见图 1.28）在 MyFirstApplet.java 上右击，选择"Run As"→"Java Applet"命令或使用 Run 菜单中的 Run 命令或单击工具栏中的 Run 按钮执行程序，从而得到结果。

图 1.28　在 Eclipse 中编写 Applet

【例题 1_7】在 Eclipse 环境下，测试例题 1_3、例题 1_4、例题 1_5。

在 Eclipse 中的 Ch1 项目的 src 下的创建类 Example1_3，该类会被包含在默认包 default package 中，在编辑区编写程序并运行程序，在 Eclipse 控制台中输入"I am a student,I study hard."并回车，就可得到图 1.29 所示的结果。

图 1.29　Eclipse 环境下测试例题 1_3

在 Eclipse 中的 Ch1 项目下创建类 Example1_4，在编辑区编写程序并运行程序，在 Example1_4.java 上右击，选择"Run As"→"Run Configurations…"命令，在弹出的"Run Configurations"对话框（见图 1.30）中的左侧列表中点选 Example1_4（见图 1.31），在对话框中选择"Arguments"页标签，在 Program arguments 中输入命令行参数 200 300，单击"Run"按钮，就可得到结果。注意，每个命令行参数之间有且仅有一个空格。

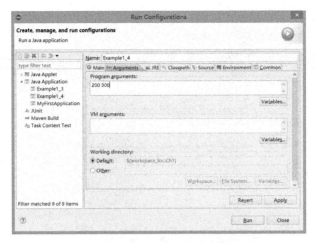

图 1.30　给程序设定命令行参数

图 1.31　Eclipse 环境下测试例题 1_4

在 Eclipse 中的 Ch1 项目下创建类 Example1_5，在编辑区编写程序并运行程序，在"输入"对话框中输入文本并单击"确定"按钮，就可在控制台输出结果，如图 1.32 所示。

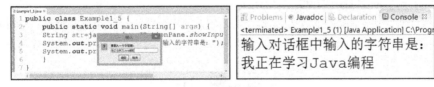

图 1.32　Eclipse 环境下测试例题 1_5

1.9　其他几个问题

1. Java 学习路线

Java 是一种编程工具，但不能简单地仅仅将 Java 视为一种编程语言，而要将其视为是一个庞大、开放的技术体系。在学习了 Java SE 后，可以为后续的学习奠定良好的基础，但不能就此断线，可以沿着 Java EE 或 Java ME 这两条技术路线继续深入学习 Java。

2. 学会查阅 JDK 开发文档

JDK API 文档（见图 1.33）中包含了 Java API 中的系统类包，在每个类包中罗列了其中的系统类及接口的定义及详细的使用方法。JDK 文档是编程人员不可或缺的编程手册。

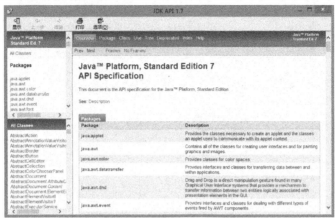

图1.33 JDK API 文档

用好开发文档，一方面可以为程序员提高开发效率，不用事事从头做起，而是要用好系统类、接口中提供的方法；另一方面，初学者可以从开发文档中系统类、接口及的定义方法中学习 Java 编程的思想。

3．Java 其他常用的开发环境

除了 Eclipse 外，JCreator、NetBeans、MyEclipse 及 JBuilder 等都是 Java 开发的常用集成开发环境。

小　　结

本章主要介绍了 Java 的发展、特点、Java 程序的分类、Java 程序的运行机制、Java 虚拟机等，详细讨论了 Java SE 开发环境 JDK 及 Eclipse 的安装与配置方法、简单 Java Application 及 Applet 程序在 JDK 及 Eclipse 环境下的开发方法等。

通过本章的学习，要重点掌握 Java SE 开发环境的配置、Java 的跨平台特性、Java 虚拟机机制、简单 Java 程序在非集成开发环境 JDK 以及集成开发环境 Eclipse 下的开发方法等。

习　题　1

一、选择题

1. 一个 Java 程序运行从上到下的环境次序是（　　）。
   A．操作系统、Java 程序、JRE/JVM、硬件　　B．JRE/JVM、Java 程序、硬件、操作系统
   C．Java 程序、JRE/JVM、操作系统、硬件　　D．Java 程序、操作系统、JRE/JVM、硬件
2. Java Application 的 main 方法声明格式正确的是（　　）。
   A．public static int main (String args[ ])　　B．public static void main (string args[ ])
   C．public void main (String args[ ])　　D．public static void main(String args[ ])
3. 若有一个名为 FirstApplet.java 的 Java Applet 源程序，与之对应的 HTML 文件为 FirstApplet.html，则在 JDK 环境下执行该小程序的命令为（　　）。

A. java FirstApplet B. javac FirstApplet.java
C. appletviewer FirstApplet.java D. appletviewer FirstApplet.html

4. 下面 JDK 命令中，用于编译 Java 源程序的是（   ）。

A. java B. javac C. javap D. javadoc

二、上机实践题

1. 下载、安装并配置 JDK 及 Eclipse 环境。
2. 分别在 JDK 及 Eclipse 环境下，编写 Application 程序，在屏幕（控制台）输出"以下内容。

****
我很喜欢学习 Java
****

3. 分别在 JDK 及 Eclipse 环境下，编写 Applet 程序，在其中以红色、字体为宋体、斜体、字号 40 号输出（绘制）"我很喜欢学习 Java"，以绿色，字体为宋体、粗体、字号 40 号输出（绘制）"我要好好学习 Java"（效果见图 1.34）。

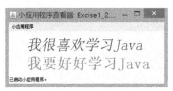

图 1.34　效果图

4. 从命令行中输入若干数字字符串，求它们的和并输出。
5. 从控制台读入若干数字字符串，求它们的和并输出。
6. 利用输入对话框 javax.swing.JOptionPane 输入若干数字字符串，求它们的和并输出。

第 2 章　Java 语法基础

【本章内容提要】
- Java Application 程序的基本输入、输出方法；
- Java 注释；
- 标识符与关键字；
- Java 数据类型；
- 运算符与表达式；
- Java 语句。
- Java 程序流程控制。

Java 语法借鉴了 C、C++的语法，但同时摈弃了 C、C++中较难的诸如指针等语法点。学习时要注意对比，主要掌握 Java 中独有的一些语法点。

2.1　Java Application 程序的基本输入、输出方法

java.lang.System 类代表系统，系统级的很多属性和控制方法都放置在该类的内部。System 类内部包含 in、out 和 err 三个成员变量，分别代表标准输入流（标准输入设备一般指键盘），标准输出流和标准错误输出流（标准输出设备一般指显示器）。

在 Java 中，涉及输入、输出，都与输入、输出流有关。具体内容在后面章节介绍，此处只介绍标准输入、输出方法。

2.1.1　Java 向显示器（控制台）输出数据

- System.out.println(参数)或 System.out.print(参数)，参数可以是字符串、变量或表达式等，前者输出数据后换行，后者输出数据后不换行。
- System.out.printf("格式控制部分", 表达式 1, 表达式 2, …, 表达式 n)。JDK1.5 之后，Java 中增加了与 C 语言类似的 printf 方法，用于灵活输出各种类型的数据并控制其输出格式。其中，格式控制部分可以是%d、% c、% f、%s 等分别表示输出整型、字符型、浮点型、字符串等；也可以像 C 语言一样，以%md、% m.nf 的形式控制数据在控制台输出的位数。

【例题 2_1】 基本输出方法测试。

```
public class Ch2_1 {
    public static void main(String[] args) {
        int a=2,b=3;
        System.out.println(a);
```

```
        System.out.println("a="+a);
        System.out.println(a+b);
        System.out.println("a+b="+(a+b));
        System.out.println("println方法测试");
        System.out.print("print方法测试");
        System.out.println();
        System.out.print("print方法测试\n");
        System.out.printf("%3d\n%c\n%4.1f",20,'A',12.34);
    }
}
```

程序运行结果如图 2.1 所示。

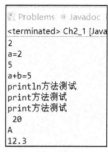

分析：程序分别向控制台输出了整型变量 a，表达式 a+b，字符串等的值。可以用"+"表示字符串的连接，如"a="+a 中"a="原样输出，紧接着输出变量 a 的值。可以用 System.out.println();的办法换行，也可以'\n'转义字符来换行。

图 2.1　数据输出

2.1.2　Java 从键盘读入数据

- System.in.read()，返回输入数值的 ASCII 码,,该值为 0～255 范围内的 int 字节值。若返回值为-1，说明没有读取到任何字节，读取工作结束。
- System.in.read(byte[] b)，读入多个字节到缓冲区字节数组 b 中，返回值是读入的字节数。

【例题 2_2】基本输入方法测试。
```
import java.io.IOException;
public class Ch2_2 {
    public static void main(String[] args) {
        int n=0;
        try {
            System.out.println("请输入若干英文字符并按回车键结束输入:");
            while ((n= System.in.read()) != -1) {      //读入
                System.out.print((char) n);            //输出
            }
        } catch (IOException e) {
            System.out.println(e.toString());
        }
    }
}
```

程序运行结果如图 2.2 所示。　　　　　　　　图 2.2　数据的读入与输出

说明：程序中 System.in.read()方法执行过程可能引发 IOException 异常，因此，要用 try{},catch{}结构对异常进行进行捕获和处理。同时，程序要用 import java.io.IOException;导入 IOException 类，程序的执行结果是输入什么英文字符序列并回车后，就会原样输出这个字符序列。将程序改写成如下形式，运行结果不变。
```
import java.io.IOException;
public class Ch2_2 {
    public static void main(String[] args) {
        int n=0;
        byte b[]=new byte[1000];              //创建字节数组
        System.out.println("从键盘输入若干英文字符并按回车键结束输入");
        try {
```

```
            n= System.in.read(b);         //读取输入的字符并存放在字节数组b中
        } catch (IOException e) {
            e.printStackTrace();
        }
        for (int i=0;i<n;i++){            //遍历输出数组元素
            System.out.print((char)b[i]);
        }
    }
}
```

2.1.3 用 Scanner 灵活读入数据

从 JDK1.5 开始，Java 类库中增加的 Scanner 类允许通过控制台从键盘以各种灵活的方式读取数据。首先按如下方式创建 Scanner 的对象，注意其构造方法参数是 System.in。

```
Scanner in = new Scanner (System.in);
```

Scanner 类的常用方法如下：
- String nextLine( )：读取输入的下一行内容。
- int nextInt( )：读取输入的下一个整数。
- double nextDouble( ) ：读取输入的下一个浮点数。
- boolean nextBoolean( ) ：读取输入的下一个布尔字符。
……

【例题 2_3】 用 Scanner 读取不同类型的数据。

```java
import java.util.Scanner;
public class Ch2_3 {
    public static void main(String[] args) {
        System.out.println("请键入一行文本: ");
        Scanner in=new Scanner(System.in);           //创建Scanner对象
        String s=in.nextLine();                      //读取一行
        System.out.println(s);                       //输出s
        System.out.println("请键入不同类型的数字: ");
        int i=in.nextInt();                          //读取整型数据
        float f=in.nextFloat();                      //读取单精度浮点型数据
        double d=in.nextDouble();                    //读取双精度浮点型数据
        boolean b=in.nextBoolean();                  //读取布尔型数据
        System.out.printf("%d\t%f\t%10.8f\t",i,f,d);//以一定的格式输出
        System.out.println( b);
    }
}
```

程序运行结果如图 2.3 所示。

分析：程序中分别用 Scanner 的 nextXXX()方法按不同类型读入键盘输入的数据，输出时可以用 System.out.printf()的格式控制部分灵活控制输出格式。'\t'是代表水平制表符的转义字符，即移到下一个 Tab 位置。

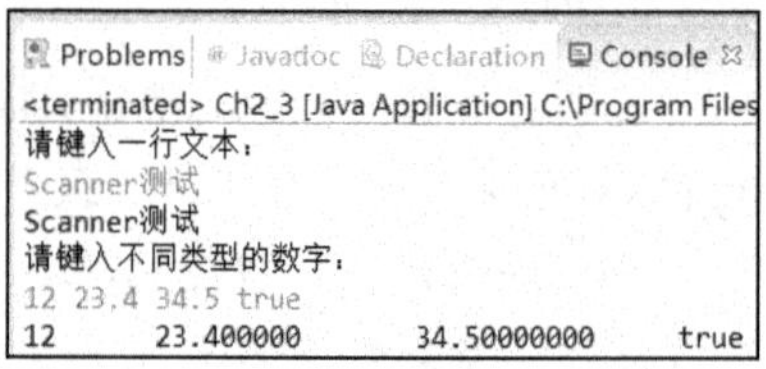

图 2.3 Scanner 读入各种类型数据

2.2 Java 注释

在程序中加注释是提高程序可读性和增强可维护性的重要方式。Java 中的注释包括以下三种类型：

- 单行注释：//注释内容。
- 多行注释（块注释）：/*注释内容*/。
- 文档注释：/**注释内容*/。

注释部分不参加程序编译。

注意：注释可用在调试程序过程中，如在编写程序时，其中某段程序暂时不用，就可以将其注释起来，在需要这段程序时再去掉注释即可。

2.3 标识符与关键字

2.3.1 标识符

Java 语言使用 Unicode 字符集。Unicode 字符集以 0~65 535 进行编码，可以表示 65 536 个字符，其中，包括了 ASCII 表中的字符及各语言中的部分字符等。

【例题 2_4】 显示 Unicode 字符集中的汉字字符。

```java
public class Ch2_4 {
    public static void main(String[] args) {
        int num=0;
        System.out.println("Unicode字符集中的汉字有: ");
        for(char c='\u4E00';c<='\u9FA5';c++){
            if(num%20==0){                    //每行输出20个字符
                System.out.println();         //换行
            }
            System.out.print(c+" ");
            num++;
        }
    }
}
```

分析：'\u4E00'~'\u9FA5'是 Unicode 字符集中汉字字符的十六进制表示的起止范围，每输出 20 个汉字字符就换行。

与其他编程语言一样，Java 标识符也是用以标识类名、成员变量名、成员方法名、数组名或文件名等的符合以下过规则的字符序列：

- 标识符由字母、下画线、美元符及数字组成，长度不限。
- 标识符的首字母不能是数字。
- 标识符不能是 Java 关键字。同时，也不能是 true、false、null。
- 标识符区分大小写。

注意：实际应用中，标识符定义要"见名知意"，尤其在定义类名等时，避免多个类名的冲突以及含义模糊。如定义学生类，应将其定义为 Student，不要简写为 Stu 或 xuesheng 等。必要时，还可以用多个单词连接为一个标识符，如 UserName、user_name 等。

HelloWorld、DataClass、_983、$bS5_c7 等都是合法的标识符，class、DataClass#、98.3、Hell World 等都是非法标识符。

Java 中，一些标识符的约定俗成的命名规则（非强制）如下：

- 类名：首字母大写，通常由多个单词合成一个类名，要求每个单词的首字母也要大写，如 class HelloWorldApp 等。
- 方法名：若由多个单词合成，则第一个单词首字母小写，中间的每个单词的首字母大写，如 setStudentName、isButtonPressed 等。
- 包名：包名为全小写的名词，中间可由点分隔开，如 Java 系统类包的名称 java.awt.event，自定义包名 org.mypackage1 等。
- 接口名：命名规则与类名相同，如 interface Shape；
- 变量名：变量名一般小写，如 name、sex、age、address、user_name 等。
- 常量名：基本数据类型的常量名为全大写，如果是由多个单词构成，可以用下画线隔开，例如：int YEAR、int WEEK_OF_MONTH。

定义规范的标识符是养成良好编程习惯的重要组成部分，在初学时，应通过查阅 JDK 文档以及阅读大量优秀的例程等途径逐渐培养。

2.3.2 关键字

和其他语言一样，一些被赋予特定含义并做专门用途的单词被称为关键字。Java 中也有许多保留关键字，如 public、static、class 等，这些保留关键字不能当作标识符使用。Java 中，所有 Java 关键字都是全小写的。

下面是 Java 中的关键字：

abstract assert boolean break byte case catch char class continue default do double else extends final finally float for if implements import instanceof int interface long native new package private protected public return short static strictfp super switch synchronized this throw throws transient try void volatile while

以上关键字在后续章节中会逐步展开介绍和应用，无须记住它们，要在应用中逐渐掌握。注意：true，false，null 不属于 Java 关键字。

2.4 Java 数据类型

2.4.1 基本数据类型

如图 2.4 所示，Java 的基本数据类型包括：boolean（布尔型）、byte（字节型）、short（短整型）、int（整型）、long（长整型）、char（字符型）、float（单精度浮点数型）、double（双精度浮点数型）等 8 种。

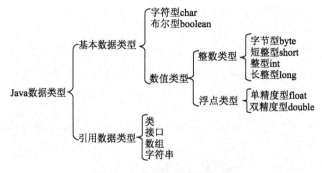

图 2.4　Java 基本数据类型

与其他编程语言一样，Java 中的不同简单数据类型决定了该类型在内存中的存储方式和所占空间大小、数据的表达范围、可参与的运算不同等。

2.4.2 常量与变量

1．布尔型常量与变量

布尔型常量取值：true 或 false。

boolean 是定义布尔型变量的关键字如 "boolean b,b1=true,b2=false;"。

【例题 2_5】 定义布尔型变量并输出其值。

```
public class Ch2_5 {
    public static void main(String[] args) {
    boolean b1 = true,b2 = false;
    System.out.println("b1 的值是:" + b1);
    System.out.println("b2 的值是:" + b2);
    }
}
```

2．整型常量与变量

（1）int 型

int 型常量有十进制、八进制、十六进制三种表示法。

- 十进制：十进制整数常量，如 456，-123 等。
- 八进制：以数字 0 开头的八进制整数，如 023，-067 等。
- 十六进制：以 0x 或 0X 开头的十六进制整数，如 0x1a3，0XB23，-0x45 等。

使用关键字 int 来声明 int 型变量，声明时也可以赋给初值。如：

int x= 123,y=9898,z;

对于 int 型变量，内存分配给 4 个字节，占 32 位。故最大的整型常量是 2 147 483 647。该值由 Integer.MAX_VALUE 表示，最小的整型常量是-2 147 483 648，该值由 Integer.MIN_VALUE 表示，Integer 是 Java 中提供的 int 类型的包装类，MAX_VALUE、MIN_VALUE 是 Integer 类的常量。

（2）byte 型

Java 中不存在 byte 型常量的表示法，若要表示 byte 型常量值，可以用强制类型转换如(byte)12。也可以把一定范围内的 int 型常量赋值给 byte 型变量。

使用关键字 byte 来声明 byte 型变量如：

byte a= -10,b=45;

对于 byte 型变量，内存分配给 1 个字节，占 8 位。

（3）short 型

和 byte 型类似，Java 中也不存在 short 型常量的表示法，若要表示 byte 型常量值，可以用强制类型转换如(short)21。也可以把一定范围内的 int 型常量赋值给 short 型变量。

使用关键字 short 来声明 short 型变量如：

short c=10,d=23;

对于 short 型变量，内存分配给 2 个字节，占 16 位。

（4）long 型

long 型常量：long 型常量用后缀 L 或 l 来表示，例如 678L（十进制）、0567l（八进制）、0x3edL

（十六进制）。

使用关键字 long 来声明 long 型变量，如

`long a1=24432L,a2=0x3edL;`

对于 long 型变量，内存分配给 8 个字节，占 64 位。

【例题 2_6】 测试输出各种整数类型数据的取值范围。

```
public class Ch2_6{
    public static void main(String[] args) {
    System.out.println("long 的最大值：  "+Long.MAX_VALUE);
    System.out.println("long 的最小值：  "+Long.MIN_VALUE);
    System.out.println("int 的最大值: "+Integer.MAX_VALUE);
    System.out.println("int 的最小值: "+Integer.MIN_VALUE);
    System.out.println("short 的最大值: "+Short.MAX_VALUE);
    System.out.println("short 的最小值: "+Short.MIN_VALUE);
    System.out.println("byte 的最大值: "+Byte.MAX_VALUE);
    System.out.println("byte 的最小值: "+Byte.MIN_VALUE);
    }
}
```

程序运行结果图 2.5 所示。

分析：程序中 Long、Integer、Short、Byte 等是 Java 对基本数据类型 long、int、short、byte 等的"包装类"，这些"包装类"位于 java.lang 包当中，是对基本数据类型的类封装，在这些"包装类"中提供了相关常量以及多个方法，用于在基本数据类型和 String 类型之间互相转换以及处理相应基本数据类型时非常有用的其他一些方法等。

图 2.5 各种整数类型数据的取值范围

程序中的 MAX_VALUE、MIN_VALUE 就是这些"包装类"中提供的常量。

【例题 2_7】 定义整型变量并输出其值。

```
public class Ch2_7{
    public static void main(String[] args) {
        byte b1 = -122;
        short s1 = -32768;
        int i1 = 1544567543;
        long l1 = 85747576576576543444L;
        System.out.println("b1 的值是:" + b1);
        System.out.println("s1 的值是:" + s1);
        System.out.println("i1 的值是:" + i1);
        System.out.println("l1 的值是:" + l1);
        int i2 = 20;           //十进制
        int i3 = 020;          //八进制
        int i4 = 0x20;         //十六进制
        System.out.println("i1 的值是:" + i2);
        System.out.println("i2 的值是:" + i3);
        System.out.println("i3 的值是:" + i4);
    }
}
```

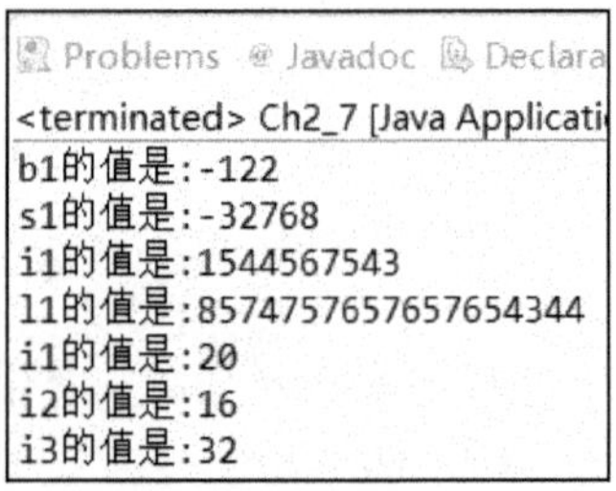

程序运行结果图 2.6 所示。

图 2.6 整型变量值的输出

注意：当给该例题的各个变量赋值时超出该变量数值表示范围时，编译系统会报错。

3．字符型常量与变量

Java 中的字符采用 Unicode 字符集的编码方案，是 16 位的无符号整数，占 2 个字节，表示的字符为 0 ~ 65 535。

字符型常量：对于可输入字符，用单引号扩起的 Unicode 表中的一个字符来表示，如 'a'、'啊' 等。

对于不可输入字符，常采用转义字符表示。'\n'表示换行；'\r'表示回车；'\t'表示 Tab 键；'\b'表示退格；'\t'表示水平制表位；'\''表示单引号；'\"'表示双引号；'\\'表示反斜线等。'\ddd'表示用三位八进制数代表的 ASCII 字符，从'\000'到'\377'，可表示 256 个 ASCII 字符；'\uxxxx'表示用四位十六进制数代表 Unicode 字符；从'\u0000'到'\uffff'，可表示所有的 Unicode 字符。

- 字符型变量：使用关键字 char 来声明 char 型变量同时可以赋予初值。如：
char ch='A', ch1='好'；
- 对于 char 型变量，内存分配给 2 个字节，占 16 位。

【例题 2_8】 字符变量测试。

```java
public class Ch2_8 {
    public static void main(String[] args) {
        char c1='好',c2='\'',c3='c';
        System.out.printf("%3c%3c%3c\n",c1,c2,c3);
        int a=(int)c1;
        System.out.println("c1在ascii表中的位置编号为: "+a);
        int b=23456;
        char c4=(char)b;
        System.out.println("整数b在ascii表中的对应的字符为: "+c4);
    }
}
```

程序运行结果图 2.7 所示。

分析：程序中可以用%c 格式控制符控制字符变量的输出，%3c 是指每个字符变量输出时占 3 个字符的宽度。字符变量与整型变量之间可以相互转化。如 "int a=(int)c1;" 及 "char c4=(char)b;"。

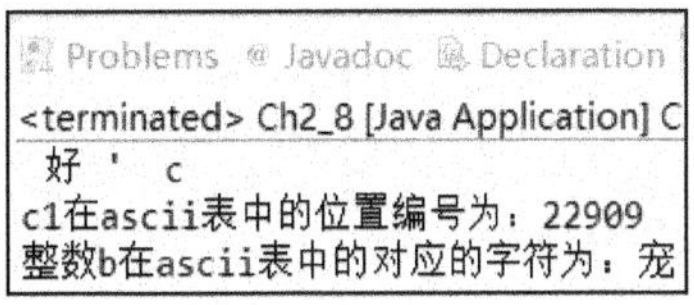

图 2.7 字符变量及其 Unicode 编码

4．浮点型常量与变量

（1）浮点型常量

浮点型被分为单精度 float 类型与双精度 double 类型。浮点型常量有两种表示形式：十进制小数形式和科学记数法形式。

- 十进制小数形式：小数点两边的数字不能同时省略且小数点不能省略。合法的 double 型浮点数如 3.14、1.0。
- 科学记数法形式：如 1.2×10^{-10} 在 Java 中表示为 1.2e-10 或 1.2E-10，这是一个 double 型的浮点数。E 或 e 的前面必须有数字且 E 或 e 后边必须是一个正、负整数（正号可省略）。由于 E 或 e 的后边必须是一个整数，那么 $1.2\times10^{-2.5}$ 该如何表示？可用 java.lang.Math 类中的方法 pow()，表示为 Math.pow(1.2,-2.5)。

数值后边加上 d 或 D 表示是 double 型的浮点数常量（d 或 D 可省略）。只有在数值后边加上 F

或 f 才表示是 float 型的浮点数常量。

注意：Java 浮点型常量默认为 double 型，如要声明一个常量为 float 型，则一定要在数字后面加 f 或 F 。

如：`double d = 12345.6;`   //正确
　　`float f = 12.3f;`   //必须加 f 否则会出错

（2）浮点型变量
- 使用关键字 float 来声明 float 型变量。对于 float 型变量，内存分配给 4 个字节，占 32 位。
- 使用关键字 double 来声明 double 型变量，对于 double 型变量，内存分配给 8 个字节，占 64 位。

2.4.3　数据类型转换

Java 语言中，数据类型的转换有"自动类型转换"和"强制类型转换"两种方式。

Java 中数据的基本类型（不包括逻辑类型）按精度从"低"到"高"排列：byte→short→char→int→long→float→double。

当把低精度的变量的值赋给高精度的变量时，系统自动完成数据类型的转换。

高精度的数据类型转换为低精度的数据类型时，要加上强制转换符，但可能造成精度损失或溢出，使用时要格外注意。

强制类型转换格式：

(要转换的数据类型)(表达式);

【例题 2_9】类型转换测试。

```java
public class Ch2_9 {
    public static void main(String[] args) {
        int i=100;
        float f=123.45f;
        double d=456.45678;
        System.out.println("i="+i+" f="+f+" d="+d);
        d=f;             //自动转换
        System.out.println("d="+d);
        i=(int)f;        //强制转换
        System.out.println("i="+i);
    }
}
```

程序运行结果图 2.8 所示。

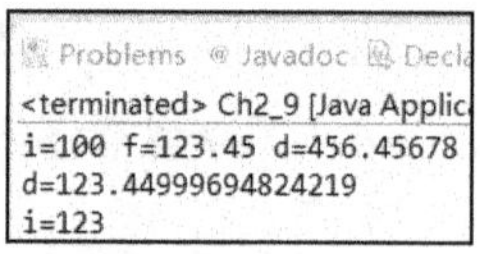

图 2.8　类型转换结果

2.4.4　引用数据类型

Java 引用数据类型主要包括类、数组、字符串、接口等。这些引用类型将在第 3 章以后逐步展开讲解。

2.5　运算符与表达式

与其他高级编程语言一样，Java 中表达各种运算的符号称为运算符，运算符的运算对象称为操作数。Java 也提供了算术运算符、逻辑运算符、关系运算符、位运算符等。用运算符连接操作数形成的式子称为表达式，表达式具有一定的值。

2.5.1 算术运算符及其表达式

（1）加、减、乘、除和求余运算符：+，-，*，/，%

都是二目运算符。其中，"+""–"运算符不仅可以用于加、减运算，而且可以作为正数和负数的前缀。

Java 中对"+"运算符进行了重载，除了用于算术加法运算外，还可用于完成字符串的连接操作。"+"运算符两侧的操作数中只要有一个是字符串(String)类型，系统会自动将另一个操作数转换为字符串然后再进行连接。

如 int c = 12; System.out.println("c=" + c);中是将字符串"c="和整型值 12 连接成一个完整的字符串进行输出。

"/"用于整型表示取整，如 7/2 结果为 3。"/"用于 float、double 表示实数相除，如 7.0/2 结果为 3.5。例如下面的语句：
int a=7,b=2; float c; c=a / b;

c 的结果是 3.0。因此若要使 a/b 按实数除法进行，可用强制类型转换：c = (float)a /b，即先将 a 的类型转换成 float 类型，然后"/"将按实数相除进行，c 的结果为 3.5；。

"%"用于整型表示取余数，如 15%2 结果为 1，(–15) % 2 结果为–1，15 % (–2) 结果为 1、(–15) % (–2) 结果为–1。"%"用于 float、double 表示实数取余。如 15.2%5 的结果为 0.1999999999999993。

（2）自增、自减运算符：++，--

是单目运算符，可以放在操作元之前，也可以放在操作元之后。操作元必须是一个整型或浮点型变量。作用是使变量的值增 1 或减 1，如：

++x（--x）表示在使用 x 之前，先使 x 的值增（减）1。

x++（x--）表示在使用 x 之后，使 x 的值增（减）1。

【例题 2_10】自加、自减运算测试。
```
public class Ch2_10 {
    public static void main(String arg[]) {
        int a=20,b=20;
        int x=b++;
        System.out.print("x="+x);
        System.out.println(" b="+b);
        x=++b;
        System.out.print("x="+x);
        System.out.println(" b="+b);
        x=--a;
        System.out.print("x="+x);
        System.out.println(" a="+a);
        x=a--;
        System.out.print("x="+x);
        System.out.println(" a="+a);
    }
}
```
程序运行结果图 2.9 所示。

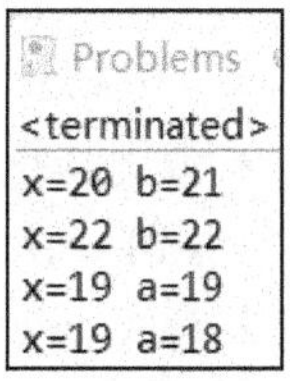

```
Problems
<terminated>
x=20 b=21
x=22 b=22
x=19 a=19
x=19 a=18
```

图 2.9 类型转换结果

2.5.2 关系运算符

关系运算符用于比较两个操作数，运算结果是布尔类型的值 true 或 false。所有关系运算符都是二目运算符。Java 中共有六种关系运算符：>（大于）、 >=（大于等于）、 <（小于）、 <=（小于等于）、 !=（不等于）、 ==（等于）。前四种优先级相同，且高于后面的两种。

Java 中，任何类型的数据（包括基本数据类型和引用类型）都可以通过==或!=来比较是否相等或不等。只有 char、byte、short、int、long、float、double 类型才能用于>、>=、<、<=这四种关系运算。

【例题 2_11】关系运算测试。

```
public class Ch2_11 {
    public static void main(String[] args) {
        int a=100,b=10;
        boolean boo=true;
        System.out.println(a==b);
        System.out.println(a>b);
        System.out.println(a<=b);
        System.out.println(a!=b);
        System.out.println(a>b==boo);
    }
}
```

程序运行结果图 2.10 所示。

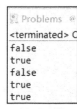

图 2.10　关系运算测试结果

2.5.3 逻辑运算符

逻辑运算符包括 &&（与），||（或），!（非）。其中，&&、||为二目运算符，实现逻辑与、逻辑或；! 为单目运算符，实现逻辑非。

逻辑运算符的操作元必须是 boolean 型数据，逻辑运算符可以用来连接关系表达式。

例如：(4>1)&&(13>16)的结果为 false；(5<3)||true 的结果为 true；!(5!=3)的结果为 false 等。

2.5.4 位运算符

Java 中的位运算符有：~（按位求反）、&（与）、|（或）、^（异或）、>>（保留符号的右移）、>>>（不保留符号的右移）、<<（左移）。

位运算符是对操作数按其在计算机内部的二进制表示按位进行操作。参与运算的操作数只能是 int、long 类型，其他类型的数据要参与位运算要转换成这两种类型。

- 按位求反运算符~是单目运算符，对操作数的二进制数据的每一个二进制位都取反，即 1 变成 0，而 0 变成 1。如~1001 结果是 0110。
- 与运算符&是双目运算符，参与运算的两个操作数，相应的二进制数位进行与运算即 x & 0 = 0, x & 1 = x（x 是 0 或 1）。
- 或运算符|是双目运算符,参与运算的两个操作数,相应的二进制数位进行或运算即 x | 0 = x, x | 1 = 1(x 是 0 或 1)。
- 异或运算符^是双目运算符，参与运算的两个操作数，相应的二进制数位进行异或运算即不相同则结果为 1，相同则结果为 0。
- 保留符号位的右移运算符>>是将一个操作数的各个二进制位全部向右移若干位，左边空出的位全部用最高位的符号位来填充。在移位之前，自动先对所要移动的位的个数进行除以 32（或 64）取余数的运算，然后再进行右移。即若 a 是 int 型，则 a >> n 就是 a >>（n%32），若 a 是 long 型，则 a >> n 就是 a >>（n % 64）。向右移一位相当于整除 2，用右移实现除法运算速度要比通常的除法运算速度要快。

- 不保留符号位的右移运算符>>>与>>不同的是，右移后左边空出的位用 0 填充。同样在移位之前，自动先对所要移动的位的个数进行除以 32（或 64）取余数的运算，然后再进行右移。即若 a 是 int 型，则 a >>> n 就是 a >>>（n%32），若 a 是 long 型，则 a >>> n 就是 a >>>（n % 64）。

- 左移运算符<<是将一个操作数的所有二进制位向左移若干位，右边空出的位填 0。同样，在移位之前，自动先对所要移动的位的个数进行除以 32（或 64）取余数的运算，然后再进行左移。即若 a 是 int 型，则 a<< n 就是 a <<（n%32），若 a 是 long 型，则 a << n 就是 a <<（n % 64）。在不产生溢出的情况下，左移一位相当于乘 2，用左移运算比通常的乘法运算速度要快。

【例题 2_12】位运算测试。

```
public class Ch2_12 {
    public static void main(String[] args) {
        int a=2,b=8;
        System.out.println("a<<2="+(a<<2));
        System.out.println("b>>3="+(b>>3));
        System.out.println("b>>>3="+(b>>>3));
        System.out.println("~a="+(~a));
        System.out.println("a&b="+(a&b));
        System.out.println("a|b="+(a|b));
        System.out.println("a^b="+(a^b));
    }
}
```

程序运行结果图 2.11 所示。

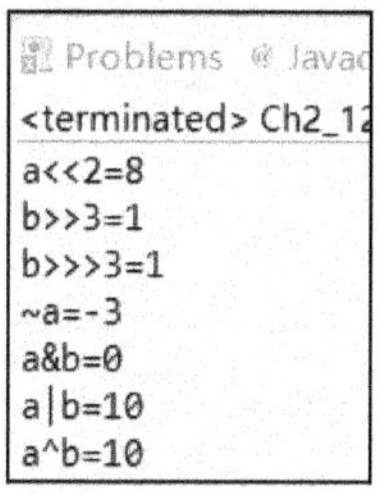

图 2.11　位运算测试结果

2.5.5　赋值运算符及其表达式

赋值运算符"="是二目运算符，用于将赋值符右边的操作数的值赋给左边的变量，左面的操作元必须是变量，不能是常量或表达式。赋值运算符的优先级较低，结合方向右到左。赋值表达式的值就是"="左面变量的值。若赋值运算符两边的类型不一致，且右边操作数类型不能自动转换到左边操作数的类型时，则需要进行强制类型转换。例如：

```
int a,b; float f=12.3f;
a=b=100;        //合法
f=a;            //合法
b=f;            //非法
b=(int)f;       //合法
```

2.5.6　复合赋值运算符及其表达式

Java 中提供了+=，-=，*=，/=，%=，&=，|=，^=，>>=，<<=，>>>=等符合赋值运算符。都是双目运算符。例如：
```
int a=10,b=20;
```

则 a+=b;与 a=a+b;是等价的。

2.5.7 条件运算符及其表达式

条件运算符?:是三目运算符，条件表达式的格式是：e1?e2:e3。

其中 e1 是一个布尔表达式，若 e1 的值是 true，则整个条件表达式的值为 e2 的值，否则整个条件表达式的值为 e3 的值。如：

```
int a=10,b=20;
int max;
max=(a>b)?a:b;   //求 a 与 b 的较大者
```

2.5.8 instanceof 运算符及其表达式

instanceof 运算符是二目运算符，其表达式的格式为

对象名 instanceof 类名

左面的操作元是一个对象；右面是一个类。当左面的对象是右面的类或子类创建的对象时，该运算符运算的结果是 true ，否则是 false。

当各种运算符同时出现在一个混合表达式中时，就涉及优先级问题，如逻辑! 运算优先于算数运算；算数运算优先于关系运算，算数*、/运算优先于算数+、-；关系运算中>、>=、<及<=优先于!=、==运算；关系运算优先于逻辑&&、||运算；逻辑&&、||运算优先于赋值运算等。在编程实践中，无须死记这些优先级，在需要提高某个运算优先级时，可以通过使用"()"运算符的方法实现。

2.6 Java 语句

Java 语言里的语句可分为以下 6 类：

- 表达式语句。表达式语句是指在一个表达式的最后加上一个分号而构成的一个语句。分号是语句不可缺少的部分。例如：
```
int x=100,y=10;
boolean z=((float)x/y)== ((double)x/y);
```
- 复合语句。复合语句是指用 "{" 和 "}" 把一些语句括起来构成的语句，一个复合语句也称作一个代码块。例如：
```
{
    int a=3, b=4;
    int sum=a+5*b;
    System.out.println(sum);
}
```
- 控制语句。包括顺序语句、条件分支语句和循环语句；流程跳转语句如 break 语句、continue 语句、return 语句等。
- 方法调用语句。方法调用语句泛指由调用方法而产生的语句，其中包括类、对象中调用类中方法所产生的行为等。例如系统输出方法调用 System.out.println("Hello");
- package 与 import 语句。package 为包声明语句，import 语句是导入包的语句，其各自用法将在第 3 章中详细介绍。
- 空语句。 一个分号也是一条语句，被称作空语句。

2.7 Java 程序流程控制

2.7.1 顺序结构

顺序结构是在程序执行时,根据程序中语句的书写顺序依次执行语句形成的结构,它是编程中最简单、基本的结构。

2.7.2 选择结构

1. if 语句

if 语句包括三种形式:
- 单分支 if 语句,其格式为
```
if(条件表达式){
    语句块
}
```
- 双分支 if 语句,其格式为
```
if(条件表达式){
    语句块 1
  }
  else{
    语句块 2
  }
```
- 多分支 if 语句,其格式为
```
if(条件表达式 1){语句块 1}
else if(条件表达式 2){语句块 2}
else if(条件表达式 3){语句块 3}
    ...
else if(条件表达式 n-1){语句块 n-1}
else{语句块 n}
```
- if 语句的嵌套

所谓 if 语句的嵌套,是指在 if 语句中又包含一个或多个 if 语句的情况。一般形式如下:
```
if(条件表达式 1) {
    if(条件表达式 2)
     {语句 1}
        else{语句 2}
    }
    else {
      if(条件表达式 3)
     {语句 3}
        else{语句 4}
    }
```

【例题 2_13】判断一个年份是否闰年。
```
import java.util.Scanner;
public class Ch2_13 {
    /*判断一个年份是否闰年的规则是该年份能被 400
     * 整除或者能被 4 整除,但不能被 100 整除*/
    public static void main(String[] args) {
```

```
    Scanner in=new Scanner(System.in);
    System.out.println("请输入一个年份: ");
    int year=in.nextInt();
    if(year%400==0||year%4==0&&year%100!=0)
        System.out.println(year+"是闰年");
    else
        System.out.println(year+"不是闰年");
    }
}
```

【例题 2_14】 判断某年某月有多少天。

```
public class Ch12_13{
    public static void main(String[] args) {
    String str1=javax.swing.JOptionPane.showInputDialog("请输入年份");
    String str2=javax.swing.JOptionPane.showInputDialog("请输入月份");
                                       //输入对话框
    int year=Integer.parseInt(str1);
    int month=Integer.parseInt(str2);     //将字符串转化为 int
    if(month<=0||month>12){
        javax.swing.JOptionPane.showMessageDialog(null,"月份应在1~12之间！");
        //消息对话框
        return;                           //跳出主函数
    }
    int day;
    if(month==1||month==3||month==5||month==7||month==8||month==10||month==12){
        day=31;
    }else if(month==4||month==6||month==9||month==11){
        day=30;
    }else{
        day=((year%4==0&&year%100!=0)||year%400==0)?29:28;
        //根据条件判定 year 是否闰年来决定 2 月份的天数
    }
    javax.swing.JOptionPane.showMessageDialog(null,year+" 年 "+month+" 月有 "+day+"天");
    }
}
```

若要判断 2000 年 2 月份有多少天，程序运行结果如图 2.12（a）、（b）、（c）所示。假定输入的月份不在 1~12 之间，则会弹出如图 2.12（d）的消息对话框。

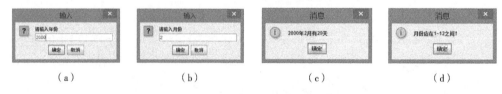

（a）　　　　　　（b）　　　　　　（c）　　　　　　（d）

图 2.12　例题 2_14 运行结果

本例中用到了 javax.swing 包中的 JOptionPane 类调用了其静态方法 showInputDialog 产生输入对话框，调用了其静态方法 showMessageDialog 产生消息对话框。具体用法在后面章节中还会深入介绍。

【例题 2_15】 输出以下分段函数的结果值

$$y = \begin{cases} -1 & \text{当 } x < 0 \\ 0 & \text{当 } x = 0 \\ 1 & \text{当 } x > 0 \end{cases}$$

```java
import java.util.Scanner;
public class Ch2_15 {
    public static void main(String[] args) {
        double x;
        int y=0;
        Scanner in=new Scanner(System.in);
        System.out.println("请输入x: ");
        x=in.nextDouble();           //读入 x
        if(x<0)  y=-1;
        if(x==0) y=0;
        if(x>0)  y=1;
        System.out.println("y 的取值为: "+y);
    }
}
```

Ch2_15 中的
```java
if(x<0)  y=-1;
if(x==0) y=0;
if(x>0)  y=1;
```
也可替换为
```java
if(x<0)   y=-1;
else{
    if(x==0)  y=0;
    else  y=1;
}
```
或替换为
```java
if (x>=0) {
    if (x>0) y=1;
        else y=0;
}
else  y=-1;
```

【例题 2_16】 求解一元二次方程 $ax^2+bx+c=0$（$a \neq 0$）的根。

```java
import java.util.Scanner;
import javax.swing.JOptionPane;
public class Ch2_16 {
    public static void main(String args[]){
        double a,b,c,delta,x1,x2,realpart,imagpart;
        String s=JOptionPane.showInputDialog("请输入a,b,c:");
        //在输入对话框中输入a,b,c的值，a,b,c之间以空格分开
        Scanner scan=new Scanner(s);//扫描并解析出a,b,c
        a=scan.nextDouble();
        b=scan.nextDouble();
        c=scan.nextDouble();
        if(Math.abs(a)<=Math.pow(10,-6))
```

```
                System.out.print("该方程不是一个一元二次方程");
            else{
                delta=b*b-4*a*c;
                if(Math.abs(delta)<=Math.pow(10, -6)){
                    System.out.println("该方程有两个相等实根: ");
                    System.out.printf("x1=x2=%8.4f",-b/(2*a));
                }
                else if(delta>Math.pow(10, -6)){
                    x1=(-b+Math.sqrt(delta))/(2*a);
                    x2=(-b+Math.sqrt(delta))/(2*a);
                    System.out.println("该方程有两个不等实根: ");
                    System.out.printf("x1=%8.4f\tx2=%8.4f",x1,x2);
                }
                else{
                    realpart=-b/(2*a);
                    imagpart=Math.sqrt(-delta)/(2*a);
                    System.out.println("该方程有两个共轭复根: ");
                    System.out.printf("%8.4f+%8.4fi\n",realpart,imagpart);
                    System.out.printf("%8.4f+%8.4fi\n",realpart,imagpart);
                }
            }
        }
    }
```

程序运行结果如图2.13（a）、（b）、（c）所示。

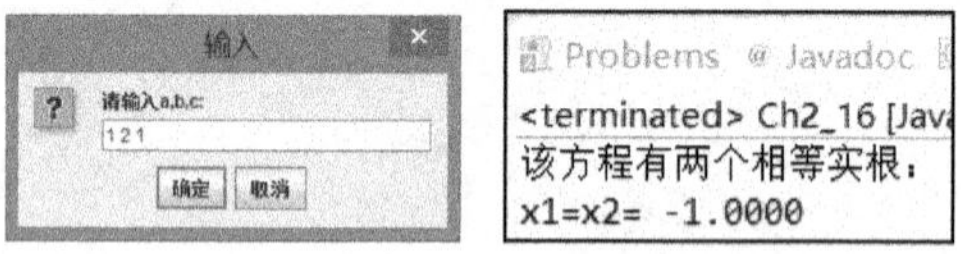

（a）有相等实根

 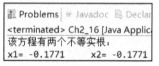

（b）有不等实根

（c）有两个共轭复根

图2.13 程序运行结果

说明：运用 JOptionPane.showInputDialog()方法可以弹出一个输入对话框，在其中可以输入文本，以字符串形式返回。Scanner 类提供了解析字符串的各种灵活的方法，如 nextDouble()，可以将以特定分隔符如空格等分割开的特定数值。在判定一个数是否为 0 时，用数学类 Math 类调用其求绝对值的静态方法 abs()以及求解幂次的静态方法 pow()，如 Math.abs(a)<=Math.pow(10,-6)表示$|a|\leq 10^{-6}$。

当然，本例也可以通过从控制台读入 a、b、c。
```
Scanner scan=new Scanner(System.in);
System.out.print("请输入 a 的值: ");a=scan.nextDouble();
System.out.print("请输入 b 的值: ");b=scan.nextDouble();
System.out.print("请输入 c 的值: ");c=scan.nextDouble();
```

2．switch 语句

用于实现多分支选择结构。其语句格式为：
```
switch(表达式)
{
   case 常量值1:
        若干语句
        break;
   case  常量值2:
        若干语句
         break;
    ...
   case 常量值n:
        若干语句
         break;
   default:
        若干语句
}
```

注意：

- switch 语句中表达式的值可以是 byte、short、int 或 char 类型。
- 常量值1到常量值n 必须也是整型或字符型。
- switch 语句首先计算表达式的值，若表达式的值和某个 case 后面的常量值（case 后的常量值必须互不相同），就执行该 case 里的若干语句直到碰到 break 语句为止；若没有一个常量值与表达式的值相同，则执行 default 后面的若干个语句。
- default 是可有可无的，若它不存在，并且所有的常量值都和表达式的值都不相同，那么 switch 语句就不会进行任何处理。
- break 语句的作用是跳出该 switch 结构体，并且继续执行 switch 结构体以后的语句。

【例题 2_17】 将5分制成绩转化为百分制成绩，假定用字符'A' ~ 'E'或'a' ~ 'e'分别表示百分制 90~100、80~89、70~79、60~69、60 分以下的成绩。
```
import java.io.IOException;
public class  Ch2_17 {
   public static void main(String[] args) {
     char score=' ';
     System.out.println("请输入 5 分制成绩: ");
     try {
          score=(char)System.in.read();//读入成绩,调用 read()方法时要进行异常处理
v}  catch (IOException e) {
          e.printStackTrace();
     }
     switch(score)
```

```
            {
            case 'A':
            case 'a':System.out.println("您的分数在 90~100 之间");break;
            case 'B':
            case 'b':System.out.println("您的分数在 80~89 之间");break;
            case 'C':
            case 'c':System.out.println("您的分数在 70~79 之间");break;
            case 'D':
            case 'd':System.out.println("您的分数在 60~69 之间");break;
            case 'E':
            case 'e':System.out.println("很遗憾！您的分数在 60 分以下");break;
            default:
                System.out.println("输入的字符有误，请输入'A'～'E'或'a'～'e'之间的字符！");
            }
        }
    }
```

2.7.3 循环结构

Java 所提供的循环结构语句有 for、while 及 do-while 三种。

1．while 循环

while 语句语法格式：
`while(条件表达式){循环体;}`

- while 是关键字，条件表达式可以是任意的布尔表达式。
- while 语句是先判断条件表达式为真时，则执行循环体，当条件表达式为假时，程序流程就转移到循环体后的语句执行。
- 循环体可以是单条语句，也可以是复合语句块。如果循环体是单条语句，则可以省略大括号。
- 循环体或条件表达式中至少应该有这样的操作，它的执行会改变 while(条件表达式)中条件表达式的值，否则 while 语句就会永远执行下去，不能终止，成为一个死循环。

2．do-while 循环

do-while 语句语法格式：`do { 循环体;}while(条件表达式);`

- do 和 while 是关键字，条件表达式可以是任意的布尔表达式。
- do-while 是无条件地先执行一次循环体，再来判断条件表达式是否为真，若为真，则执行循环体，否则跳出 do-while 循环，执行循环体后的语句。
- 循环体可以是单条语句，也可以是复合语句块。如果循环体是单条语句，则可以省略大括号。
- 循环体或条件表达式中至少应该有这样的操作，它的执行会改变 while(条件表达式);中条件表达式的值，否则 while 语句就会永远执行下去，不能终止，成为一个死循环。
- do-while 结构中 while 语句的后面有个分号";"。do-while 循环至少执行一次循环体，而 while 的循环体可能一次都不能执行。

3．for 循环

for 循环的语法格式： `for(表达式1;表达式2;表达式3){循环体;}`

- for 是关键字，三个表达式之间用分号隔开。

- 表达式 1 完成循环变量初始化等工作；条件表达式 2 可以是任意的布尔表达式，用来判断循环是否继续；表达式 3 用来改变循环变量的值。
- 循环体可以是单条语句，也可以是复合语句块。如果循环体是单条语句，则可以省略大括号。
- for 语句的三个表达式都是可选的即可以全为空，但 ";" 不能省略。但若表达式 2 也为空，则表示当前循环是一个无限循环，需要在循环体中书写另外的跳转语句方能终止循环，否则，系统会陷入"死"循环。

4．循环嵌套

循环嵌套是指循环体内包含其他循环语句，while 循环、do-while 循环与 for 循环都可以实现自身嵌套，也可以相互嵌套。

【例题 2_18】分别用 3 种不同的循环求解 1～100 之间整数的和。

```
//用while循环实现如下:
public class Ch2_18_1 {
    public static void main(String[] args) {
        int sum=0,i=1;
        while(i<=100){
            sum+=i;
            i++;
        }
        System.out.println("1-100之间所有整数的和为:\nsum="+sum);
    }
}
//用do-while循环实现如下:
public class Ch2_18_2 {
    public static void main(String[] args) {
        int sum=0,i=1;
        do{
            sum+=i;
            i++;
        }
        while(i<=100);
        System.out.println("1-100之间所有整数的和为:\nsum="+sum);
    }
}
//用for循环实现如下:
public class Ch12_18_3 {
    public static void main(String[] args) {
        int sum=0;
        for(int i=1;i<=100;i++){
            sum+=i;
        }
        System.out.println("1-100之间所有整数的和为:\nsum="+sum);
    }
}
```

2.7.4 流程跳转语句

Java 语言支持三种跳转语句：break 语句、continue 语句和 return 语句，通过这些语句，可把控制转移到程序的其他部分。

（1）break 语句

break 语句是用关键 break 加上分号构成的语句。break 语句的作用是使程序的流程从 switch 结构或循环结构的代码块中跳出。当 break 语句出现在 switch 结构中时，则跳出 switch 结构并执行 switch 结构之后的语句；当 break 语句出现在循环体中时（尤其是嵌套循环结构），遇到 break 语句，则只跳出当前包含 break 语句的循环体，转到当前循环体后的语句执行。

（2）continue 语句

continue 语句只能出现在循环体中。表示跳出本次循环，立即开始下一次循环。

（3）return 语句

退出当前方法，跳到上层调用方法。如果当前方法的返回类型不是 void，则用 return 表达式;的方式将返回值带到调用处。

【例题 2_19】 输出 1～1 000 之间能够被 3 和 5 整除的所有整数，且每行输出 20 个整数。

```
public class Ch2_19 {
    public static void main(String[] args) {
        int count=0;                          //计数变量
        for(int i=1;i<=1000;i++)
          if(i%3==0&&i%5==0)
          {  count++;
             System.out.printf("%4d",i);
             if(count%20==0)
                System.out.println();         //换行
          }
    }
}
```

也可以将程序改成下面形式：
```
public class Ch2_19{
    public static void main(String[] args) {
        int count=0;                          //计数变量
        for(int i=1;i<=1000;i++)
          if(i%3!=0||i%5!=0)
          continue;
          else{count++;
             System.out.printf("%4d",i);
             if(count%20==0)
                System.out.println();         //换行
          }
    }
}
```

【例题 2_20】 输出 100～200 之间的所有素数，且要求每行输出 10 个素数。
```
public class Ch2_20 {
    public static void main(String[] args) {
      int n,k,i,count=0;
        for(n=101;n<=200;n=n+2)  { // n从100变化到200,对每个奇数进行判定
            k=(int)Math.sqrt(n);
    //Math 类调用其静态方法 sqrt 对 n 进行开平方,但其返回值是 double ,将其强制转化为 int 型
            for  (i=2;i<=k;i++)
```

```
                    if (n%i==0) break;          // 如果n被i整除，终止内循环
                    if (i>=k+1)                 // 表示n未被i整除
                    {
                        System.out.printf("%4d",n);   // 输出素数
                        count++;                       // count用来统计素数个数
                    }
                    if(count%10==0)
                        System.out.println();         // m累计到10的倍数，换行
                }
            }
        }
```

程序执行结果如图2.14所示。

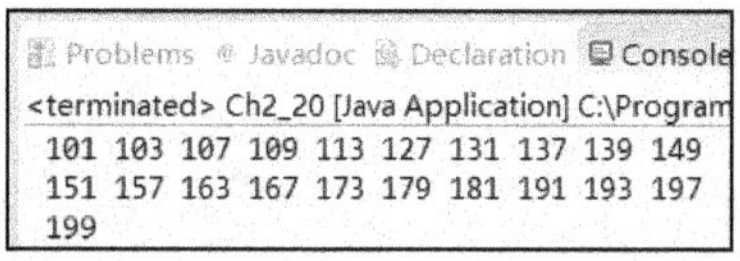

图 2.14　100～200之间的所有素数

小　　结

本章介绍了Java Application程序的基本输入、输出方法；Java程序的注释；Java标识符与关键字；Java的基本数据类型、常量与变量的定义与用法；Java的运算符与表达式的应用方法；Java程序的控制结构的应用等。

由于Java的基本语法部分与C语言语法类似，有C语言程序设计基础的读者可以复习性地学习本章内容，同时比较与C语法的不同之处。

习　　题　2

一、选择题

1. 下面那些标识符是正确的（　　）。
   A. Example1_2　　　B. hello world　　　C. 123$temp　　　D. class
2. 下面不属于Java关键字的是（　　）。
   A. class　　　　　　B. interface　　　　C. Class　　　　　D. enum
3. 下面（　　）是Java中的一个整型常量。
   A. 35.d　　　　　　B. -20　　　　　　　C. 1,234　　　　　D. "123"
4. 下列关于自动类型转换的说法中，正确的一个是（　　）。
   A. int 类型数据可以自动转换为char类型数据
   B. char 类型数据可以被自动转换为int类型数据
   C. boolean 类型数据不可以做自动类型转换，但是可以做强制转换
   D. long 类型数据可以被自动转换为short类型数据
5. 下面不属于java位运算符的是（　　）。
   A. &　　　　　　　B. ^　　　　　　　　C. |　　　　　　　D. !
6. 若设 int a=3,b=5,c=0,则表达式 c=a++---b 的值是（　　）。
   A. 0　　　　　　　B. -1　　　　　　　　C. 1　　　　　　　D. 2
7. 设各个变量的定义如下，选项的值为true 的有（　　）。
   int a=3,b=3; boolean bool=true;

A. (++a==b)||bool　　B. ++a==b++　　C. ++a==b　　D. (++a==b)&bool

8. 语句 System.out.println(5^2)输出的结果为（　　）。

　A. 6　　　　　　B. 7　　　　　　C. 10　　　　　　D. 25

9. 下列程序段执行结果是（　　）。

```
int i=1, j=8;
do {
   if(i++>--j)
   continue;
   }
while(i<4);
System.out.println(i+j);
```

　A. 10　　　　　B. 11　　　　　C. 9　　　　　D. 8

二、上机实践题

1. 编写程序：输出所有"水仙花数"（所谓"水仙花数"是指一个 3 位数，其各位数字立方和等于该数本身。例如：153 就是一个"水仙花数"。

2. 编写程序通过求解 $\pi/4 \approx 1-1/3+1/5-1/7+1/9-1/11+\cdots$ 的值，直到发现某一项的绝对值小于 10^{-6} 为止，最终得到 π 的近似值。

3. "猜数字"游戏：利用(int)(Math.random()*100)+1;得到一个 1～100 之间的随机整数，要求程序通过控制台读入用户猜测的数字，若与程序生成的随机数相等，则输出"恭喜你，答对了！"，若大于该随机数，则输出"错误!太大，请重猜:"，若小于该随机数，则输出"错误!太小，请重猜:"，同时要求统计用户猜对该随机数的次数并输出。

4. 求 $a+aa+aaa+\cdots+\underbrace{aaa\cdots aaaa}_{n\uparrow a}$ 的和，其中 a 为 1 至 9 之中的一个数，如 5+55+555+…前 10 项之和。项数 n 由程序指定。

第 3 章　Java 面向对象程序设计基础

【本章内容提要】
- 面向对象概述；
- Java 类与对象；
- 方法重载；
- this 关键字；
- static 与静态成员；
- 参数传递；
- 类的关联与依赖关系；
- package 与 import 关键字；
- JavaBean；
- jar 命令的用法。

3.1　面向对象概述

面向对象是一种程序分析、设计和编程的思想和方法集合，是当前最为常用的程序设计和实现方法。

3.1.1　类与对象的概念

- 类是面向对象思想中的核心概念之一，是面向对象程序的基本组成单位即一个面向对象程序是由若干个类组成的。类是对相同或相似的现实事物共同属性和行为的抽象和封装。例如汽车有很多种包括轿车、卡车和公共汽车等，但它们存在着许多共同点如具有车牌号、排气量等属性；都具有加速、刹车等行为，因此，就可以定义一个汽车类来封装和描述这些共同的属性和行为。
- 如《Java 编程思想》一书中所提到的"一切皆是对象"，在面向对象思想中，每个现实事物都可以被视为对象。类仅仅封装和描述了多个具有相同或相似的现实事物的共同属性和行为（共性）。但每个具体的现实事物的属性和行为又各有不同。因此，应给一个类实例化出多个不同的对象来反映各个对象的"个性"。例如：有一个学生类，其中封装了学生的学号、姓名、性别、年龄等属性信息及显示学生信息的行为。那么，如何区分两个不同的学生呢？这就需要将类进行实例化（创建对象），给每个对象赋以不同的属性值和不同的输出学生信息的行为实现。

- 类是对一组对象共性的抽象与封装，是创建对象的"模板"；对象是对类的实例化。

3.1.2 面向对象的主要特性

1．抽象

抽象是对具体事物一般化的过程，即对具有特定属性的对象进行概括，从中归纳出这一类对象的共性，并从共同性的角度描述共有的状态和行为特征。

现实问题空间中的一切事物，都具有名称、静态属性信息及动态行为信息等，而将这些映射成程序求解空间时，就可以将现实事物以"类"的形式进行表示，用类名代表现实事物的名称，用类的成员变量代表其静态属性，用类的成员方法代表其动态行为。

2．封装

封装是面向对象的重要特性之一。它有两个含义：其一是指把对象的属性和行为看成一个密不可分的整体，将这两者"封装"在一个不可分割的独立单位（对象）中；其二是实现"信息隐蔽"，即把不需要让外界知道的信息隐藏起来，有些对象的属性及行为允许外界用户知道或使用，但不允许更改，而另一些属性或行为，则不允许外界知晓，或只允许使用对象的功能，而尽可能隐蔽对象的功能实现细节。

3．继承

继承是面向对象方法中的特性之一。面向对象方法中允许类之间存在"父子"继承关系。子类可以继承父类允许被继承的属性和行为，同时，子类可以在继承的基础上增加它自己独有的属性和行为。反过来说，父类可以派生出多个子类。

面向对象程序设计中的继承机制，大大增强了程序代码的可复用性，提高了软件的开发效率，降低了程序产生错误的可能性，也为程序的修改扩充提供了便利。

若一个子类只允许继承一个父类，称为单继承；若允许继承多个父类，称为多继承。Java 只支持单继承。但通过 Java 接口的方式可以弥补子类不能享用多个父类的成员的不足。

4．多态

多态是面向对象程序设计的又一个重要特征。在面向对象的程序设计中，多态一般是指不同的处理可以用相同的名字，以产生不同的结果。

Java 中的多态主要体现在两个方面：其一是"方法重载"即同一个类中允许定义多个同名方法，编译时依据方法参数的不同进行匹配调用，它是一种静态多态（编译时静态绑定）；其二是与继承相关的"方法覆盖"，如当子类继承了父类并覆盖（重写）了父类的某个方法，那么可以用父类对象调用父类中的该方法，也可以用子类的上转型对象调用子类覆盖过的该方法，产生不同的行为。这种多态是一种动态多态（运行时动态绑定）。

3.2　Java 类与对象

3.2.1　Java 类的定义

Java 类定义的一般格式：

```
[修饰符] class 类名 [ extends 父类] [implements  接口名]
{
```

成员变量定义部分；
成员方法定义部分；
}

"[修饰符] class 类名 [extends 父类名] [implements 接口名]"是类的声明部分。其中，class 是定义类的关键字，extends 是子类继承父类的关键字，implements 是表示实现接口的关键字。类名、父类名、接口名必须是合法的 Java 标识符。"[]"表示可选。两个大括号以及之间的内容是类体。类体主要包括成员变量和成员方法定义两部分。

（1）成员变量

类中的成员变量用以对类所代表的现实事物静态状态属性进行描述。一个类的成员变量可以是 Java 简单数据类型变量，也可以是其他类的对象、数组、字符串或接口变量等。

其一般定义格式为

[修饰符] 变量类型 变量名[=变量初值]；

（2）成员方法

类中的成员方法用以对类所代表的现实事物动态行为属性进行刻画。成员方法的一般定义格式为：

[修饰符] 返回值类型 方法名(形式参数列表)
　{
　　方法体
　}

【例题 3_1】 定义一个"圆"类，类中包含三个成员变量半径、周长、面积，用于求解圆周长和圆面积的两个成员方法。

```
public class Circle{                //定义圆类
/*成员变量定义*/
  double radius;                    //定义圆半径
  double length,area;               //定义圆面积
/*成员方法定义*/
  public double getLength(){        //求解圆周长的方法
    return 2*3.14*radius;
  }
  public double getArea(){          //求解圆面积的方法
    return 3.14*radius*radius;
  }
}
```

此时，只是定义了圆类，但还没有对其创建对象，无法使用和操作类中的成员变量和方法。

3.2.2 构造方法与对象的创建

1. 构造方法

构造方法的作用是创建对象，并对对象的成员变量进行初始化。构造方法是 Java 类中特殊的方法。它具有以下特征：

- 构造方法的方法名与类名相同。
- 构造方法没有返回值类型。
- 在给类创建对象时，构造方法被系统自动调用。
- 构造方法的主要功能是完成类对象的初始化工作。

- 若在一个类中没有定义构造方法,则在给类创建对象时,系统提供默认构造方法;反之,若果自定义了构造方法,系统则不再提供默认构造方法。
- Java 支持构造方法重载,即允许在一个类中定义若干构造方法,但一定要保证这些构造方法的参数不同。
- 默认构造方法没有参数。

2. 对象的创建

定义类的目的在于使用它,对其创建对象是使用类的前提。类是创建对象的模板,对象是类的实例化。一个类可以创建多个对象。

类对象的声明格式:类名 对象名;

类对象的创建格式:new 构造方法;

【例题 3_2】对例题 3_1 定义的圆类创建对象并操作其成员变量和方法。

```java
class Circle{//定义圆类
   double radius;//定义圆半径
   double length,area; //定义圆面积
   public double getLength(){    //求解圆周长的方法
      return 2*3.14*radius;
   }
   public double getArea(){      //求解圆面积的方法
     return 3.14*radius*radius;
   }
}
public class Ch3_2 {
    public static void main(String[] args) {
     Circle circle ;                //声明 Circle 类对象 circle
     circle=new Circle();           //用 Circle 类的默认构造方法创建对象
     circle.radius=3.0;             //对象用"."引用运算符引用自己的成员变量并给其赋值
     System.out.println("圆半径 radius=: "+circle.radius);
     System.out.println("圆周长 length=: "+circle.getLength());
     System.out.println("圆面积为 area=: "+circle.getArea());
                                    //对象用"."引用运算符调用自己的成员方法
    }
}
```

程序运行结果如图 3.1 所示。

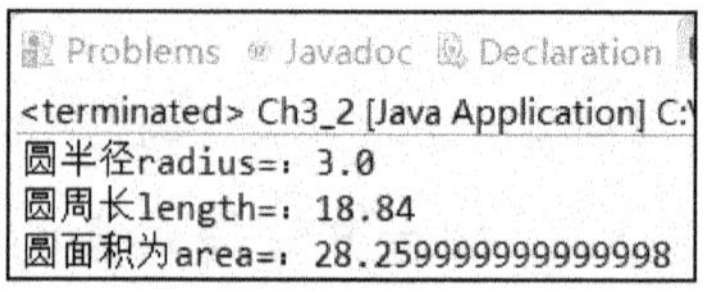

上面程序中,使用 Circle 类的默认构造方法创建对象。一般地,我们需要在类中自定义构造方法。那么,如果在 Circle 类定义中增加自定义构造方法后,情况会怎样呢?

图 3.1 圆类对象的输出

【例题 3_3】在圆类中自定义构造方法。

```java
package myproject.ch3_3;
public class Circle {
  double radius;
  double length,area;
  public Circle(){}                      //定义无参构造方法
  public Circle(double radius){          //定义含参构造方法
  this.radius = radius;
  }
```

```java
    public double getLength(){
      return 2*3.14*radius;
    }
    public double getArea(){
      return 3.14*radius*radius;
    }
}
package myproject.ch3_3;
public class Ch3_3 {
    public static void main(String[] args) {
      Circle circle1=new Circle();        //用 Circle 类的无参构造方法创建对象
      circle1.radius=3.0;
      System.out.println("circle1 的圆周长为: "+circle1.getLength());
      System.out.println("circle1 的圆面积为: "+circle1.getArea());
      Circle circle2=new Circle(6.0);     //用 Circle 类的含参构造方法创建对象
      System.out.println("circle2 的圆周长为: "+circle2.getLength());
      System.out.println("circle2 的圆面积为: "+circle2.getArea());
    }
}
```

程序运行结果如图 3.2 所示。

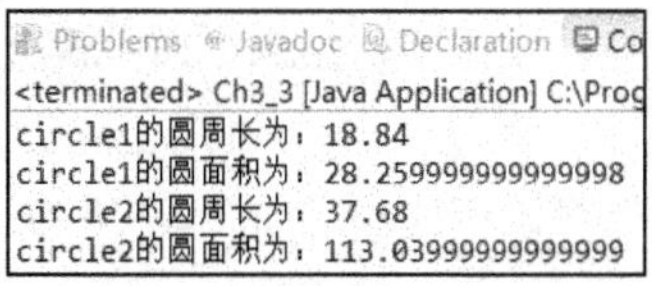

说明：

本例中，当类中定义了无参构造方法 public Circle(){}后并用其给 Circle 类创建对象时，Java 不再提供默认构造方法；无参构造方法和含参构造方法形成了构造方法重载。

图 3.2　例题 3_3 运行结果

为了避免与例题 3_2 的 Circle 类名冲突，也为了方便管理各个类，本例中用 package myproject.ch3_3;定义了"包"，并将类 Circle 与类 Ch3_3 在 myproject.ch3_3 包中单独定义。将多个类包含在包中单独定义，而不要将它们定义在同一源文件中是一个良好的编程习惯。

3.2.3　对象声明及创建的内存模型

一个类对象被声明后，并不给该对象分配内存空间，只有创建了该对象后，才给该对象分配内存空间。

【**例题 3_4**】声明并创建 Person 类对象 zhangsan、lisi 并输出各自的相关信息，比较对象声明和创建时对象引用的变化。

```java
package myproject.ch3_4;
public class Person { //人员类
    /*成员变量定义*/
    String name;
    String sex;
    int age;
    /*构造方法*/
    public Person(String name, String sex, int age){
       this.name = name;
       this.sex = sex;
       this.age = age;
    }
    /*用于显示人员信息的成员方法**/
```

```java
    void showInfo(){
      System.out.println("姓名 name="+name+" 性别 sex="+sex+" 年龄 age="+age);
    }
}
package myproject.ch3_4;
public class Ch3_4{
    public static void main(String[] args) {
        Person zhangsan;                            //声明对象 zhangsan
        //System.out.println("zhangsan 对象的引用"+zhangsan);
        zhangsan=new Person("张三","男",18);       //创建对象 zhangsan
        System.out.println("zhangsan 对象的引用"+zhangsan);
        zhangsan.showInfo();
        zhangsan.age=20;                            //修改 zhangsan 的年龄
        zhangsan.showInfo();
        Person lisi=new Person("李四","男",19);    //声明并创建 lisi 对象
        System.out.println("lisi 对象的引用"+lisi);
        lisi.showInfo();
    }
}
```

程序运行结果如图 3.3 所示。

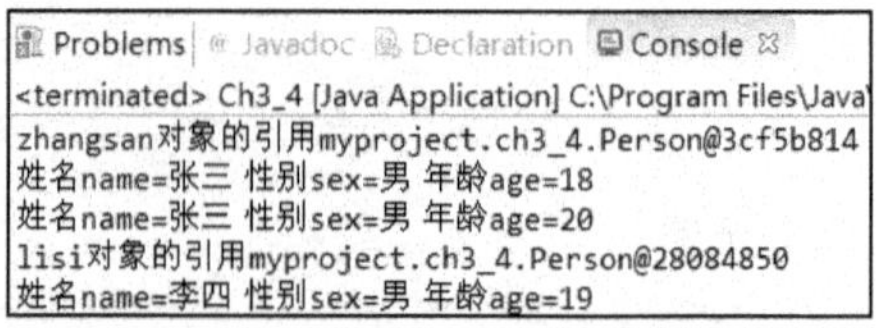

分析：一个类可以创建出多个对象，如本例中 Person 类可以创建出 zhangsan 和 lisi 对象，不同对象创建后在内存中会分配不同的地址空间，所谓某个对象的引用是指分配给该对象的地址空间的首地址，实质上就是指向该对象实体的存储空间的指针（但请注意，Java 在语法层面并不存在"指针"这一概念）。

图 3.3 例题 3_4 运行结果

程序的输出结果显示：zhangsan 和 lisi 对象的引用不同即分配给它们的地址不同。

程序中有被注释起来的"//System.out.println("zhangsan 对象的引用"+zhangsan);"语句，若去掉该语句的注释符并运行程序，会报错（见图 3.4）。这是因为，该语句之前只声明了 zhangsan 对象，但没有创建对象，系统不会为 zhangsan 分配存储空间，此时 zhangsan 并没有确定的引用值。

若将"Person zhangsan;"改成"Person zhangsan=null;"，再运行程序则没有问题。

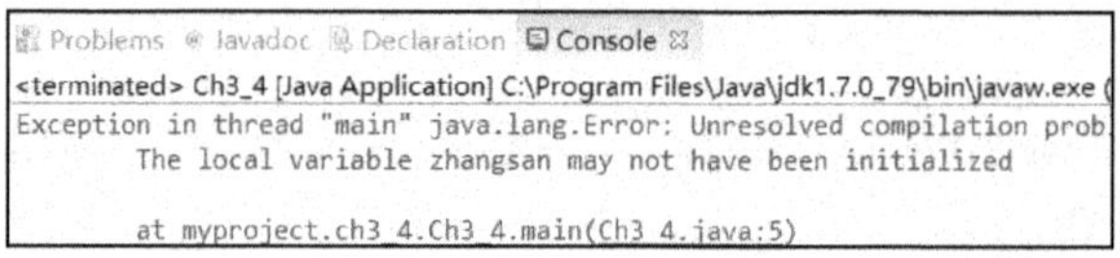

图 3.4 出错信息

3.2.4 对象间的赋值

【例题 3_5】 对象间赋值测试。

```java
package myproject.ch3_5;
public class Person {
    String name;
    String sex;
    int age;
}
package myproject.ch3_5;
```

```
public class Ch3_5{
    public static void main(String[] args) {
        Person zhangsan=new Person();     //用默认构造方法创建对象
        zhangsan.age=20;                  //给 zhangsan 的 age 成员变量赋值
        System.out.println("zhangsan对象的引用"+zhangsan);
        System.out.println("zhangsan的年龄age="+zhangsan.age);
        Person lisi=new Person();
        lisi.age=18;
        System.out.println("lisi对象的引用"+lisi);
        System.out.println("lisi的年龄age="+lisi.age);
        zhangsan=lisi;                    //对象的赋值
        System.out.println("zhangsan对象的引用"+zhangsan);
        System.out.println("lisi对象的引用"+lisi);
        System.out.println(zhangsan.age);
        System.out.println(lisi.age);
    }
}
```

程序运行结果如图 3.5 所示。

分析：当执行“zhangsan=lisi;”时，实际上使得zhangsan对象和lisi对象的引用值变得一样了，因此，无论输出zhangsan.age还是lisi.age，都是取得原来lisi对象实体中的age值。

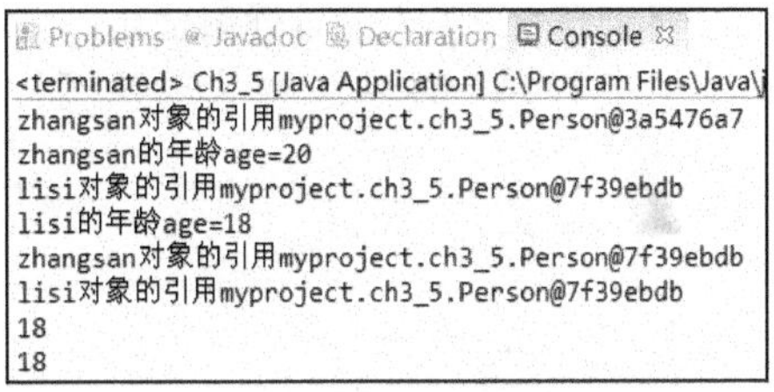

```
Problems  Javadoc  Declaration  Console
<terminated> Ch3_5 [Java Application] C:\Program Files\Java\j
zhangsan对象的引用myproject.ch3_5.Person@3a5476a7
zhangsan的年龄age=20
lisi对象的引用myproject.ch3_5.Person@7f39ebdb
lisi的年龄age=18
zhangsan对象的引用myproject.ch3_5.Person@7f39ebdb
lisi对象的引用myproject.ch3_5.Person@7f39ebdb
18
18
```

图 3.5　对象间赋值结果

3.3 方法重载

在同一个 Java 类中，允许多个方法使用同一个名字，但方法的参数列表不同，完成的功能也不同。参数列表不同表现在：参数的个数不同、类型不同或顺序不同等。方法重载是 Java 实现多态性的方式之一。

当调用这些重载的同名的方法时，Java 根据参数类型和参数的数目来确定到底调用哪一个方法。返回值类型可以相同也可以不相同。注意：无法以返回值类型作为重载函数的区分标准。

Java 系统类中有很多方法重载现象，如 java.io 包中的文件字节输入流类 FileInputStream 的重载构造方法，FileInputStream 重载的 read()方法如图 3.6 所示。

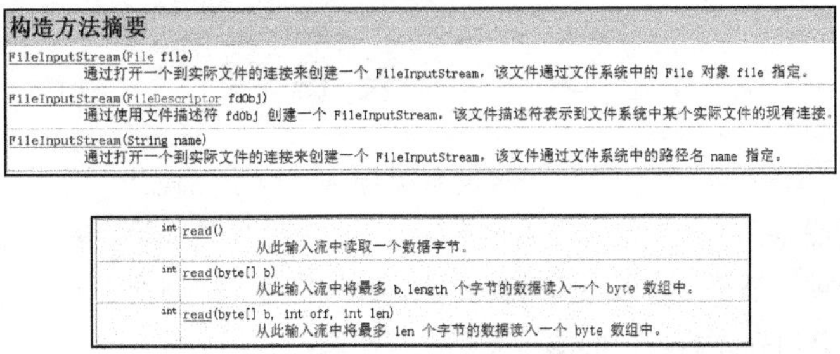

构造方法摘要
FileInputStream(File file) 通过打开一个到实际文件的连接来创建一个 FileInputStream，该文件通过文件系统中的 File 对象 file 指定。
FileInputStream(FileDescriptor fdObj) 通过使用文件描述符 fdObj 创建一个 FileInputStream，该文件描述符表示到文件系统中某个实际文件的现有连接。
FileInputStream(String name) 通过打开一个到实际文件的连接来创建一个 FileInputStream，该文件通过文件系统中的路径名 name 指定。

int	read() 从此输入流中读取一个数据字节。
int	read(byte[] b) 从此输入流中将最多 b.length 个字节的数据读入一个 byte 数组中。
int	read(byte[] b, int off, int len) 从此输入流中将最多 len 个字节的数据读入一个 byte 数组中。

图 3.6　java.io. FileInputStream 类的重载构造方法和 read()方法

【例题 3_6】方法重载测试。

```
package myproject.ch3_6;
public class A {
```

```java
    int abs(int a){              //自定义求绝对值方法
        return a>0?a:-a;
    }
    double abs(double a){        //自定义求绝对值方法
        return a>0?a:-a;
    }
}
package myproject.ch3_6;
public class Ch3_6{
    public static void main(String[] args) {
        A a = new A();
        System.out.println("78.8 的绝对值为: "+a.abs(78.8));
        System.out.println("-23 的绝对值为: "+a.abs(-23));
    }
}
```

程序运行结果如图 3.7 所示。

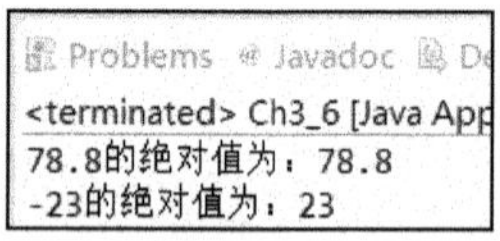

图 3.7　例题 3_6 运行结果

【例题 3_7】 方法重载测试 2。

```java
package myproject.ch3_7;
public class A {
    public int add(int x,int y){
        return x+y;
    }
    public int add(int x,int y,int z){
        return x+y+z;
    }
    public float add(float x,float y){
        return x+y;
    }
}
package myproject.ch3_7;
public class Ch3_7 {
    public static void main(String[] args) {
        A a = new A();
        System.out.println(a.add(20,30));
        System.out.println(a.add(3,4,5));
        System.out.println(a.add(20.0f,30.0f));
    }
}
```

程序运行结果如图 3.8 所示。

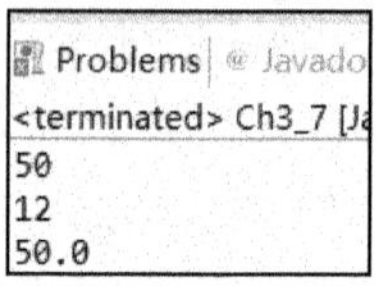

图 3.8　例题 3_7 运行结果

3.4　this 关键字

this 的含义是"当前类对象本身"。this 的常见用法有以下三种：
- 当类的成员变量和方法的局部变量同名时，用"this.成员变量"表示成员变量。

方法的局部变量是指在方法内定义的变量，包括方法形参。局部变量的作用域仅限于所定义的方法内，不能在其所在方法以外的地方对它进行访问。
- 当 this 出现在类的类成员方法中时，this 代表包含该成员方法的类对象本身。
- 当 this 出现在类的构造方法中时，可以以"this(参数列表)"的形式调用类中其他重载的构造方法。

【例题 3_8】 this 关键字测试 1。
```java
package myproject.ch3_8;
public class Person{
  String name;
  String sex;
  int age;
  public Person(String name, String sex, int age) {
    //构造方法形参与 Person 的成员变量同名
    this.name = name;
    this.sex = sex;
    this.age = age;
  }
  public void sayHello(String s){
     System.out.println(s);
  }
}
package myproject.ch3_8;
public class Ch3_8 {
   public static void main(String[] args) {
     Person zs=new Person("张三","男",18);
     System.out.println("zs 的姓名: "+zs.name+"\n"+"zs 的性别: "
+zs.sex+"\n"+"zs 的年龄: "+zs.age);
     zs.sayHello("您好! ");
   }
}
```
程序运行结果如图 3.9 所示。

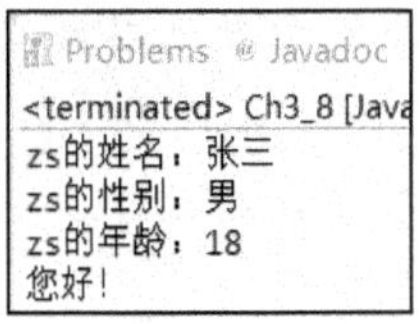

图 3.9 例题 3_8 运行结果

说明：本例中，由于类 Person 的成员变量与构造方法中的形式参数同名，为了区分两者，在成员变量前加 this.引导。

【例题 3_9】 this 关键字测试 2。
```java
package myproject.ch3_9;
public class A {
    public A f()                    //f 方法的返回值为 A 类对象的引用
    {
      return this;                  //返回 A 类对象引用
    }
}
package myproject.ch3_9;
public class Ch3_9 {
    public static void main(String[] args) {
      A a = new A();
      System.out.println("a 的引用: "+a.f());//通过 f()方法得到对象 a 的引用
      System.out.println("a 的引用: "+a);    //通过 A 类对象 a 直接得到对象 a 的引用
    }
}
```
程序运行结果如图 3.10 所示。

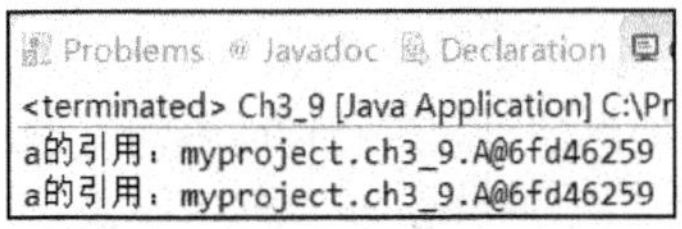

说明：this 代表当前类对象本身，本例中 this 包含在 A 类的成员方法 f()方法中，那么 this 就表示对象 a 本身。

图 3.10 例题 3_9 的运行结果

【例题 3_10】 this 调用其他构造方法。
```
package myproject.ch3_10;
public class A {
    A(String str) {                        //含参构造方法
        System.out.println(str);
    }
    A() {                                  //无参构造方法
        this("调用含参的构造方法！");        //调用含参构造方法
        System.out.println("我是无参构造方法。");
    }
}
package myproject.ch3_10;
public class Ch3_10 {
    public static void main(String[] args) {
        new A();                           //用无参构造方法给类 A 创建对象
    }
}
```
程序运行结果如图 3.11 所示。

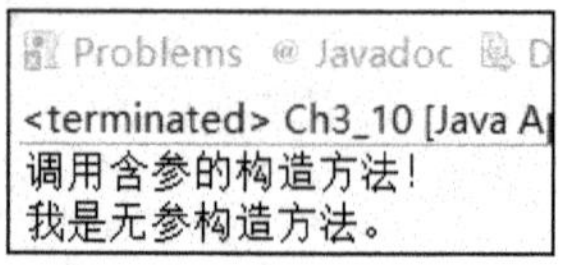

说明：Java 允许构造方法重载，本例无参构造方法中 this ("调用含参的构造方法！");表示调用另一个重载的构造方法 A(String str)。this 调用本类中的其他重载构造方法时，调用语句要放在该构造方法的第一行。

图 3.11　例题 3_10 运行结果

3.5　static 与静态成员

static 关键字可以修饰成员变量、成员方法、代码块以及内部类。

3.5.1　静态变量

若在类的成员变量前面加 static 修饰，则该成员变量被称为静态变量或类变量。没有被 static 修饰的变量，叫实例变量。

静态变量和实例变量的区别如下：

（1）内存分配不同

Java 虚拟机只为静态变量分配一次内存，在加载类的过程中完成静态变量的内存空间分配。此后，对该静态变量的操作实际上在操作同一个内存空间，即所有对象共享该静态变量的存储空间。

对于实例变量，每给类创建一个对象，都会为实例变量分配一次内存，且占用不同的内存空间。不同对象对实例变量的操作由于是针对不同的内存空间，所以互不影响。

（2）访问方式不同

对于实例变量，只能先给类创建对象后，用"对象名.实例变量"的形式对其进行访问。

对于静态变量，应通过类名访问静态变量即"类名.静态变量"。也可以用"对象名.静态变量"的形式对其进行访问，但是这种方式是不推荐使用的。

【例题 3_11】 静态变量测试。
```
package myproject.ch3_11;
public class A {
```

```
        int a;              //实例变量
        static int b;       //静态变量（类变量）
}
package myproject.ch3_11;
public class Ch3_11 {
    public static void main(String[] args) {
        A a1=new A();
        A a2=new A();
        //创建A的对象a1,a2
        a1.a=100;
        a2.a=300;
        A.b=1000;
        System.out.println("a1 的 a 值为: "+a1.a);
        System.out.println("a2 的 a 值为: "+a2.a);
        System.out.println("静态变量b的值为: "+A.b);
        System.out.println(a1.b);
        System.out.println(a2.b);
    }
}
```

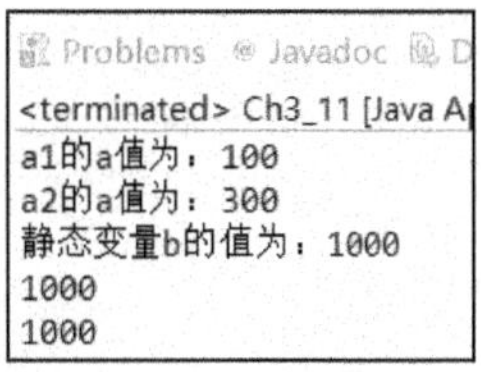

图 3.12　例题 3_11 运行结果

程序运行结果如图 3.12 所示。

分析：对于类 A 的实例变量 a，由 A 的对象 a1，a2 去引用和赋值；对于静态变量 b，则是通过类名 A.b 的方式进行访问和赋值。结果显示，A.b、a1.b、a2.b 访问到的 b 值都是 1000，而 a1.a 和 a2.a 却各不相同，这就说明静态变量的空间是共享的，而给不同对象的实例变量 a 分配不同的空间。程序中会有警告信息（见图 3.13），这是提示静态变量应以静态的方式进行访问即 A.b 的方式。

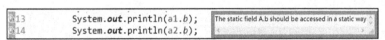

图 3.13　警告信息

3.5.2　静态方法

若在类的成员方法前面加 static 修饰，则该成员方法被称为静态方法或类方法；没有被 static 修饰的方法，叫实例方法。

对于实例方法，只能先给类创建对象后，以"对象名.实例方法"的形式进行调用。

对于静态方法，应以"类名.静态方法"的形式进行调用。也可以用"类对象.静态方法"的形式进行调用，但是这种方式是不推荐使用的。

静态方法的特点：

- 在静态方法里只能直接访问类中其他的静态成员(包括变量和方法)，而不能直接访问类中的非静态成员。这是因为，对于非静态的方法和变量，需要先创建类的实例后才可使用。
- 静态方法不能以任何方式引用 this 和 super 关键字。因为静态方法在使用前是不需要创建任何对象的，当静态方法被调用时，this 所引用的对象根本就没有产生。
- 子类只能继承、重载、隐藏父类的静态方法，子类不能重写父类的静态方法，也不能把父类不是静态的方法重写成静态的方法。
- 非静态方法可以访问静态成员。

注意：Java 中的 main 方法为什么是静态的？

Java 是一个纯面向对象的语言，而 main 也是在一个类里，若不是静态的，则在没有对象时，main 是不能调用的，这样一来程序就没有入口了，所以 main 方法一定要是静态的，而对象的创建必须是 JVM 执行主方法后才创建，而 main 就是主方法，只有 main 执行了才开始创建对象。

【例题 3_12】 静态成员测试。

```
package myproject.ch12;
public class StaticTest {
   int a;
   static int b;
   void f1(){
      a=100;
      b=200;                //非静态方法访问静态变量
      System.out.println("实例方法 f1()");
      System.out.println("a="+a);
      System.out.println("b="+b);
   }
   static void f2(){
      // a=100;              //非法，静态方法中不能访问非静态成员变量
      b=300;                 //合法，静态方法中可以直接访问静态变量
      System.out.println("静态方法 f2()");
      System.out.println("b="+b);
   }
}
package myproject.ch12;
public class Ch3_12 {
 public static void main(String[] args) {
     StaticTest test=new StaticTest();
     test.f1();
     StaticTest.f2();     //用类名直接调用静态方法
     //test.f2();
     //可以用这种方法调用静态方法，但不建议这样做
    }
}
```

程序运行结果如图 3.14 所示。

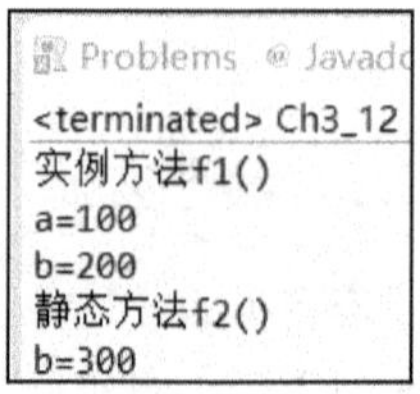

图 3.14 静态成员测试结果

3.5.3 静态代码块

类中可以包含以下形式的静态代码块：

```
class 类名
{
static
{ 代码块;}
…;
}
```

若在类中包含多个静态代码块，JVM 将按其出现的先后次序依次执行它们，每个静态代码块只被执行一次。

【例题 3_13】 静态代码块的执行。

```
package myproject.ch3_13;
public class StaticCodeTest {
    public StaticCodeTest (){
```

```java
    System.out.println("构造方法");
  }
  static{//静态代码块 1
    System.out.println("静态代码块 1");
  }
  static {//静态代码块 2
    System.out.println("静态代码块 2");
  }

  static {//静态代码块 3
    System.out.println("静态代码块 3");
  }
}
package myproject.ch3_13;
public class Ch3_13 {
  public static void main(String args[]){
    System.out.println("主方法");
    new StaticCodeTest();    //创建对象
    new StaticCodeTest();    //创建对象
  }
}
```

程序运行结果如图 3.15 所示。 图 3.15 静态代码块测试结果

分析:本例中先执行,输出"主方法";第一次创建了 StaticCodeTest 的对象,JVM 按照 StaticCode Test 包含的 3 个静态代码块定义的先后次序依次执行它们,再调用构造方法;第二次创建 StaticCodeTest 的对象,只调用了构造方法,静态代码块并不执行,这是因为每个静态代码块只被执行一次。

将例题 3_13 改写成下面形式:

```java
package myproject.ch3_13;
public class StaticCodeTest {
  public StaticCodeTest (){//构造方法
    System.out.println("构造方法");
  }
  static{//静态代码块 1
    System.out.println("静态代码块 1");
  }
  static {//静态代码块 2
    System.out.println("静态代码块 2");
  }
  public static void main(String args[]){// 主方法
   System.out.println("主方法");
   new StaticCodeTest();       //创建对象
   new StaticCodeTest();       //创建对象
  }
  static {//静态代码块 3
    System.out.println("静态代码块 3");
  }
}
```

这段程序的运行结果如图 3.16 所示。 图 3.16 例题 3_13 改写后的运行结果

请观察图 3.16 的结果,进一步总结静态代码块的执行特点。

3.6 参 数 传 递

Java 在参数传递时,也有按值传递和引用类型参数传递两种方式。

基本数据类型参数都是按值传递,每次传递参数时,把参数的原始数值复制一份并把复制出来的数值传递到方法内部,在方法内部修改时,则修改的是复制出来的值,而原始的值不发生改变。

引用类型参数(如类对象、数组名等)传递指每次传递参数时,把参数的引用(在内存中的存储地址)传递到方法内部,在方法内部通过引用(存储地址)改变对应存储区域的内容。当方法内部修改了参数的值以后,参数原始的值发生改变。

值得注意的是:String 对象、String 常量在传递前后,其值并不发生改变。

【例题 3_14】 基本数据类型的参数传递。

```
package myproject.ch3_14;
class Calculate{
    public int sum(int x,int y){
        return 2*x+y;
    }
}
package myproject.ch3_14;
public class Ch3_14 {
    public static void main(String[] args) {
        int a=100,b=200,result;
        Calculate cal=new Calculate();
        System.out.println("参数传递前 a="+a+" b="+b);
        result=cal.sum(a,b);          //传递实参
        System.out.println("result="+result);
        System.out.println("参数传递后 a="+a+" b="+b);
    }
}
```

程序运行结果如图 3.17 所示。

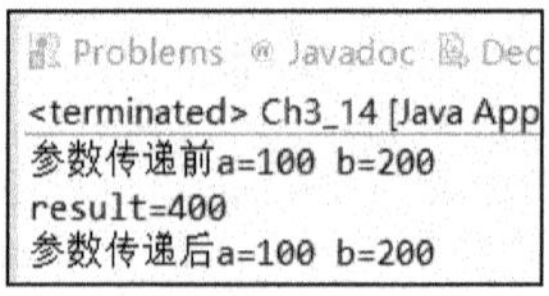

说明:实参 a,b 在传递给形参 x,y 前后的输出值并没有发生改变。

图 3.17 基本数据类型参数传值结果

【例题 3_15】 引用类型的参数传递——数组传递的例子。

```
package myproject.ch3_15;
public class A {
    void change(int[] a){          //整型的一维数组 a 做形参
        a[0]= a[0]*10;
    }
}
package myproject.ch3_15;
public class Ch3_15 {
    public static void main(String[] args) {
        int[] arr={1,2,3,4,5};     //整型的一维数组 arr 静态初始化
        System.out.println("参数传递前 arr[0]="+arr[0]);
        A a=new A();
        a.change(arr);             //以数组名 arr 为实参
        System.out.println("参数传递后 arr[0]="+arr[0]);
    }
}
```

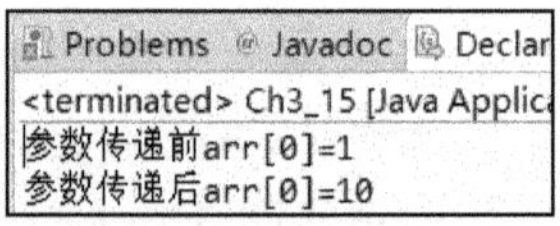

程序运行结果如图 3.18 所示。

图 3.18 数组参数传递结果

程序运行结果显示，在参数传递前后，arr[0]的结果发生了变化。

【例题 3_16】 引用类型的参数传递——类对象传递的例子。

```
package myproject.ch3_16;
public class A{
    int x;
}
package myproject.ch3_16;
public class B {
    void fun(A aa){           //以A类的对象a作为形参
        aa.x = aa.x + 1;      //改变aa.x的值
    }
}
package myproject.ch3_16;
public class Ch3_16 {
    public static void main(String[] args) {
        A a = new A();                  //创建A的对象
        a.x = 10;                       //给A类的成员变量x赋值
        System.out.println("参数传递前a.x=" + a.x);    //输出A类的成员变量x的值
        B b = new B();                  //创建B的对象
        b.fun(a);                       //调用B的f方法并传递实参，实参为A的对象a
        System.out.println("参数传递后a.x=" + a.x);    //输出A类的成员变量x的值
    }
}
```

程序运行结果如图 3.19 所示。

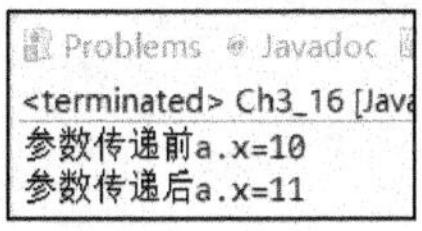

图 3.19 对象作为参数传递测试结果

【例题 3_17】 引用类型的参数传递——类对象传递的例子。

```
package myproject.ch3_17;
public class A {
    public void test(String str1,String str2){    //以str1,str2做形参
        str1="def";                               //String常量
        str2=new String("DEF");                   //String对象
    }
}
package myproject.ch3_17;
public class Ch3_17 {
    public static void main(String[] args) {
        String s1="abc";                    //String常量s1
        String s2=new String("ABC");        //String对象s2
        A a=new A();                        //创建A的对象
        System.out.println("参数传递前: ");
        System.out.println("s1="+s1);
        System.out.println("s2="+s2);
        a.test(s1, s2);                     //调用test方法，以s1,s2作为实参
        System.out.println("参数传递后: ");
        System.out.println("s1="+s1);
        System.out.println("s2="+s2);
    }
}
```

程序运行结果如图 3.20 所示。

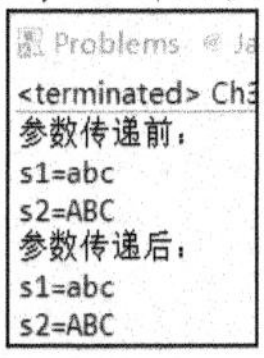

图 3.20 String 常量和 String 对象的参数传递结果

3.7　类的关联与依赖关系

在面向对象编程中，依赖与关联是类与类之间的常见关系。下面，只是讨论 Java 类之间关联与依赖关系的常见表现形式，并不讨论它们严格的定义。

1. 关联关系的常见表现形式

若在一个类 A 中声明另一个类 B 的对象作为 A 的成员变量，则 A 与 B 之间存在关联关系，称 A 关联于 B。如：

```
class A{
    private B b;            //声明B类的对象b作为A的成员变量
    ...
}
class B{
    ...
}
```

2. 依赖关系的常见表现形式

若一个类 A 的某个方法的形参或返回值是其他类 B 的对象，则 A 和 B 之间存在依赖关系，称 A 依赖于 B。如：

```
class A{
public void f(B b){
    ...
    }
}
class B{
    ...
}
```

【例题 3_18】学生选课问题：定义课程类 Course，学生类 Student，学生要选课，则在学生类和课程类间存在关联和依赖关系。

```
package myproject.ch3_18;
public class Course {                       //课程类
    private String couseno;                 //课程号
    private String coursename;              //课程名
    public Course(String couseno, String coursename) {//构造方法
        this.couseno = couseno;
        this.coursename = coursename;
    }
    public String getCouseno() {            //获得课程号的方法
        return couseno;
    }
    public String getCoursename() {         //获得课程名的方法
        return coursename;
    }
}
package myproject.ch3_18;
public class Student {                      //学生类
    private String stuno;                   //学号
    private String stuname;                 //姓名
```

```java
        private Course course;                    //以课程类对象作为学生类的成员变量
        public Student(String stuno, String stuname) {//构造方法
            this.stuno = stuno;
            this.stuname = stuname;
        }
        public String getStuno() {                //获得学号的方法
            return stuno;
        }
        public String getStuname() {              //获得姓名的方法
            return stuname;
        }
        public void selectCourse(Course course){//给所选课程赋值的方法
        //以课程类对象作为方法形参
            this.course=course;
        }
        public Course getSelectedCourse(){        //获得所选课程的方法
            return course;
        }
    }
    package myproject.ch3_18;
    public class Ch3_18{
        public static void main(String[] args) {
            Course course=new Course("008","Java 程序设计");  //创建课程类对象
            Student stu=new Student("20150003","李胜利");      //创建学生类对象
            stu.selectCourse(course);         //调用选课方法，以课程类对象 course 作为实参
            Course c=stu.getSelectedCourse();//获得所选课程
            System.out.println("学生信息: ");
            System.out.println("学号: "+stu.getStuno()+"姓名: "+stu.getStuname());
            System.out.println("该生所选课程:");
            System.out.println("课程编号: "+c.getCouseno()+", 课程名称: "+c.get
    Coursename());
        }
    }
```

程序运行结果如图 3.21 所示。

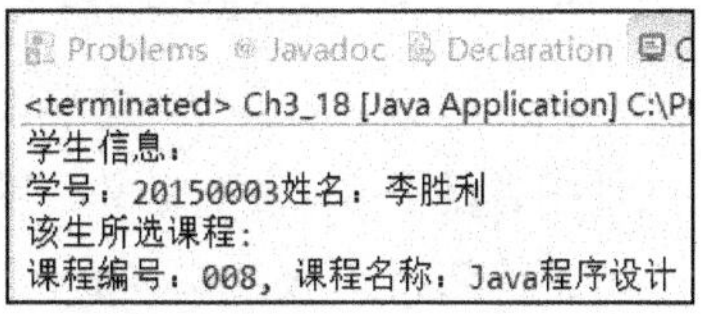

分析：学生类 Student 中定义了 private Course course，即以课程类 Course 的对象作为其成员变量，因此，Student 类关联于 Course 类。学生类 Student 的 public void selectCourse(Course course)方法形参是课程类 Course 的对象，因此，Student 类依赖于 Course 类。

图 3.21 例题 3_18 运行结果

3.8 package 与 import 关键字

3.8.1 Java 中的包

就像 Windows 操作系统中的文件夹可以分门别类地管理一组文件或子文件夹一样，Java 中，包用来方便管理一组相关的类和接口，有系统包和用户自定义包两类。在物理存储上，包对应着一个文件夹。

包的作用主要有：
- 划分类名空间：包是一种名字空间机制，同一包中的类（包括接口）名不能重名，不同包中的类名可以重名，解决了类和接口的命名冲突问题。
- 控制类之间的访问：包是一个访问域，对包中的类有保护作用。例如，若类被 public 修饰，则该类不仅可供同一包中的类访问，也可以被其他包中的类访问。若类声明无修饰符，则该类仅供同一包中的类访问等。
- 分门别类地组织 Java 应用中的一组功能相关的类或接口。

Java 系统包主要有 JDK 提供的系统类和接口包以及其他扩展类和接口包。JDK 提供的常用系统类和接口包主要有：
- java.lang：包含一些 Java 的基础核心类如所有 Java 类的共同父类 Object；简单数据类型的包装类 Integer、Long 、Boolean、Character、Float 和 Double 等；封装最常用的数学方法如平方根、绝对值、正余弦的 Math 类；用于字符串操作的 String，StringBuffer 类等，该包在使用时是由系统自动导入的，因此，不需要显式用 import 关键字进行导入。
- java.awt (abstract window toolkits)其中的类和接口被称为"重量级组件"，是早期 Java 用来构建 GUI 可视程序的类与接口集合。
- javax.swing：一个扩展类包，其中的类和接口被称为"轻量级组件"，是继 awt 之后推出的构建 Java GUI 可视程序的类和接口集合。
- java.util：Java 实用工具和接口的集合类，如时间、日历类、字符串解析类、集合框架类等。
- java.io：提供了 Java 输入、输出流类和接口的集合。
- java.sql：提供了 Java 数据库操作的类和接口的集合。
- java.net：提供了网络编程相关的操作和接口，如 URL、 InetAddress、Socket、ServerSocket、DatagramSocket、DatagramPacket 等。
- java.applet 包：是用来实现在 Internet 浏览器中运行 Java Applet 的工具类，其中包含一个非常有用的类 java.applet. Applet。

3.8.2 用 package 关键字自定义包

Java 运用 package 关键字定义包，且 package 语句必须作为 Java 源文件的第一条语句。若源文件中省略了 package 定义，则该源文件被默认为包含在默认类包下。

其语法格式为：package 包名；

包名的定义一般以层次结构出现如：package 标识符 1.标识符 2.标识符 3；而且包名中的标识符应该全部小写如 myproject.ch3_18。

3.8.3 用 import 关键字导入包

为了能够使用某一个包的类和接口，我们需要在 Java 程序中使用 import 导入该包。在 Java 源文件中 import 语句应位于 package 语句之后，所有类的定义之前。

【例题 3_19】在 com.mysoft.pack1 包中定义类 Calculate，该类中拥有计算两个 double 型数据的和、差、积、商的方法，在另一个包 com.mysoft.pack2 中定义类 packageTest，在其主方法中创建 Calculate 对象并测试其各个方法。

1．在 JDK 环境下

首先，在记事本中编写如下程序：
```java
// Calculate.java
package com.mysoft.pack1;
public class Calculate {
    public double add(double x, double y) { return x+y; }
    public double subtract(double x, double y) { return x-y; }
    public double multiply(double x, double y) { return x*y; }
    public double divide(double x, double y) { return x/y; }
}
// packageTest.java
package com.mysoft.pack2;
import com.mysoft.pack1.Calculate;
public class PackageTest {
    public static void main(String[] args) {
        double a=20.0,b=10.0;
        Calculate cal=new Calculate();
        System.out.println("a+b="+cal.add(a,b));
        System.out.println("a-b="+cal.subtract(a,b));
        System.out.println("a*b="+cal.multiply(a,b));
        System.out.println("a/b="+cal.divide(a,b));
    }
}
```

先将这两个源程序保存在 E 盘根目录下。打开命令提示符窗口：将当前路径转到 E 盘根目录，输入 javac –d . Calculate.java 以及 javac –d . PackageTest.java，若编译无误，则会自动在 E 盘根目录下形成与 Calculate、PackageTest 所在包对应的文件夹结构（见图 3.22）。这是因为，javac 命令的 –d 参数就表示生成与源程序包名对应的文件夹结构，"."表示当前目录。注意：在 javac 与–d，–d 与"."，"."与源文件名之间都分别有且仅有一个空格。

  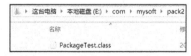

图 3.22 用 javac –d .命令形成的包结构

继续在命令提示符下输入 java com.mysoft.pack2.PackageTest 并回车执行程序得到图 3.23 所示的测试结果。

图 3.23 JDK 环境下对 Java 包的测试结果

2．在 Eclipse 环境下

在 Eclipse 环境下新建带包结构的程序相对很容易。

以上面例题为例，首先在 Eclipse 的 Package Explorer 中当前项目 Ch3 的 src 上右击，选择"new"→"Package"命令，在弹出的对话框（见图 3.24）中输入包名 com.mysoft.pack1，单击"Finish"按钮完成包的新建。

在 com.mysoft.pack1 上右击，选择"new"→"Class"命令，在弹出的对话框（见图 3.25 所示）中输入类名 Calculate 单击"Finish"按钮完成类的新建。此时，在 Eclipse 编辑区会生成该类简单框架代码，在其中继续编辑完整的 Calculate 源程序即可。采用同样的办法，新建 com.mysoft.pack2 包，并在该包中新建 PackageTest.java，形成图 3.26 所示的项目结构。

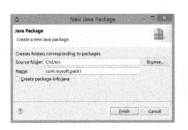
图 3.24　新建包

图 3.25　新建类

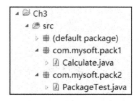
图 3.26　项目结构

打开 Eclipse 当前工作空间文件夹（E:\Java Project），可以在 Ch3 的 src 文件夹下观察到与包结构对应的文件夹结构。在 pack1、pack2 中分别存放着 Calculate.java 及 PackageTest.java 文件；打开 Ch3 的 bin 文件夹，也可以观察到与包结构对应的文件夹结构，但不同的是，bin 文件夹中的 pack1、pack2 中分别存放着字节码文件 Calculate.class 及 PackageTest.class（见图 3.27）。

图 3.27　工作空间下对应于包的目录结构

3.9　JavaBean

在 Java 应用开发中，尤其是 Java EE 应用开发中，一种特殊的类—JavaBean 的使用非常频繁。

JavaBean 有广义和狭义的两种形式。广义的 JavaBean 是指普通的 Java 类；狭义的 JavaBean 是指严格按照以下规范编写的 Java 类。

- 需要定义一个无参的构造方法；
- 需要定义若干成员变量，且访问权限一般为 private；
- 对于每个成员变量，都需要一对 getter( )、setter( )方法。成员变量的读写，通过 getter( )和 setter( )方法进行。

狭义 JavaBean 中的各个成员变量一般被声明为 private 的，外部其他类只能通过其各自的通过 getter( )和 setter( )方法来访问这些成员变量，很好地实现了信息隐藏，体现了类的封装性。

【例题 3_20】 定义一个封装学生属性（学号、姓名、性别、年龄、成绩、婚否）的 JavaBean，并通过 getter()、setter()方法读写这些属性。

```java
package myproject.ch3_20;
public class Student{                //一个 JavaBean 类 Student
    private String stuno;            //学号
    private String stuname;          //姓名
    private String stusex;           //性别
    private int age;                 //年龄
    private double score[];          //成绩数组
    private boolean married;         //婚否
    public Student(){}               //无参构造方法
    /*以下是各个成员变量的 getter(),setter()方法*/
    public String getStuno() {
      return stuno;
    }
    public void setStuno(String stuno) {
      this.stuno = stuno;
    }
    public String getStuname() {
      return stuname;
    }
    public void setStuname(String stuname) {
      this.stuname = stuname;
    }
    public String getStusex() {
      return stusex;
    }
    public void setStusex(String stusex) {
      this.stusex = stusex;
    }
    public int getAge() {
      return age;
    }
    public void setAge(int age) {
      this.age = age;
    }
    public double[] getScore() {
      return score;
    }
    public void setScore(double[] score) {
      this.score = score;
    }
    public boolean isMarried() {
      return married;
    }
    public void setMarried(boolean married) {
      this.married = married;
    }
}
package myproject.ch3_20;
public class Ch3_19 {
    public static void main(String[] args) {
```

```java
        Student zs=new Student();              //创建 Student 对象
        /*以下是调用各个 setter()方法给各个 JavaBean 成员变量赋值*/
        zs.setStuno("20150009");
        zs.setStuname("张三");
        zs.setStusex("男");
        zs.setAge(19);
        double score[]={88.5,79.5,94,88,95};
        zs.setScore(score);
        zs.setMarried(false);
        /*以下是调用各个 getter()方法获取各个 JavaBean 成员变量值并输出*/
        System.out.println("学号: "+zs.getStuno());
        System.out.println("姓名: "+zs.getStuname());
        System.out.println("性别: "+zs.getStusex());
        double score1[]=zs.getScore();         //返回值是一个 double 数组
        System.out.print("成绩: ");
        for(double s:score1 )                  //遍历数组
            System.out.printf("%4.1f ",s);
        System.out.println();
        System.out.println("婚否: "+zs.isMarried());
    }
}
```

程序运行结果如图 3.28 所示。

在 Eclipse 环境下，给一个 JavaBean 的各个成员变量增加 getter()、setter()方法非常方便，只需在产生这些方法的位置右击，选择 "Source" → "Generate Getters and Setters" 命令，在弹出的对话框（见图 3.29）中勾选需要生成 getter 和 setter 方法的属性，单击 "OK" 按钮即可。

图 3.28　操作 JavaBean 的结果　　　　图 3.29　给 JavaBean 生成 getter()、setter()方法

在 Student 这个 JavaBean 的定义中，需要注意的是对于基类型为 double 型的数组类型属性 score[]，其 getter 方法的返回值是 double 型数组，其 setter 方法的形参也是 double 型数组；对于布尔型属性 married，用 isMarried()方法表示其 "getter()" 方法。

3.10　jar 命令的用法

在 JDK 安装路径的\bin 目录下有一个 jar.exe 文件，是 JDK 提供的常用命令之一。利用该命令，可以用来对大量的.class 文件进行压缩，存为以.jar 为扩展名的文件，主要用途是扩展系统类库和打包发布应用程序。

在命令提示符下输入 jar 命令，可以观察到该命令的各种参数及其作用，如图 3.30 所示。

3.10.1 将应用程序打包为 jar 文件

【例题 3_20】将一个简单的窗体程序打包发布为可执行的 jar 文件。

（1）编写源文件如下：

```java
import javax.swing.JFrame;
class MyFrame extends JFrame{
    public MyFrame(){
        this.setTitle("被打包发布的窗口程序");
//窗体标题栏标题
        this.setSize(500,300);    //窗体大小
        this.setVisible(true);    //设置窗体可见
    }
}
public class JarTest1{
    public static void main(String[] args) {
        new MyFrame();           //创建窗体对象
    }
}
```

图 3.30 jar 命令参数

（2）保存源文件并进行编译，假定将源文件保存在 E:\下并进行编译。

（3）在记事本中编写一个清单文件，清单文件内容为

```
Manifest-Version: 1.0
Main-Class: JarTest1
Created-By: 1.7.0_79
```

注意：Manifest-Version:与 1.0，Main-Class:与 JarTest1，Created-By:与 1.7.0_79 之间分别必须有且只有一个空格，第一行的行前不可以有空行，行与行之间不能有空行。其中 Manifest-Version 表明了清单文件的版本；Main-Class 指明了主类；Created-By 表示生成该 jar 文件的 JDK 版本号。

将该清单文件也保存在 E:\下，命名为 manifest1.mf。

（4）在命令提示符中，将路径转到 E:\，在命令提示符下输入：jar cvfm window.jar manifest1.mf JarTest1.class MyFrame.class 并回车（见图 3.31）。

此时，在 E:\下生成一个 window.jar，双击该文件或在命令提示符下输入：java –jar window.jar 并回车，可以运行程序（见图 3.32）。

图 3.31 将程序打包成可执行的 jar 包

图 3.32 可执行 jar 包执行结果

说明：jar 命令中参数数 c 表示要创建一个 jar 文件，v 表示生成详细的报表，并输出至标准设备，f 表示指定 jar 文件的文件名，m 表示指定 manifest.mf 文件。

jar 文件本质上是一个压缩文件,可以用压缩软件打开 jar 文件(见图 3.33),可以看出,该 jar 文件中包含了源程序生成的字节码文件。在命令提示符下键入:jar –tf window.jar 并回车也可以用查看 jar 文件的内容(见图 3.34)。

图 3.33 jar 文件中包含的字节码文件　　　　图 3.34 用 jar 命令查看 jar 文件内容

注意:若在压缩软件中设置了关联.jar 文件,则所有的 jar 文件会显示为本压缩软件图标。

3.10.2 生成 jar 文件扩展类库

可以使用 jar.exe 把自己编写的一些实用类的字节码文件压缩成一个 jar 文件,然后其存放到 JDK 安装路径的 jre\lib\ext 文件夹中。这样,其他程序就可以使用这个 jar 文件中的类来创建对象了。

【例题 3_21】定义自己的实用类 jar 文件并将其添加到 JDK 的 jre\lib\ext 目录下。

(1)编写源文件如下:
```
//A.java
class A{
  public void outPut(){
     System.out.println("我的实用类 1");
    }
}
//B.java
class B{
  public void outPut(){
     System.out.println("我的实用类 2");
    }
}
```

(2)保存源文件并进行编译,假定将源文件保存在 E:\下并进行编译。

(3)编写清单文件如下:
```
Manifest-Version: 1.0
Class: A B
Created-By: 1.7.0_79
```
将该清单文件也保存在 E:\下,命名为 manifest2.mf。

(4)在命令提示符下键入:jar cvfm use.jar manifest2.mf A.class B.class 并回车(如图 3.34 所示)。则在 E:\下生成一个 use.jar。

(5)将 use.jar 复制到 jre\lib\ext 下。

(6)编写一个测试程序:
```
public class JarTest2 {
    public static void main(String[] args) {
     A a=new A();
     a.outPut();
     B b=new B();
     b.outPut();
     }
}
```

将该程序保存到与 A.java、B.java 不同的路径下，如 F:\ 下并编译，此时，可能发生找不到 A，B 的编译错误，可以在环境变量 classpath 中增加 "C:\Program Files\Java\jre7\lib\ext\use.jar;"。

再次编译、执行 JarTest2 即可，如图 3.35 所示。

图 3.35　使用自定义的扩展 jar 文件

3.10.3　Eclipse 环境下的文件打包

【例题 3_22】在 Eclipse 环境下生成 .jar 文件。

（1）在 Eclipse 中编写程序如下：

```java
import javax.swing.JFrame;
class MyFrame extends JFrame{
    public MyFrame(){
        this.setTitle("被打包发布的窗口程序");      //窗体标题栏标题
        this.setSize(500,300);                    //窗体大小
        this.setVisible(true);                    //设置窗体可见
    }
}
public class JarTest1{
    public static void main(String[] args) {
        new MyFrame();                            //创建窗体对象
    }
}
```

注意：先在 Eclipse 环境下运行一次该程序，否则会导致在打包过程中无法找到该程序主类的情况（即在图 3.37 的 Launch configutation 下拉列表中不显示 JarTest1-Ch3）。

（2）在 Eclipse 的 Package Explorer 中的 Ch3 项目下的 JarTest1 上右击，选择"Export..."命令，在弹出的"Export"对话框（见图 3.36）中选择"Java"→"Runnable JAR file"命令，单击"Next"按钮，在弹出的对话框（见图 3.37）的"Launch configuration"下拉列表框中选择"JarTest1-Ch3"，单击"Export destination"右侧的"Browse"按钮，在保存对话框（见图 3.38）中选择保存文件，在"Runnable JAR File Export"对话框中单击"Finish"按钮即可。

图 3.36　导出可执行 jar 文件

图 3.37　选择主类及可执行 jar 文件存放路径

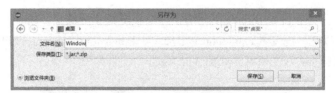

图 3.38　保存 jar 文件

若在图 3.36 所示的 "Export" 对话框中选择 "JAR File"，则可以将选择导出的文件导出为一个普通的 jar 文件。

小　　结

本章首先介绍了面向对象编程中类与对象的概念；介绍了面向对象的抽象、封装、继承以及多态等特征；重点介绍了 Java 类、成员变量、成员方法的定义格式；构造方法的作用及特点；对象的声明及创建方法；方法重载的含义及其用法；this 关键字的用法；静态成员与非静态成员的区别，静态代码块的使用方法；方法的参数传递；类之间的关联与依赖关系；包的定义及其使用方法；JavaBean 的定义及其使用；jar 命令的用法等内容。

习　题　3

一、选择题

1. 构造方法何时被调用？（　　）
   A. 类定义时。　　　　　　　　　B. 使用对象的属性时。
   C. 使用对象的方法时。　　　　　D. 对象被创建时。
2. 如果希望方法直接通过类名称访问，在定义时要使用的修饰符是（　　）。
   A. static　　　B. final　　　C. abstract　　　D. this
3. 函数重载是指（　　）。
   A. 两个或两个以上的方法取相同的方法名，但形参的个数或类型不同
   B. 两个以上的方法取相同的名字和具有相同的参数个数，但形参的类型可以不同
   C. 两个以上的方法名字不同，但形参的个数或类型相同
   D. 两个以上的方法取相同的方法名，并且方法的返回类型相同。
4. 下面 Java 系统类包在使用时不需要用 import 显式导入的是（　　）
   A. java.io　　　B. java.net　　　C. java.util　　　D. java.lang

二、上机实践题

1. 定义人员类 Person。要求：

（1）定义私有的成员变量 String name（姓名），String sex（性别），age（年龄），address（户籍所在地）；

（2）定义一个不含参的构造方法以及一个以各个成员变量为形参的含参构造方法；

（3）给每个成员变量添加 setter() 与 getter() 方法，用于给各个成员变量赋值和取值；

（4）定义成员方法 showInfo()，用于输出人员信息或返回人员信息；

（5）在主方法 main()中创建 Person 类对象 zhangsan，lisi，并测试 setter()、getter()、showInfo() 方法。

2. 设计一个复数类。要求：

（1）复数类包括实部 realPart 和虚部 imagePart 两个成员变量。

（2）定义一个无参构造方法，一个以 realPart、imagePart 为参数的含参构造方法。

（3）定义一个用于输出复数的方法 "public String outputComplexNumber();"。

（4）定义一个成员方法 ComplexNumber add(ComplexNumber anotherComplexNumber)，用于求解两个复数之和。

（5）在主方法 main()中创建两个 ComplexNumber 的对象 complexnumber1、complexnumber2，调用 add 方法对 complexnumber1、complexnumber2 求和，并通过 outputComplexNumber()方法分别输出 complexnumber1，complexnumber2 以及求和结果。(如 complexnumber1 =5-9i，complexnumber2=7+13i，complexnumber1 + complexnumber2=12+4i）。

3. 学生选课问题：

（1）定义一个课程类 Course，包含成员变量课程编号 String couseno、课程名称 String coursename；一个含参构造方法 public Course(String couseno, String coursename); 两个成员方法 public String getCouseno()，public String getCoursename()，用于获取课程编号与课程名称。

（2）定义一个学生类 Student，包括成员变量学号 stuno、姓名 stuname、所选课程 course；一个含参构造方法 public Student(String stuno, String stuname)；成员方法 public String getStuno()，public String getStuname()，表示选课的方法 public void selectCourse(Course course)，返回所选课程的方法 public Course getSelectedCourse()。

（3）创建 Course 类对象 course，Student 类对象 stu，stu 调用 selectCourse 方法进行选课，调用 getSelectedCourse()方法得到所选课程信息，输出学生及其所选课程信息。

4. 上机运行下面程序，观察结果，总结 Java 参数传递机制。

```
package xiti3;
class Point{
    int x;
    int y;
    public Point(int x, int y) {
        this.x = x;
        this.y = y;
    }
}
public class ParamTransferDemo {
    public  void test(int a,int[] arr,String str1,String str2,Point p){
        a=0;
        arr[0]=1000;
        str1="def";
        str2=new String("DEF");
        p.x=100;
        p.y=200;
    }
    public static void main(String[] args) {
 ParamTransferDemo demo=new ParamTransferDemo ();
```

```java
            int n = 10;
            int[] a = {10,20,30};
            String s1="abc";
            String s2=new String("ABC");
            Point point=new Point(1,2);
            System.out.println("参数传递前: ");
            System.out.println("n="+n);
            System.out.println("a[0]="+a[0]);
            System.out.println("s1="+s1);
            System.out.println("s2="+s2);
            System.out.println("x="+point.x+",y="+point.y);
            demo.test(n,a,s1,s2,point);
            System.out.println("参数传递后: ");
            System.out.println("n="+n);
            System.out.println("a[0]="+a[0]);
            System.out.println("s1="+s1);
            System.out.println("s2="+s2);
            System.out.println("x="+point.x+",y="+point.y);
        }
    }
```

5. 上机练习 JDK 中 jar 命令的用法，在 Eclipse 环境练习 jar 包的生成方法。

第 4 章　Java 继承与多态

【本章内容提要】
- Java 继承；
- 权限修饰符；
- 子类继承性；
- 变量覆盖与方法重写；
- super 的用法；
- 对象的上下转型；
- instanceof 关键字；
- 多态；
- final 修饰符。

4.1　Java 继 承

继承机制是面向对象编程中实现代码复用的重要机制之一，是对现实世界"子承父业"在编程思想中的应用。

继承是在已有类的基础上构建新的类。已有的类称为超类、父类或基类，产生的新类称为子类或派生类。父类派生出子类，或子类继承父类，以达到让子类不必一切从头做起，在子类中只新定义自己独有的属性和方法，与父类相同的属性和方法从父类中继承得到即可。

在 Java 中，表示继承的关键字是 extends。Java 只支持单继承，即一个子类只能有一个父类。但一个父类可以派生出多个子类。

【例题 4_1】　简单的继承测试。

```
package myproject.ch4_1;
public class Father{
    int a=10;
    public int getA() {
      return a;
    }
}
package myproject.ch4_1;
public  class Son extends Father{      //Son 继承 Father
    int b;                              //定义新属性
    public int getB() {                 //定义新方法
```

```
        return b;
    }
}
package myproject.ch4_1;
public class Ch4_1 {
    public static void main(String[] args) {
        Father father=new Father();    //创建父类对象
        System.out.println("父类的 a="+father.a);
        System.out.println("父类的 a="+father.getA());
        Son son=new Son();             //创建子类对象
        son.a=30;                      //修改继承来的 a 值
        son.b=20;                      //给自己的新属性赋值
        System.out.println("子类的 a="+son.a);
        System.out.println("子类的 a="+son.getA());
        System.out.println("子类的 b="+son.b);
        System.out.println("子类的 b="+son.getB());
    }
}
```

程序运行结果如图 4.1 所示。

分析：子类 Son 在继承了父类 Father 的成员变量 a 和成员方法 getA() 的基础上，定义了新的成员变量 b 和成员方法 getB()。因此，Son 具有 a、b 两个成员变量和 getA()、getB()两个成员方法，体现了"继承中有发展"。同时，对于继承来父类中的成员变量和方法，如能够直接满足子类的需要，则无须在子类中重复定义它们，体现了代码重用的思想。

图 4.1　类继承测试结果

4.2　权限修饰符

Java 的权限修饰符有 public 、protected 、缺省（即什么都不写）、 private，可以用来修饰类的成员变量以及成员方法，用来控制这些成员变量或成员方法的访问权限。

只可以用 public 和缺省权限来修饰类(内部类除外)，不能用 protected 和 private 修饰类。public 类可以被任意其他类访问，而缺省权限修饰的类只可以被同一个包中的其他类访问。

访问限制修饰符按访问权限从高到低的排列顺序依次为 public、protected、缺省、private。

【例题 4_2】权限修饰符测试。

```
package pack1;
public class AccessTest1 {
    /*定义被不同权限修饰符修饰的成员变量*/
    public String s1="public 修饰的变量";
    protected String s2="protected 修饰的变量";
    String s3="缺省权限修饰的变量";
    private String s4="private 修饰的变量";
    /*定义被不同权限修饰符修饰的成员方法*/
    public void f1(){
        System.out.println("public 修饰的方法");
    }
    protected void f2(){
        System.out.println("proteced 修饰的方法");
```

```java
        }
        void f3(){
            System.out.println("缺省权限修饰的方法");
        }
        private void f4(){
            System.out.println("private修饰的方法");
        }
        public static void main(String[] args){
            AccessTest1 test=new AccessTest1();
            System.out.println("在同一个类中，可以访问下列修饰符修饰的方法、变量: " );
            System.out.println(test.s1);
            System.out.println(test.s2);
            System.out.println(test.s3);
            System.out.println(test.s4);
            test.f1();
            test.f2();
            test.f3();
            test.f4();
        }
}
```

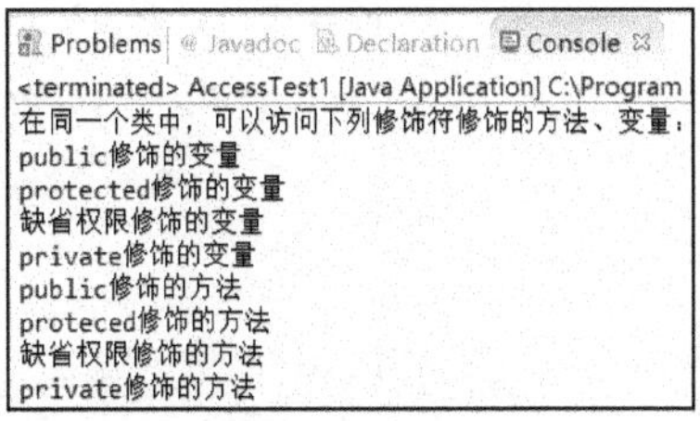

程序运行结果如图 4.2 所示。

图 4.2 例题 4_2 运行结果

说明：在一个类中，任何访问权限修饰符修饰的变量和方法都可以被访问。

继续编写程序如下：

```java
package pack1;
public class AccessTest2 {
    public static void main(String[] args) {
        AccessTest1 test=new AccessTest1();      //创建 AccessTest1 对象
        System.out.println(test.s1);
        System.out.println(test.s2);
        System.out.println(test.s3);
        //System.out.println(test.s4);            //s4 不能被访问
        test.f1();
        test.f2();
        test.f3();
        //test.f4();                              //f4() 不能被访问
    }
}
```

程序运行结果如图 4.3 所示。

图 4.3 同一个包中类间的访问权限

说明：AccessTest1 与 AccessTest2 类处在同一个包 pack1 中，在 AccessTest2 中可以访问到 AccessTest1 中除了 private 修饰的变量和方法。

继续编写程序如下：

```java
package pack2;
import pack1.AccessTest1;
public class AccessTest3 {
    public static void main(String[] args) {
```

```
        AccessTest1 test=new AccessTest1();  //创建pack1包中AccessTest1类的对象
        System.out.println(test.s1);
        //System.out.println(test.s2);        //s2不能被访问
        //System.out.println(test.s3);        //s3不能被访问
        //System.out.println(test.s4);        //s4不能被访问
        test.f1();
        //test.f2();                          //f2()不能被访问
        //test.f3();                          //f3()不能被访问
        //test.f4();                          //f4()不能被访问
    }
}
```
程序运行结果如图4.4所示。

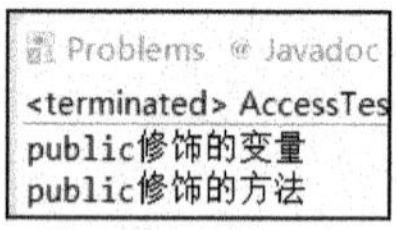

说明：AccessTest3 和 AccessTest1 类分别处在不同包 pack2 与 pack1 中，此时，AccessTest3 只能访问到 AccessTest1 中的 public 成员变量和方法。

图4.4 不同包中类间的访问权限

4.3 子类继承性

子类可以从父类中继承哪些成员变量和成员方法呢？这取决于子类与父类是否处在同一个包中以及父类中各个成员变量和成员方法的权限修饰。

1. 子类和父类在同一包中的继承性

如果子类和父类在同一个包中，那么，子类自然地继承了其父类中不是 private 的成员变量作为自己的成员变量，并且也自然地继承了父类中不是 private 的方法作为自己的方法，继承的成员变量或方法的访问权限保持不变。

【例题 4_3】子类继承性测试 1。

```
package pack1;
public class AccessTest4 extends AccessTest1 {//继承AccessTest1
    public static void main(String[] args) {
        AccessTest4 son=new AccessTest4();    //给子类AccessTest4创建对象son
        System.out.println(son.s1);
        System.out.println(son.s2);
        System.out.println(son.s3);
        //System.out.println(son.s4);         //s4不能被继承和访问
        son.f1();
        son.f2();
        son.f3();
        //son.f4();//f4()不能被继承和访问
    }
}
```
程序运行结果如图4.5所示。

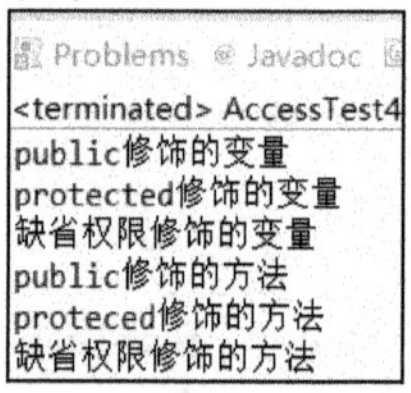

图4.5 同一个包中子类的继承性

说明：AccessTest4 与 AccessTest1 类处于同一包 pack1 中并继承了 AccessTest1，AccessTest4 可以从父类 AccessTest1 中继承来非 private 修饰的变量和方法。

2. 子类和父类不在同一包中的继承性

如果子类和父类不在同一个包中，那么，子类继承了父类的 public、protected 修饰的成员变量作为子类的成员变量，并且继承了父类的 public、protected 修饰的方法为子类的方法，继承的成员或方法的访问权限保持不变。

【例题 4_4】子类继承性测试 2。
```
package pack2;
import pack1.AccessTest1;
public class AccessTest5 extends AccessTest1 {    //继承pack1包中的AccessTest1
    public static void main(String[] args) {
        AccessTest5 son=new AccessTest5();         //给子类AccessTest5创建对象
        System.out.println(son.s1);
        System.out.println(son.s2);
        //System.out.println(son.s3);              //s3 不能被继承和访问
        //System.out.println(son.s4);              //s4 不能被继承和访问
        son.f1();
        son.f2();
        //son.f3();                                //f3() 不能被继承和访问
        //son.f4();                                //f4() 不能被继承和访问
    }
}
```

程序运行结果如图 4.6 所示。

说明：AccessTest5 与 AccessTest1 类分别处于不同包 pack2 和 pack1 中，并且 AccessTest5 继承了 AccessTest1。此时，AccessTest5 可以继承和访问 AccessTest1 中被 public、protected 修饰的变量和方法。

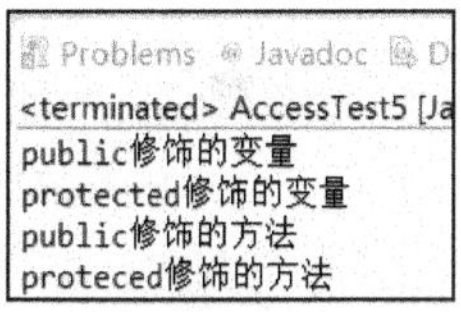

图 4.6 不同包中子类的继承性

虽然，父类中 private 修饰的成员变量不能被子类继承和直接访问，但可以被子类间接访问。

【例题 4_5】 子类间接访问父类的私有成员变量。
```
package pack1;
public class Person{                           //父类
    private String name ;                      //private成员变量
    private int age ;                          //private成员变量
    public String getName() {                  //public方法
        return name;
    }
    public void setName(String name) {         //public方法
        this.name = name;
    }
    public int getAge() {                      //public方法
        return age;
    }
    public void setAge(int age) {              //public方法
        this.age = age;
    }
}
package pack1;
public class Student extends Person{           //子类
    private String stuno ;                     //private成员变量
    public String getStuno() {                 //public方法
        return stuno;
```

```
        }
        public void setStuno(String stuno) {       //public 方法
            this.stuno = stuno;
        }
    }
    package pack1;
    public class AccessTest6{
        public static void main(String arsg[]){
            Student zhangsan=new Student()   ;
            //zhangsan.name="张三";                    //不能继承和直接访问 name
            //zhangsan.age=18;                        //不能继承和直接访问 age
            zhangsan.setName("张三") ;                //通过调用 public 方法给 name 赋值
            zhangsan.setAge(19) ;                    //通过调用 public 方法给 age 赋值
            zhangsan.setStuno("20150001"); ;
            System.out.println("姓名: "+zhangsan.getName()+ " 年龄: "+
              zhangsan.getAge() + " 学号: "+zhangsan.getStuno()) ;
            //通过调用 public 方法获得 name、age 的值
        }
    }
```

程序运行结果:"姓名：张三 年龄：19 学号：20150001"。

分析：父类 Person 中的 name 与 age 成员变量都是私有的，不能被子类 Student 继承和直接访问，但可以通过 Person 类中操作 name 与 age 成员变量的 public 修饰的 getName()和 setName()方法间接访问到。

4.4 变量覆盖与方法重写

1. 变量覆盖

子类可以定义与继承来的父类变量同名的变量，此时，子类中的同名变量会覆盖（隐藏）父类同名变量。

【例题 4_6】变量覆盖测试。

```
package myproject.ch4_6;
public class Father {
    int a=100;
}
package myproject.ch4_6;
public class Son extends Father {
    int a=200;                //定义与父类中同名变量 a
    public static void main(String arsg[]){
        Son son=new Son();
        System.out.println(son.a);
    }
}
```

程序运行结果为：200

若将子类 Son 中 int a=200;语句注释掉，再次运行程序将得到结果 100，即从父类 Father 中继承来的 a 值。

2. 方法重写

当子类从父类继承了某方法，但该方法不能满足自己的要求时，可以重写(覆盖)父类的方法。

【例题 4_7】在一个教务管理系统中，有管理员、教师和学生三种不同权限类型的用户。定义一个父类 User 类，定义其子类 Administrator、Teacher、Student 类，测试方法覆盖的用法。

```java
package myproject.ch4_7;
public class User{
  String name;
  String type;
  public void showInfo(){
      System.out.println("用户名: "+name+" 类型: "+type);
  }
  public void use(){
      System.out.println("我可以使用本系统！");
  }
}
package myproject.ch4_7;
public class Administrator extends User{
    @Override
    public void use() {//重写use()方法
      System.out.println("我可以使用本系统的所有功能！");
    }
}
package myproject.ch4_7;
public class Teacher extends User{
    @Override
    public void use() {//重写use()方法
      System.out.println("我可以使用本系统的成绩录入、修改和查看功能！");
    }
}
package myproject.ch4_7;
public class Student extends User{
    @Override
    public void use() {//重写use()方法
       System.out.println("我可以使用本系统的成绩查看功能！");
    }
}
package myproject.ch4_7;
public class Ch4_7 {
    public static void main(String[] args) {
      Administrator a=new Administrator();
      a.name="管理员";
      a.type="超级用户";
      a.showInfo();
      a.use();
      Teacher t=new Teacher();
      t.name="教师1";
      t.type="教师";
      t.showInfo();
      t.use();
```

```
        Student s=new Student();
        s.name="学生1";
        s.type="学生";
        s.showInfo();
        s.use();
    }
}
```

程序运行结果如图 4.7 所示。

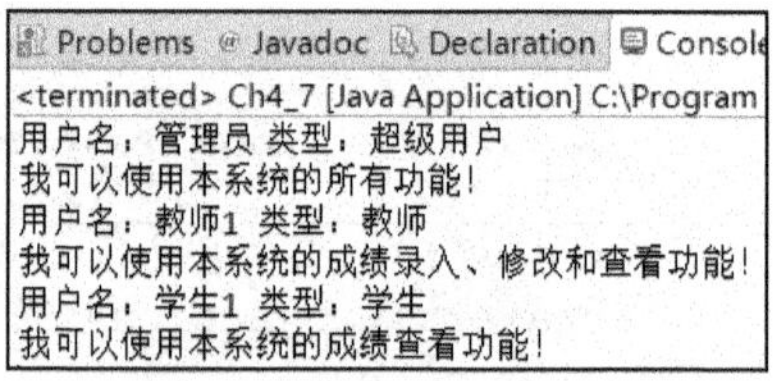

图 4.7 方法重写测试结果

分析：Adminstrator、Teacher、Student 类从 User 类中继承了 name 和 type 成员变量以及 showInfo() 和 use() 方法，但不同权限的用户对系统操作的权限不同，因此，要在各个子类中重写 use() 方法。

在方法重写过程中需要注意：

- 子类重写的方法必须与父类被重写的方法具有相同的方法名称、参数列表和相同或相容的返回值类型，否则不构成重写。在 JDK1.5 之后，允许重写方法的类型可以是父类方法的类型的子类型，即不必完全一致，比如父类方法的类型是 People，重写方法的类型可以是 Student（假设 Student 是 People 的子类）。
- 子类重写父类方法时，不能降低被重写方法的访问权限。
- 父类的静态方法不能被子类重写为非静态的方法。同样，父类中的实例方法也不能被子类重写为静态方法。
- 方法重写只针对实例方法，父类中的静态方法，子类只能隐藏、重载和继承。
- 父类中能被子类继承的实例方法，才会在子类中被重写。
- 子类重写的方法不能比父类中被重写的方法声明抛出更多的异常。

4.5 super 的用法

super 代表当前对象的直接父类对象，是当前对象的直接父类对象的引用。

1. super 语句出现在子类构造方法中

- 当用子类的构造方法创建一个子类对象时，子类的构造方法总是先调用其父类的某个构造方法。若没有给父类定义构造方法，则子类调用父类的默认构造方法，即在子类构造方法的第一句默认存在 super();语句。
- 在类的构造方法中，可显式地通过 super 语句调用其父类的构造方法。super 语句一定要是子类构造方法的第一条语句。
- 当给父类定义了构造方法（无参或含参构造方法）时，父类不再提供默认构造方法。此时要注意，当给父类定义了含参构造方法时，应同时给父类定义一个无参构造方法，以避免在子类中没有用 super 语句显式地调用父类构造方法时出现错误，这是因为，当子类不用 super 语句显式地调用父类构造方法时，子类就会默认地调用父类无参构造方法。

【例题 4_8】子类调用父类无参构造方法。

```
package myproject.ch4_8;
public class A{
    public A(){          //无参构造方法
        System.out.println("A 的无参构造方法");
    }
```

```
}
package myproject.ch4_8;
public class B extends A{
    public B() {
      //super();              //显式调用父类无参构造方法 A()
      System.out.println("B 的无参构造方法");
    }
}
package myproject.ch4_8;
public class Ch4_8 {
    public static void main(String[] args) {
      new B();                //创建子类对象
    }
}
```

程序运行结果如图 4.8 所示。

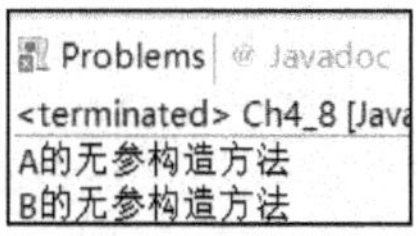

图 4.8 子类调用父类无参构造方法

分析：在用 new B();给子类 B 创建对象时，会先自动调用父类 A 的无参构造方法 A(),无论有没有 super();这条语句都不会影响程序运行结果。只不过写上 super();表示显式调用父类无参构造方法 A()。

【例题 4_9】子类调用父类含参构造方法。

```
package myproject.ch4_9;
public class A{
    String s;
    public A(){               //无参构造方法
    }
    public A(String s) {      //含参构造方法
      this.s = s;
      System.out.println("s="+s);
    }
}
package myproject.ch4_9;
class B extends A{
    String str;
    public B(String s,String str) {
      super(s);               //显式调用父类构造含参构造方法
      this.str = str;
      System.out.println("str="+str);
    }
}
package myproject.ch4_9;
public class Ch4_9 {
    public static void main(String[] args) {
      new B("aaa","bbb");     //创建子类对象
    }
}
```

程序运行结果如图 4.9 所示。

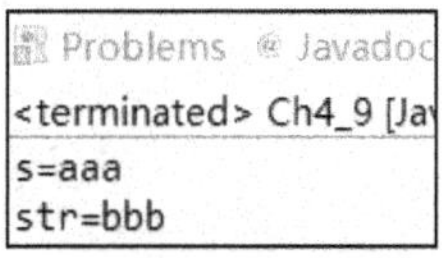

分析:本例中 B 类的含参构造方法 public B(String s,String str)中用 super(s);语句显式调用父类含参构造方法。

图 4.9 子类调用父类含参构造方法

将例题 4_9 的程序改写如下：
```java
package myproject.ch4_9;
public class A{
    String s;
}
package myproject.ch4_9;
class B extends A{
    String str;
    public B(String s,String str) {
       //super();
       this.s=s;
       this.str = str;
       System.out.println("s="+s);
       System.out.println("str="+str);
    }
}
package myproject.ch4_9;
public class Ch4_9 {
    public static void main(String[] args) {
       new B("aaa","bbb");         //创建子类对象
    }
}
```
改写后的程序运行结果与例题 4_9 运行结果相同，此时，父类 A 中并没有定义任何构造方法，那么 B 的构造方法中的 super();就表示调用父类的默认构造方法，这条语句写不写都是默认存在的。

继续将例题 4_9 的程序改写如下：
```java
package myproject.ch4_9;
public class A{
    String s;
    public A(){                    //无参构造方法
    }
    public A(String s) {           //含参构造方法
       this.s = s;
       System.out.println("s="+s);
    }
}
package myproject.ch4_9;
class B extends A{
    String str;
    public B(String s,String str) {
       this.s=s;
       this.str = str;
       System.out.println("s="+s);
       System.out.println("str="+str);
    }
}
package myproject.ch4_9;
public class Ch4_9 {
    public static void main(String[] args) {
       new B("aaa","bbb");         //创建子类对象
```

        }
    }

改写后的程序运行结果与例题 4_9 运行结果相同，此时，子类 B 的构造方法中没有用 super 显式地调用父类的某个构造方法，则默认地，子类 B 会调用父类定义的无参构造方法 public A(){}。
若将父类 A 中的无参构造方法注释掉如：
    /* public A(){//无参构造方法
    }*/
则程序编译会报错，如图 4.10 所示。

```
Exception in thread "main" java.lang.Error: Unresolved compilation problem:
    Implicit super constructor A() is undefined. Must explicitly invoke another constructor

    at myproject.ch4_9.B.<init>(B.java:4)
    at myproject.ch4_9.Ch4_9.main(Ch4_9.java:4)
```

图 4.10 父类不提供无参构造方法时的出错信息

一般地，应在父类定义含参构造方法的同时，定义父类的无参构造方法，避免以上错误的发生。

2. 用 super 引用父类变量或调用父类方法

在子类中可以采用 "super.<方法名>([实参表])" 和 "super.<父类变量名>" 的方式分别访问父类中的被子类覆盖的方法和被子类隐藏的变量。

【例题 4_10】super 调用父类被覆盖的方法测试。

```java
package myproject.ch4_10;
public class A{
    String s;
    public A() {                //无参构造方法
    }
    public A(String s) {        //含参构造方法
        this.s = s;
    }
    public void show(){
        System.out.println("s="+s);
    }
}
package myproject.ch4_10;
class B extends A{
    String str;
    public B(String s,String str) {
        super(s);               //调用父类含参构造方法
        this.str = str;
    }
    public void show(){         //重写父类 A 的 show()方法
        super.show();           //调用父类的 show()方法
        System.out.println("str="+str);
    }
}
package myproject.ch4_10;
public class Ch4_10 {
    public static void main(String[] args) {
        B b=new B("aaa","bbb");  //创建B的对象
```

```
        b.show();              //调用 B 的 show()方法
    }
}
```
本程序的运行结果与例题 4_9 运行结果相同。

【例题 4_11】super 引用父类中被隐藏的变量测试。
```
package myproject.ch4_11;
class A{
    int n;
    double result;
    public A() {}           //无参构造方法
    public double f(){      //求和方法
       for(int i=1;i<=n;i++)
           result=result+i;
       return result;
     }
}
package myproject.ch4_11;
class B extends A{
    int n=100;
    double result;
    //定义与父类 A 同名的成员变量, 父类的同名变量被隐藏
    public B(){
       super.n=n;           //给父类 A 的 n 赋值
    }
    public double f() {   //重写 f()方法, 实现求解平均值
        result=super.f()/n;//调用父类的 f()方法
     return result;
    }
}
package myproject.ch4_11;
public class Ch4_11 {
    public static void main(String[] args) {
    B b=new B();            //创建子类 B 的对象
    System.out.println("B 的 f()方法的结果: result="+b.f());
    //调用子类的 f()方法
    }
}
```
程序运行结果如图 4.11 所示。

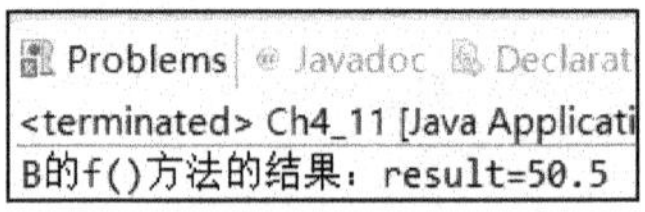

```
Problems  @ Javadoc  Declarat
<terminated> Ch4_11 [Java Applicati
B的f()方法的结果: result=50.5
```

图 4.11 例题 4_11 运行结果

4.6 对象的上下转型

1. 对象的上转型

一个父类的引用可以"指向"其子类的对象或将子类对象赋值给父类对象的现象被称为对象的"上转型"。

一般的, 若有 class A{...} class B extends A{...}, 即 B 是 A 的子类。当给子类创建对象 B b=new B();后, 可以有以下赋值方式: A a=b; 此时, 称 a 是 b 的"上转型对象"。

2. 对象的下转型

Java 也允许将父类对象赋值给子类对象，但必须进行强制类型转化：如 B 是 A 的子类，A a=new A();后，可以进行如下赋值：B b=(B)a；此时，称将父类对象 a "下转型" 为子类对象 b。

注意：上转型对象不能访问其子类对象新增加的成员变量和方法，只能访问被子类继承并覆盖的变量或重写的方法。

【例题 4_12】 对象的上下转型测试。

```
package myproject.ch4_12;
public class Father{                                //父类
    String name;
    public Father(String name) {                    //含参构造方法
        this.name = name;
    }
    public void showInfo(){
      System.out.println(name);
    }
}
package myproject.ch4_12;
public class Son1 extends Father{                   //子类 Son1
    String str1;                                    //Son1 新增的成员变量
    public Son1(String name, String str1) {         //Son1 的含参构造方法
        super(name);                                //调用父类含参构造方法
        this.str1 = str1;
    }
    public void showInfo() {                        //重写 showInfo()
      System.out.println(name+", "+str1);
    }
    public void f(){                                //Son1 新增的方法
        System.out.println("Son1 新增的方法");
    }
}
package myproject.ch4_12;
class Son2 extends Father{                          //子类 Son2
    String str2;                                    //Son2 新增的成员变量
    public Son2(String name, String str2) {         //Son2 的含参构造方法
        super(name);                                //调用父类含参构造方法
        this.str2 = str2;
    }
    public void showInfo() {                        //重写 showInfo()
         System.out.println(name+", "+str2);
    }
    public void f(){                                //Son2 新增的方法
        System.out.println("Son2 新增的方法");
    }
}
package myproject.ch4_12;
public class Ch4_12 {
    public static void main(String[] args) {
```

```
        Father f;                              //声明父类对象
        f=new Son1("子类1","子类1中新增的str1");        //f是子类Son1的上转型对象
        f.showInfo();                          //调用到Son1中重写过的showInfo()方法
        //f.str1="aaa";                        //非法，不可以访问子类新增的成员变量
        //f.f();                               //非法，不可以访问子类新增的成员方法
        f=new Son2("子类2","子类2中新增的str2");        //f是子类Son2的上转型对象
        f.showInfo();                          //调用到Son2中重写过的showInfo()方法
        Son2 s2=(Son2)f;                       //将f下转型为子类Son2的对象；
        s2.str2="***子类2中新增的str2***";
//下转型对象s2可以访问子类Son2中新增的变量str2
        s2.showInfo();
        s2.f();//下转型对象s2可以访问子类Son2中新增的f()方法
    }
}
```

程序运行结果如图4.12所示。

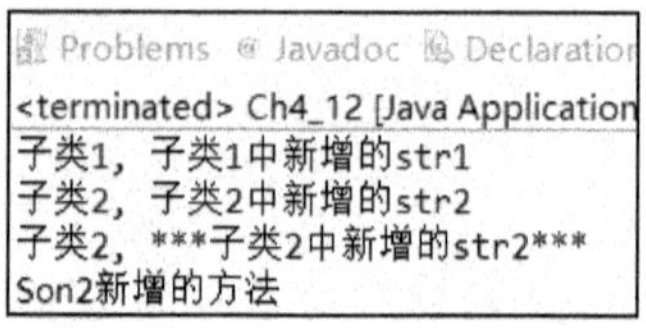

图4.12 对象上下转型测试结果

4.7 instanceof 关键字

instanceof 关键字用于判断某个对象是否是某个类的对象。基本格式：对象 instanceof 类，若该对象是某个类的对象，则结果为 true，否则为 false。

【例题4_13】 instanceof 测试。
```
package myproject.ch4_13;
class A{                                       //父类A
}
class B extends A{                             //子类B
}
public class Ch4_13 {
    public static void main(String[] args) {
        A  a=new B();                          //上转型
        System.out.println(a instanceof A);    //输出true
        System.out.println(a instanceof B);    //输出true
        A  a1=new A();                         //创建对象父类
        System.out.println(a1 instanceof A);   //输出true
        System.out.println(a1 instanceof B);   //输出false
    }
}
```
在进行对象下转型操作前，需要先用 instanceof 判断某个对象是否为某个类的对象，再进行下转型操作，以避免类型转换异常的发生。

【例题4_14】 instanceof 在对象下转型过程中的使用。
```
package myproject.ch4_14;
public class A{
    public void a_f(){
        System.out.println("a_f()输出");
    }
}
package myproject.ch4_14;
public class B extends A{
```

```java
    public void b_f(){
       System.out.println("b_f()输出");
    }
}
package myproject.ch4_14;
public class C extends A{
    public void c_f(){
       System.out.println("c_f()输出");
    }
}
package myproject.ch4_14;
public class Ch4_14 {
    public void test(A a){      //以父类 A 的对象作为方法形参
       if(a instanceof B){
          B b=(B)a;             //将 a 下转型为 B 的对象 b
          b.a_f();
          b.b_f();
       }
       if(a instanceof C){
          C c=(C)a;             //将 a 下转型为 C 的对象 c
          c.a_f();
          c.c_f();
       }
    }
    public static void main(String[] args) {
       Ch4_14 cast=new Ch4_14();
       B b1=new B();
       cast.test(b1);
       C c1=new C();
       cast.test(c1);
    }
}
```

程序运行结果如图 4.13 所示。

图 4.13 例题 4_14 运行结果

4.8 多 态

多态性是面向对象方法的重要特性之一。多态是指在一个程序中相同的名字可以表示不同的实现。Java 的多态性主要表现在以下两个方面：

- 方法重载：即可以在一个类中定义多个名字相同而实现不同的成员方法，它是一种静态多态性，或称编译时多态。编译时多态是指在编译阶段，编译器根据实参的不同来静态确定具体调用多个同名方法中的哪一个。
- 方法重写（覆盖）：即子类可以对继承自父类的某个成员方法的方法体进行重新实现，它是一种动态多态性，或称运行时多态。运行时多态是指运行时系统能根据对象状态不同来调用其相应的成员方法即动态绑定。

对于方法重载这种多态形式，可以参看第 3 章 3.3 节方法重载的具体实现和相关例题，此处不再单独举例。

【例题 4_15】 与方法重写相关的多态测试。

```java
package myproject.ch4_15;
abstract class Animal{                    //抽象类
    public abstract void run();           //抽象方法
}
class Rabbit extends Animal{
    public void run() {                   //重写 run()方法
        System.out.println("兔子跑得快! ");
    }
}
class  Tortoise extends Animal{
    public void run() {                   //重写 run()方法
        System.out.println("乌龟跑得慢! ");
    }
}
public class Ch4_15 {
    public static void main(String[] args) {
      Animal a;
      a=new Rabbit();
      a.run();                            //调用 Rabbit 类重写过的 run()方法
      a=new Tortoise();
      a.run();                            //调用 Tortoise 类重写过的 run()方法
    }
}
```

程序运行结果如图 4.14 所示。

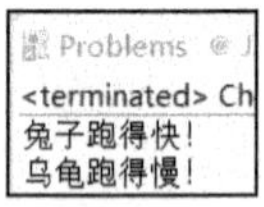

分析：本例将 Animal 类定义成了一个抽象类，其中包含了一个抽象方法 public abstract void run()，抽象方法没有方法体（关于抽象类和抽象方法的含义和用法将在第 5 章中详细介绍）。

图 4.14　例题 4_15 运行结果

当 Animal 的对象 a 分别充当不同子类对象的上转型对象并调用 run()方法时，从调用的形式上来看，程序中两次用到 a.run();语句，调用形式是一模一样的。但由于 a 在充当不同子类对象的上转型对象时，会分别调用该子类重写后的 run()方法，所以表现出不同的行为。

4.9　final 修 饰 符

final 可以修饰类、成员变量和成员方法等。
- 用 final 修饰的类不能被继承，没有子类。
- 用 final 修饰的方法不能被子类重写。
- 用 final 修饰的变量表示常量，只能被赋值一次。

继承关系的弱点是打破了封装性，这是由于子类能够访问父类的实现细节，而且能以方法重写的方式改变实现细节。出于安全的原因，Java 的一些系统类被定义为 final 类，如 java.lang.String 类是 final 类。

【例题 4_16】 final 用法测试。

```java
package myproject.ch4_16;
final class A{
    final int a=10;
```

```java
    public void f(){
        //a=30;                    //不能改变其值
        System.out.println("a="+a);
    }
}
//class B extends A{ }          //A不能被继承
class C{
    int c=100;
    final public void f(){
        System.out.println("c="+c);
    }
}
class D extends C{
    // public void f(){}         //f()方法不能被重写
}
public class Ch4_16 {
    public static void main(String[] args) {
        A a=new A();
        a.f();
        D d=new D();
        d.f();                    //调用从C类中继承来的f()方法
    }
}
```

分析例题 4_16 的运行结果，掌握 final 关键字的用法。

小 结

本章主要介绍了 Java 类继承的特点及方法；Java 权限修饰符 public、protected、缺省、private 的用法；分别讨论了子类和父类在不在同一包中子类的继承特性；变量覆盖与方法重写的含义及其各自的用法；构造方法的继承性以及 super 关键字的用法；对象上、下转型的含义及其各自用法；Java 多态的实现方法以及 final 关键字的用法等。

习 题 4

一、选择题

1. 下列关于 Java 继承的说法中不正确的是（　　）。
   A. 一个父类可以有若干个直接子类　　　B. 一个子类可以有多个直接父类
   C. Object 类是所有 Java 类的父类　　　D. 子类可以派生出子类
2. 下面类成员访问修饰符中表明"不可以被其子类访问但可以被同一包中其他类访问"的是（　　）。
   A. private　　　B. protected　　　C. 友好的　　　D. public
3. 不能直接使用 new 创建对象的类是（　　）
   A. 静态类　　　B. 抽象类　　　C. 最终类　　　D. 公有类
4. 为类定义多个名称相同、但参数的类型或个数不同的方法的做法称为（　　）。

A. 方法重载　　　　B. 方法覆写　　　　C. 方法继承　　　　D. 方法重用

二、上机实践题

1. 定义一个大学生类 UniversityStudent，包括成员变量学号 String stuno，姓名 String stuname，年龄 int stuage，学位 String degree 等；包括一个无参构造方法及一个含参构造方法 public UniversityStudent(String stuno, String stuname, int stuage, String degree)；一个用于输出大学生基本信息的成员方法 public void showInfo()。

UniversityStudent 派生出两个子类：本科生类 UnderGraduate 及研究生类 Graduate，在 UnderGraduate 新增变量专业 String specialty，在 Graduate 新增变量研究方向 String researchdirection；两个子类都各自定义含参构造方法，并在各自的构造方法中通过 super 调用父类含参构造方法，两个子类都分别重写父类的 showInfo()方法，以便输出各自的信息。

定义主类 XiTi4_2_1，在 main()方法中创建 UnderGraduate 对象 undergraduate、Graduate 对象 master、Graduate 对象 doctor 并用这些对象分别调用各自的 showInfo()方法输出相应的信息。

2. 运用对象"上转型"方法改写 XiTi4_2_1 的 main()方法，即用 UniversityStudent 对象 stu 分别调用父类及各个子类的 showInfo()方法，得到与 XiTi4_2_1 相同的输出结果。

3. 编程实现以下要求：

（1）定义一个图形类 Shape，在其中定义一个 public double getArea()方法，用于求解图形的面积。Shape 类派生出圆类 Circle、矩形类 Rectangle。

（2）Circle 类中定义一个表示圆半径的成员变量 double radius 及一个构造方法 public Circle(double radius)，重写 Shape 类的 getArea(){}方法，用于求解圆面积。

（3）Rectangle 类中定义表示矩形宽、高的成员变量 double width, height，一个构造方法 public Rectangle(double width, double height)，重写 Shape 类的 getArea(){}方法，用于求解矩形面积。

（4）定义一个柱体类 Cylinder，在其中定义表示柱体底的成员变量 Shape shape、表示柱体高的成员变量 double height；定义构造方法 public Cylinder(double height, Shape shape)，一个用于求解柱体体积的成员方法"public double getVolume();"。

（5）定义主类 XiTi4_2_3，在 main()方法中分别定义圆柱体、四棱柱体对象并输出它们各自的体积。

第 5 章　Java 抽象类与接口

【本章内容提要】
- 抽象方法与抽象类；
- Java 接口；
- Java 接口回调；
- 内部类；
- 匿名与匿名对象；
- 面向抽象（接口）编程；
- 接口的一个应用——工厂模式；
- Java 内置注解简介。

5.1　抽象方法与抽象类

abstract 关键字可以修饰类和成员方法。用 abstract 关键字来修饰一个类时，这个类被称为抽象类；用 abstract 来修饰一个成员方法时，该方法被称为抽象方法。

1. abstract 方法

用 abstract 修饰的成员方法为抽象方法。抽象方法只有方法头的声明，没有方法体的实现，如 "abstract void f();"。

2. abstract 类

一个被 abstract 关键字修饰的类被称为抽象类。抽象类具有以下特点：
- 含有抽象方法的类必须被声明为抽象类，但一个抽象类中不一定要有抽象方法。
- 抽象类必须被继承，抽象方法必须被重写。如果子类未重写抽象父类的抽象方法，则子类本身也称为一个抽象类。
- 可以给抽象类声明一个对象，但不能给抽象类创建对象。

在实际开发中，抽象类主要用来抽象出子类的一些共性行为，对这些共性行为，在抽象类中只声明方法，表示出方法能做什么，但没有必要且无法给出具体的实现，这是因为不同子类对这些方法的实现不尽相同。

【例题 5_1】　定义抽象类图形 Shape，要求其中包含求解图形周长和面积的抽象方法。定义矩形类 Rectangle、圆类 Circle 分别继承 Shape 并给求解周长和面积的方法给出方法实现，在主类中分别输出一个 Rectangle 及 Circle 类的周长和面积。

```java
package myproject.ch5_1;
public abstract class Shape{
    double length;
    double area;
    public abstract double getLength();        //求解图形周长的抽象方法
    public abstract double getArea();          //求解图形面积的抽象方法
    public void showInfo(){                    //非抽象方法
        System.out.println("我是图形类");
    }
}
package myproject.ch5_1;
public class Rectangle extends Shape{          //矩形类
    double height,width;                       //高和宽
    public Rectangle(double height, double width) {//构造方法
       this.height = height;
       this.width = width;
    }
    @Override
    public double getLength() {                //实现抽象方法
        return 2*(height+width);
    }
    @Override
    public double getArea() {                  //实现抽象方法
       return height*width;
    }
    @Override
    public void showInfo() {                   //重写showInfo()方法
        System.out.println("我是矩形类");
    }
}
package myproject.ch5_1;
public class Circle extends Shape{
    static final double PI=3.14;               //常量PI
    double radius;                             //半径
    public Circle(double radius) {             //构造方法
       this.radius = radius;
    }
    @Override
    public double getLength() {                //实现抽象方法
       return 2*PI*radius;
    }
    @Override
    public double getArea() {                  //实现抽象方法
       return PI*radius*radius;
    }
    @Override
    public void showInfo() {                   //重写showInfo()方法
        System.out.println("我是圆形类");
    }
```

```
}
package myproject.ch5_1;
public class Ch5_1 {
    public static void main(String[] args) {
        //Shape shape=new Shape();        //不能给抽象类创建对象
        Shape shape;                      //声明抽象类对象
        shape=new Rectangle(20.5,13.5);   //对象上转型
        System.out.println("矩形周长: "+shape.getLength());
        System.out.println("矩形面积: "+shape.getArea());
        shape.showInfo();
        shape=new Circle(3.0);            //对象上转型
        System.out.println("圆形周长: "+shape.getLength());
        System.out.println("圆形面积: "+shape.getArea());
        shape.showInfo();
    }
}
```

程序运行结果如图 5.1 所示。

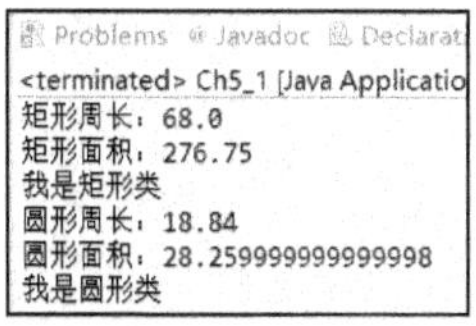

分析：本例中在定义一个图形类 Shape 时，考虑到每个图形类都应具有输出周长和面积属性以及求解周长和面积的方法。但是，由于具体图形的多样性，决定了：①无法在 Shape 类中给出具体求解某种图形的周长和面积的实现方法，因此，getLength()和 getArea()方法都是抽象方法，即只有方法声明，没有方法实现。②没有必要给出方法实现，这是因为，假定在 Shape 中给出了某种方法体，那么当诸如矩形、圆形等子类继承 Shape 后还是要重写这些方法。

图 5.1　例题 5_1 运行结果

Shape 类中的 showInfo()方法在本例中只是为了说明抽象类中可以包含非抽象方法。

5.2　Java 接口

Java 接口是 Java 中重要的引用类型。Java 只支持单继承即一个类只能有一个直接父类，而 Java 允许一个类同时实现多个接口，同时，一个接口可以同时继承多个父接口。所以，运用 Java 接口机制弥补 Java 单继承在实际应用过程中的不足。

Java 接口定义的语法格式如下：

```
[public] interface 接口名[extends 父接口名列表]
    {
        [public] [static] [final] 常量名=常量值;
        [public] [abstract] 返回值 方法名([参数列表]) [throw 异常列表];
    }
```

其中，interface 是定义接口的关键字；"[ ]"表示可选。

定义了一个接口之后，并不能直接给该接口创建实例（对象），而是需要类通过 implements 关键字来实现该接口并给接口的各个抽象方法给出方法实现。

接口具有以下主要特性：
- 接口中声明的属性默认为 public static final 的，都是常量。
- 接口中定义的方法默认为 public abstract 的，都是抽象方法。
- 一个类可以实现多个接口，类实现接口的关键字为 implements，一个接口可以被多个不同的类来实现。

- 一个类只能继承另外一个类，但能同时实现多个接口。
- Java 中不允许类的多继承，但允许接口的多继承即一个接口可以同时继承多个其他接口，并添加新的常量和抽象方法。
- 如果一个非抽象类实现某接口，则该类必须实现（重写）该接口中所有方法，而且要显式地在这些被重写的方法前加 public 进行修饰。
- 接口的权限可以是 public 或缺省的，若用 public 修饰了某个接口，则该接口可以被任何类进行实现；缺省权限的接口只能被与该接口处于同一个包中的类来实现。
- 不允许创建接口的实例，但允许定义接口类型的引用变量。

【例题 5_2】接口特性测试。

```java
package myproject.ch5_2;
interface A{                          //接口 A
    int x=100;                        //常量
    void f_a();                       //抽象方法
}
interface B{                          //接口 B
    void f_b();                       //抽象方法
}
interface C extends A,B{              //接口 C 的同时继承接口 A、B
    void f_c();                       //抽象方法
}
abstract class D{                     //抽象类 D
    int y=200;
    public abstract void f_d();       //抽象方法
}
class E implements A,B{               //类 E 实现接口 A、B
    @Override
    public void f_a() {               //实现接口 A 的抽象方法，public 必须写
      System.out.println("对接口 A 的 f_a()的实现");
    }
    @Override
    public void f_b() {               //实现接口 B 的抽象方法
      System.out.println("对接口 B 的 f_b()的实现");
    }
    public void f_e(){                //类 E 的新增方法
      System.out.println("E 类新增方法");
      System.out.println("接口 A 中 x="+x);     //接口 A 中的常量 x
    }
}
class F extends D implements C{  //类 F 继承抽象类 D 同时实现接口 C
    @Override
    public void f_a() {               //实现接口 A 的抽象方法
      System.out.println("对接口 A 的 f_a()的实现");
    }
    @Override
    public void f_b() {               //实现接口 B 的抽象方法
      System.out.println("对接口 B 的 f_b()的实现");
    }
```

```java
        @Override
        public void f_c() {            //实现接口C的抽象方法
          System.out.println("对接口C的f_c()的实现");
        }
        @Override
        public void f_d() {            //实现抽象类C的抽象方法
          System.out.println("对抽象类D的f_d()的实现");
        }
        public void f_f(){             //类F的新增方法
          System.out.println("F类自定义方法");
          System.out.println("接口A中x="+x);//接口A中的常量x
          y=500;                       //y是变量，可以重新赋值
          System.out.println("抽象类D中y="+y);
        }
}
public class Ch5_2{                    //主类
    public static void main(String args[]){
      System.out.println("接口A中常量x:"+A.x);
      //获取x值，用"接口名.常量名"的形式访问接口常量
      E e=new E();                     //创建E类对象
      e.f_a();
      e.f_b();
      e.f_e();
      F f=new F();                     //创建F类对象
      f.f_a();
      f.f_b();
      f.f_c();
      f.f_d();
      f.f_f();
    }
}
```

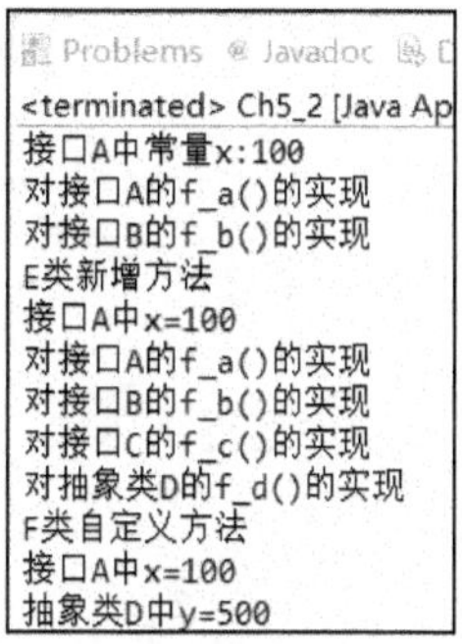

程序运行结果如图 5.2 所示。

图 5.2 接口特性测试结果

说明：接口成员只有常量和抽象方法。例如，接口 A 中的 int x=100;看似变量，但事实上会默认为 public static final int x=100;所以它是一个常量；接口 A 中的 void f_a();会被默认为 public abstract void f_a();所以，该方法是抽象方法。其他接口也是一样。

注意：接口 A 中的 void f_a();前可以缺省 public 修饰符，但当一个类实现该方法时，public 不能缺省。

5.3 Java 接口回调

1. 接口变量

对于一个类 A 而言，可以声明其对象如 A a; 我们称 a 是类 A 声明的对象。那么，若定义了一个接口 A,然后声明 A a; 我们称声明了 A 的一个接口变量,这是因为接口本身不能创建实例(对象)，不能将 a 称作接口对象。

2. 接口回调的含义

接口回调是指可以把实现了某一接口的类创建的对象赋值给该接口声明的接口变量。那么,

就可以用该接口变量调用被类实现过的接口方法,这一过程被称为接口回调。

【例题5_3】接口回调测试。

```java
package myproject.ch5_3;
interface Person{                          //定义接口
   void showInfo();                        //抽象方法
}
class Student implements Person{           //Student类实现Person接口
   public void showInfo(){                 //实现showInfo()方法
      System.out.println("我是一名学生");
   }
}
class Teacher implements Person{           //Teacher类实现Person接口
   public void showInfo(){                 //实现showInfo()方法
      System.out.println("我是一名教师");
   }
}
public class Ch5_3 {
   public static void main(String args[]){
      Person person;                       //声明接口变量
      person=new Student();                //接口变量中存放Student类对象的引用
      person.showInfo();                   //接口回调
      person=new Teacher();                //接口变量中存放Teacher类对象的引用
      person.showInfo();                   //接口回调
   }
}
```

程序运行结果如图5.3所示。

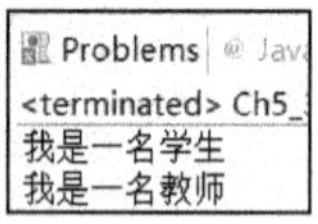

图5.3 接口回调测试结果

5.4 内 部 类

1. 内部类定义及特点

Java允许一个类的内部可以包含(嵌套)另一个类,称这个被包含的类为内部类(Inner Class),把包含内部类的类称之为外部类(Outer Class)。

内部类特点:
- 内部类对象能够访问其所在外部类的全部属性,包括私有属性。
- 内部类能够被隐藏起来,不被同一包中的其他类所见。

【例题5_4】内部类测试。

```java
package myproject.ch5_4;
class Outer {                    //外部类
   private int m=10;
   Inner inner;                  //声明内部类对象
   public Outer(){               //外部类构造方法
      inner=new Inner();         //创建内部类对象
   }
   class Inner{                  //内部类
      int n=2;
      public void f(){
```

```
            System.out.println("Outer 的 m="+m);
            n=m*n;
            System.out.println("n="+n);
        }
    }                                      //内部类定义结束
}                                          //外部类定义结束
public class Ch5_4 {
    public static void main( String[] args) {
        Outer outer=new Outer();     //创建外部类对象
        outer.inner.f();             //调用内部类方法 f(),注意层次调用格式
        System.out.println("Inner 的 n="+outer.inner.n);//访问内部类变量 n
        System.out.println("另一种创建 Inner 对象的方式: ");
        Outer.Inner inner1=new Outer().new Inner();//也可以以这种格式创建内部类对象
        inner1.f();
        System.out.println("Inner 的 n="+inner1.n);//访问内部类变量 n
    }
}
```

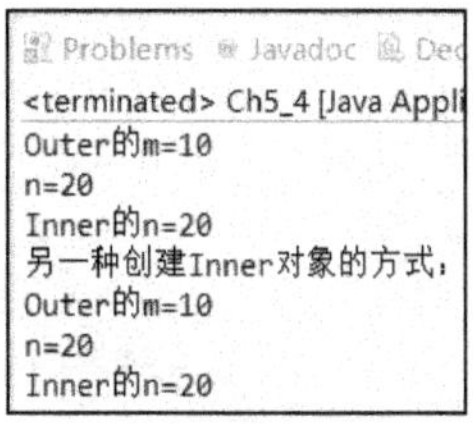

程序运行结果如图 5.4 所示。

分析：在用 Outer outer=new Outer();创建外部类对象时调用了外部类构造方法 Outer()，在该构造方法体中用 inner=new Inner();创建了内部类对象。用"外部类对象.内部类对象.内部类方法（变量）"的格式访问内部类成员。与其等价的方式是直接创建内部类对象"Outer.Inner inner1=new Outer().new Inner();"，然后用"内部类对象.内部类方法（变量）"的格式访问内部类成员。

图 5.4 内部类测试结果

执行结果显示内部类可以访问外部类的 private int m。

注意：含有内部类的源程序被编译后会生成一个外部类名$内部类名.class 的字节码文件，如本例编译后除了生成 Ch5_4.class、Outer.class 外，还会生成一个 Outer$Inner.class。

2. 静态内部类和非静态内部类

内部类可以被分为静态内部类和非静态内部类，静态内部类是用 static 修饰的内部类。非静态内部类中可以访问其外部类的所有变量，而静态内部类不能直接访问外部类的非静态变量。

【**例题 5_5**】静态内部类测试。
```
package myproject.ch5_5;
class Outer {                    //外部类
    private int m=10;
    Inner1 inner1;
    Inner2 inner2;               //声明内部类对象
    public Outer(){              //外部类构造方法
        inner1=new Inner1();
        inner2=new Inner2();     //创建内部类对象
    }
    class Inner1{                //内部类
        int n=2;
        public void f(){
            System.out.println("Outer 的 m="+m);
            n=m*n;
            System.out.println("n="+n);
```

```java
        }
    }
    static class Inner2{                    //静态内部类
        int n=2;
        public void f() {
            //System.out.println("Outer 的 m="+m);//非法,不能直接访问外部类的非静态变量m
            //n=m*n;                            // 非法,不能直接访问外部类的非静态变量m
            /*以上被注释掉的两条语句可以改写成下面两条语句**/
            System.out.println("Outer 的 m="+new Outer().m);
            //外部类对象.m 的格式访问外部类变量
            n=new Outer().m*n;
            System.out.println("n="+n);
        }
    }
}
public class Ch5_5 {
    public static void main( String[] args) {
        Outer outer=new Outer();
        outer.inner1.f();
        outer.inner2.f();
        System.out.println("Inner1 的 n="+outer.inner1.n);//访问内部类 Inner1 的
                                                          //变量 n
        System.out.println("Inner2 的 n="+outer.inner2.n);//访问内部类 Inner2 的
                                                          //变量 n
    }
}
```

程序运行结果如图 5.5 所示。

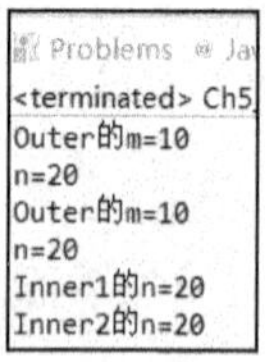

分析：首先，只有内部类才可以被修饰为 static 的。由于 Inner2 被修饰为静态的，所以它不可以直接访问外部类 Outer 的非静态变量 m，程序中被注释掉的语句是非法的，静态内部类必须通过其外部类的一个对象才能访问其外部类的非静态变量 m。

图 5.5 例题 5_5 运行结果

若将程序 "private int m=10;" 中的 m 定义为静态的，即 "private static int m=10;"，则 System.out.println("Outer 的 m="+m);n=m*n;是合法的，或将 "System.out.println("Outer 的 m="+new Outer().m);n=new Outer().m*n;" 改写成 "System.out.println("Outer 的 m="+Outer.m);n=Outer.m*n;" 也是可以的。

为了进一步了解静态内部类的用法，将例题 5_5 做如下改写：

```java
package myproject.ch5_5;
class Outer {                          //外部类
    private int m=10;
    public Outer(){}
    static class Inner2{               //静态内部类
        int n=2;
        public void f() {
            System.out.println("Outer 的 m="+new Outer().m);
            n=new Outer().m*n;
            System.out.println("n="+n);
        }
```

```
    }
}
public class Ch5_5 {
  public static void main( String[] args) {
    Outer.Inner2 inner2=new Outer.Inner2();//创建静态内部类对象,注意这种格式
    inner2.f();
    System.out.println("Inner2 的 n="+inner2.n);//访问内部类 Inner2 的变量 n
  }
}
```

改写后的程序运行结果如图 5.6 所示。

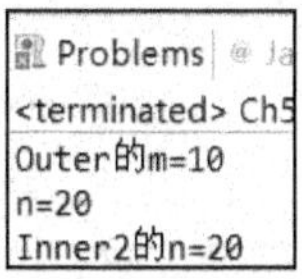

继续改写程序如下:

```
package myproject.ch5_5;
class Outer {                          //外部类
  private int m=10;
  public Outer(){}
  static class Inner2{                 //静态内部类
    static int n=2;                    //静态变量
    public static void f() {           //静态方法
      System.out.println("Outer 的 m="+new Outer().m);
      n=new Outer().m*n;
      System.out.println("n="+n);
    }
  }
}
public class Ch5_5 {
  public static void main( String[] args) {
    Outer.Inner2.f();                  //调用内部类 Inner2 的静态非法 f()
    System.out.println("Inner2 的 n="+Outer.Inner2.n);//访问内部类 Inner2 的静态变量 n
  }
}
```

图 5.6 静态内部类测试结果

改写后的程序执行结果也如图 5.6 所示,但由于静态成员可以被类名直接引用,所以本段程序中用外部类 Outer 直接引用了其静态类 Inner,而 Inner 又直接引用了它自己的静态成员变量 n 和方法 f()。

5.5 匿名类与匿名对象

Java 允许以下格式给创建一个匿名类的匿名对象:
给一个抽象父类创建匿名子类对象,格式如下:
new 抽象父类构造方法（[参数列表]）
{
  …//匿名子类的类体
}

这段代码的作用是给抽象父类创建一个匿名子类对象,通常这个匿名子类对象用于充当某个方法的实参。注意:一般地,由于抽象类不能创建实例(对象),所以不允许直接用 new 抽象父类构造方法([参数列表])的形式给一个抽象类创建对象,只有在上述情况下,new 运算符后可以

紧跟抽象类的构造方法。

- 给实现过某接口的匿名类创建匿名对象,格式如下:

```
new 接口名( )
{
    ...//实现过该接口的匿名类体
}
```

这段代码相当于创建出一个实现过该接口的匿名类对象,通常这个匿名类对象用于充当某个方法的实参。注意:一般地,由于不能给一个接口创建对象,因此,不允许直接用 new 接口名( )的形式给一个接口创建对象,只有在上述情况下,new 运算符后可以紧跟接口名。

【例题 5_6】编程模拟当计算机接入不同类型的移动存储设备,如移动硬盘、U 盘等时,输出计算机当前接入的设备名称。

```
package myproject.ch5_6;
/*定义 Disk 抽象类,包含显示 Disk 名称的抽象方法**/
abstract class Disk {
    public abstract String showDiskName();
}
/*HardDisk 是 Disk 的子类**/
class HardDisk extends Disk{
    public String showDiskName() {
        return "移动硬盘";
    }
}
/*Computer 类使用某种 Disk**/
class Computer{
  public void useDisk(Disk disk){     //以抽象类对象作为方法形参
     System.out.println(disk.showDiskName()+"连接了计算机...");
  }
}
/*主类**/
public class Ch5_6{
    public static void main(String args[]){
    Computer computer=new Computer();
    HardDisk hdisk=new HardDisk();
    computer.useDisk(hdisk);          //以 Disk 的有名子类 HardDisk 的有名对象为实参
    computer.useDisk(new Disk(){      //以 Disk 的匿名子类的匿名对象为实参
        public String showDiskName() {
            return "U 盘";
        }
    });
    }
}
```

程序运行结果如图 5.7 所示。

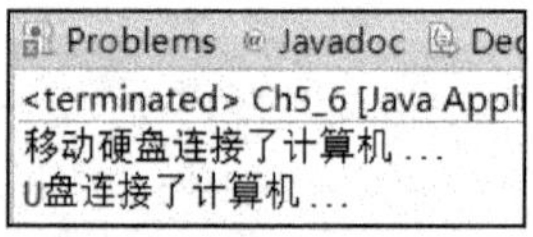

图 5.7 例题 5_6 运行结果

说明:程序中

```
new Disk(){
    public String showDiskName() {
        return "U 盘";
    }
}
```

的作用是给抽象类 Disk 的一个匿名子类创建一个匿名对象。

该做法等价于：创建 Disk 的有名子类 UDisk 的有名对象 udisk 作为 compputer.useDisk()方法的实参即增加如下 UDisk 类定义：
```
class UDisk extends Disk{
public String showDiskName() {
    return "U盘";
    }
}//这是一个有名类，其类名为 UDisk
```
在主类的主方法中增加：
```
UDisk udisk=new UDisk();              //给 UDisk 创建有名对象 udisk
computer.useDisk(udisk);              //以有名类对象 udisk 作为方法实参
```
即可以将例题 5_6 改写为以下形式：
```
package myproject.ch5_6;
/*定义 Disk 抽象类，包含显示 Disk 名称的抽象方法**/
abstract class Disk {
    public abstract String showDiskName();
}
/*HardDisk 是 Disk 的子类**/
class HardDisk extends Disk{
    public String showDiskName() {
        return "移动硬盘";
    }
}
class UDisk extends Disk{                //定义 Disk 的有名子类 UDisk
public String showDiskName() {
    return "U盘";
    }
}
/*Computer 类使用某种 Disk**/
class Computer{
  public void useDisk(Disk disk){        //以抽象类对象作为方法形参
     System.out.println(disk.showDiskName()+"连接了计算机...");
  }
}
/*主类**/
public class Ch5_6{
    public static void main(String args[]){
    Computer computer=new Computer();
    HardDisk hdisk=new HardDisk();
    computer.useDisk(hdisk);             //以 Disk 的有名子类 HardDisk 的有名对象为实参
    UDisk udisk=new UDisk();             //给 UDisk 的有名子类 UDisk 创建有名对象 udisk
    computer.useDisk(udisk);             //以有名类对象 udisk 作为方法实参
    }
}
```
其运行结果是不变的，如图 5.7 所示。

若将例题 5_6 中的抽象类 Disk 的定义换成接口，程序变为如下形式：
```
package myproject.ch5_6;
interface Disk {
   String showDiskName();
```

```java
}
class HardDisk implements Disk{
   public String showDiskName() {
       return "移动硬盘";
   }
}
class Computer{
   public void useDisk(Disk disk){
       System.out.println(disk.showDiskName()+"连接了计算机...");
   }
}
public class Ch5_6{
   public static void main(String args[]){
       Computer computer=new Computer();
       HardDisk hdisk=new HardDisk();
       computer.useDisk(hdisk);//以实现过接口Disk的有名类HardDisk的有名对象为实参
       computer.useDisk(new Disk(){//以实现过Disk接口的匿名类的匿名对象为实参
          public String showDiskName() {
              return "U盘";
          }
       });
   }
}
```

其运行结果是不变的，如图5.7所示。当然，也可以将上述程序继续改写为如下形式：

```java
package myproject.ch5_6;
interface Disk {
   String showDiskName();
}
class HardDisk implements Disk{
   public String showDiskName() {
       return "移动硬盘";
   }
}
class UDisk implements Disk{// 定义实现过Disk接口的有名类UDisk
public String showDiskName() {
       return "U盘";
   }
}
class Computer{
   public void useDisk(Disk disk){
       System.out.println(disk.showDiskName()+"连接了计算机... ");
   }
}
public class Ch5_6{
   public static void main(String args[]){
       Computer computer=new Computer();
       HardDisk hdisk=new HardDisk();
       computer.useDisk(hdisk);//以实现过接口Disk的有名类HardDisk的有名对象为实参
       UDisk udisk=new UDisk();//给有名类UDisk创建有名对象udisk
       computer.useDisk(udisk);//以有名对象udisk作为方法实参
   }
}
```

其运行结果是不变的，如图 5.7 所示。

5.6 面向抽象（接口）编程

"开闭原则"是面向对象设计的基本原则之一。所谓"开闭原则"就是让设计的系统应当对扩展开放，对修改关闭。在设计系统时，应当首先考虑到用户需求的变化，将应对用户变化的部分设计为对扩展开放，而设计的核心部分应不随应用需求的变化而持续被修改，即这部分是对修改饰关闭的。

实现"开闭原则"的重要途径就是面向抽象（接口）编程。

所谓面向抽象（接口）编程，是指当设计某种重要的类时，不让该类面向具体的类，而是面向抽象类（或接口），即让该类依赖或关联于抽象类对象（或接口变量），而不是具体类对象。

【例题 5_7】求解不同图形的面积。

```
package myproject.ch5_7;
/*图形类是一个抽象类，包含求解图形面积的抽象方法**/
abstract class Shape {
    public abstract double getArea();
}
/*矩形类是图形类的子类**/
class Rectangle extends Shape {
    double height,width,area;              //定义矩形的高、宽和面积
    public Rectangle(double height, double width) {
        this.height = height;
        this.width = width;
    }
    public double getArea() {              //实现抽象方法
        return height*width;
    }
}
/*三角类是图形类的子类**/
class Triangle extends Shape{
    double a,b,c,area;                     //定义三角形的三边和面积
    public Triangle(double a, double b, double c) {
        this.a = a;
        this.b = b;
        this.c = c;
    }
    public double getArea() {              //实现抽象方法
        double p=(a+b+c)/2;
        return Math.sqrt(p*(p-a)*(p-b)*(p-c));
    }
}
/*定义显示某种图形面积的类**/
class ShapeArea{
  public void computerArea(Shape shape){   //依赖于抽象类 Shape
    System.out.println(shape.getArea());
  }
}
```

```
/*主类**/
public class Ch5_7{
    public static void main(String args[]){
        ShapeArea show=new ShapeArea();
        Rectangle rect=new Rectangle(10.5,20.0);
        Triangle tri=new Triangle(3.0,4.0,5.0);
        System.out.println("矩形面积为: ");
        show.computerArea(rect);
        System.out.println("三角形面积为: ");
        show.computerArea(tri);
    }
}
```

这段程序实现了 ShapeArea 类面向（依赖于）抽象类 Shape 编程。程序运行结果如图 5.8 所示。

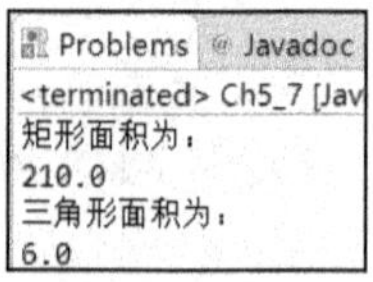

图 5.8　例题 5_7 的运行结果

分析：图形类的种类很多，也许当前只需要求解矩形、三角形的面积，以后还会需要求解梯形及其他多边形的面积，这就表示应用需求会不断变化。这些类应是"对扩展开放的"，即需要求解梯形面积时，就新定义一个梯形类就可以了。

程序中的 ShapeArea 类是本例中的"核心类"，若让该类依赖于一个具体类如 Rectangle 或 Triangle，即让 public void computerArea()方法的形参是 Rectangle 或 Triangle 对象，当有新的需求如定义了一个梯形类并需要求解梯形类面积时，就需要修改 public void computerArea()的形参，使之成为梯形类对象。这样的做法违背了"对修改关闭"的原则。

因此，需要定义一个抽象类 Shape，让 ShapeArea 类依赖于抽象类即让 public void computerArea()方法的形参是抽象类 Shape 的对象。此时，不管应用需求如何变化，Shape 类都无须修改，实现了"对修改关闭"。

注意：程序中 ShapeArea 类的 public void computerArea(Shape shape)方法形参是抽象类对象 Shape shape，show.computerArea(rect); show.computerArea(tri);两句中的实参分别是 Shape 的子类 Rectangle 和 Triangle 的对象，在 computerArea 方法执行时，实际上 shape 分别充当了 rect 和 tri 的上转型对象，会分别调用被 Rectangle 和 Triangle 重写过的 getArea()方法。

改写例题 5_7 的程序如下：
```
package myproject.ch5_7;
interface Shape {                          //接口
    double getArea();
}
class Rectangle implements Shape {         //实现 Shape 接口
    double height,width,area;              //定义矩形的高、宽和面积
    public Rectangle(double height, double width) {
        this.height = height;
        this.width = width;
    }
    public double getArea() {              //实现抽象方法
        return height*width;
    }
}
```

```java
class Triangle implements Shape{                //实现 Shape 接口
    double a,b,c,area;                          //定义三角形的三边和面积
    public Triangle(double a, double b, double c) {
        this.a = a;
        this.b = b;
        this.c = c;
    }
    public double getArea() {                   //实现抽象方法
        double p=(a+b+c)/2;
        return Math.sqrt(p*(p-a)*(p-b)*(p-c));
    }
}
/*定义显示某种图形面积的类**/
class ShapeArea{
  public void computerArea(Shape shape){        //依赖于接口 Shape
    System.out.println(shape.getArea());
  }
}
public class Ch5_7{
    public static void main(String args[]){
        ShapeArea show=new ShapeArea();
        Rectangle rect=new Rectangle(10.5,20.0);
        Triangle tri=new Triangle(3.0,4.0,5.0);
        System.out.println("矩形面积为: ");
        show.computerArea(rect);
        System.out.println("三角形面积为: ");
        show.computerArea(tri);
    }
}
```

这段程序实现了 ShapeArea 类面向（依赖于）接口 Shape 编程。程序运行不变，如图 5.8 所示。

分析：上述程序使得核心类 ShapeArea 依赖于接口 Shape，程序中 ShapeArea 类的 public void computerArea(Shape shape) 方法形参是接口变量 Shape shape，"show.computerArea(rect);" "show.computerArea(tri);"两句中的实参分别是实现过接口 Shape 的类 Rectangle 和 Triangle 的对象，在 computerArea 方法执行时，实际上是通过接口回调机制分别调用被 Rectangle 和 Triangle 重写过的 getArea()方法。

5.7 接口的一个应用——工厂模式

工厂模式是面向对象编程的基本模式之一，工厂模式属于创建型模式。"工厂"的主要任务是根据具体需求给不同类创建对象。当应用程序中调用方程序需要给被调用方程序创建对象时，并不直接去创建被调用方类的对象，而是通过一个"工厂类"来给被调用方的类创建对象。这样一来，就降低了调用方和被调用方的耦合性。

工厂模式专门负责将大量有共同接口的类实例化。工厂模式可以动态决定将哪一个类实例化，而不必事先知道每次要实例化那一个类。

【例题 5_8】工厂模式测试。
```java
package myproject.ch5_8;
interface Storage{                              //存储器接口
```

```java
        void read();                              //读存储器的方法
        void write();                             //写存储器的方法
    }
    class UsbDisk implements Storage{             //UsbDisk类，被调用方
        @Override
        public void read() {                      //实现read()方法
            System.out.println("正在读取UsbDisk...");
        }
        @Override
        public void write() {                     //实现write()方法
            System.out.println("正在将数据写入UsbDisk...");
        }
    }
    class HardDisk implements Storage{            //HardDisk类，被调用方
        @Override
        public void read() {                      //实现read()方法
            System.out.println("正在读取HardDisk...");
        }
        @Override
        public void write() {                     //实现write()方法
            System.out.println("正在将数据写入HardDisk...");
        }
    }
    class Factory{                                //工厂类
        public Storage getObject(String name){    //返回创建的对象
            if(name.equals("UsbDisk"))
                return new UsbDisk();
            else if(name.equals("HardDisk"))
                return new HardDisk();
            else
                throw new IllegalArgumentException("非法字符串");
        }
    }
    public class Ch5_8 {                          //主类，调用方
        public static void main(String[] args) {
            Storage storage=null;                 //声明接口变量
            storage=new Factory().getObject("UsbDisk");   //调用工厂类方法
            storage.read();
            storage.write();
            storage=new Factory().getObject("HardDisk");  //调用工厂类方法
            storage.read();
            storage.write();
        }
    }
```

程序运行结果如图 5.9 所示。

分析：程序中调用方 Ch5_8 并不直接创建被调用方 UsbDisk 以及 HardDisk 类的对象，而是通过工厂类创建对象，从而降低了调用方和被调用方的耦合性。这种应用在 Java 的轻型框架 Spring 中会用到。但工厂模式中调用方却直接创建了工厂类的对象，即调用方和工厂，类

图 5.9 工厂模式测试结果

存在耦合性，工厂类与各个被调用方也存在耦合性，在 Spring 框架中，通过依赖注入、控制反转以及将要产生对象的字符串如本例中的"UsbDisk"、"HardDisk"等写入 Spring 配置文件等方法进一步解决了此问题。

5.8　Java 内置注解简介

注解（Annotation）是 Java SE 5.0 以上版本新增加的功能。注解是一种"元数据"（用来描述数据的一种数据）。注解可以被添加到程序单元上，用以对程序单元进行解释和说明。Java 开发和部署工具可以读取并处理这些注解。

注解在程序中表现为一些特殊标记，采用能被 Java 编译器进行检查、验证的格式，存储有关源代码的补充性描述信息。这些标记可以在编译、类加载、运行时被读取，并执行相应的处理。注解不影响程序的执行，无论增加、删除注解，程序的执行都不受任何影响。Java 注解包括内置的标准注解、元注解及自定义注解的功能。此处只简介 Java 的 3 个内置注解。

Java SE 5.0 的 java.lang 包中预定义了三个内置注解：@Override、@Deprecated 以及 @SuppressWarnings。

1. @Override

@Override 注解只能用于方法上，是一个限定重写方法的注解类型，用来指明被注解的方法必须是重写父类中的方法。编译器在编译源程序时会检查用@Override 注解的方法是否重写父类的方法。若不是，则发出警告。

【例题 5_9】@Override 注解测试。
```
package myproject.ch5_9;
class A{
    public void f(){
        System.out.println("父类的输出");
    }
}
class B extends A{
    public void F(){ //不构成重写
        System.out.println("子类的输出");
    }
}
public class Ch5_9 {
    public static void main(String[] args) {
        A a=new B();//上转型
        a.f();
    }
}
```
程序运行结果如图 5.10 所示。

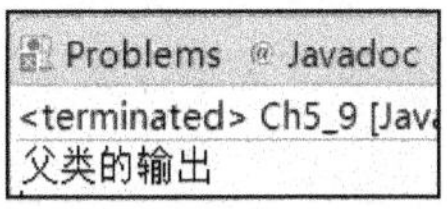

分析：程序本来的意图是子类 B 重写父类 A 的 f()方法，当执行 a.f();时输出"子类的输出"这句话，但结果却是"父类的输出"。这是因为子类 B 中在重写父类方法 f()时，错误地将方法名写成了大写 F，这就不能构成重写。为了避免此类错误的发生，可以在重写方法前加上注解@Override，这样一来，一旦发生重写错误，编译系统会报错，如图 5.11 所示。

图 5.10　例题 5_8 的结果

图 5.11 @Override 测试结果

2. @Deprecated

@Deprecated 是用来标记已过时的成员的注解类型，用来指明被注解的方法或类是一个过时的方法或类，不建议再使用它。当编译器遇到被注解为 Deprecated 的方法或类时，就会发出警告。

【例题 5_10】@Deprecated 注解测试。

```
package myproject.ch5_10;
class A{
    @Deprecated
    public void f(){
        System.out.println("Hello");
    }
}
public class Ch5_10 {
    public static void main(String[] args) {
        A a=new A();              //上转型
        a.f();
    }
}
```

分析：在类 A 的 f()方法前加了@Deprecated 注解，若在 Eclipse 环境下，发现方法名 f 上加了一横线，表示是一个过时的、建议不再使用的方法，但并不影响程序的编译和执行。在 JDK 环境下，编译该程序时，会有更加详细的提示信息，如图 5.12 所示。

图 5.12 @Deprecated 注解在 JDK 环境下的测试结果

@Deprecated 也可以加在类前面，不是一个已经过时，建议不再使用的类。如下面代码：

```
@Deprecated
class A{
    public void f(){
        System.out.println("Hello");
    }
}
public class Example2 {
    public static void main(String[] args) {
        A a=new A();              //上转型
        a.f();
    }
}
```

3. @SuppressWarnings

@SuppressWarnings 是抑制编译器警告的注解类型，用来指明被@SuppressWarnings 注解的方法、变量或类在编译时如果有警告信息，则阻止发出警告。如下所示的程序，会有警告信息。

【例题 5_11】@SuppressWarnings 注解测试。

```
package myproject.ch5_11;
import java.util.ArrayList;
import java.util.List;
class A{
    public void f(){
        List list=new ArrayList();
        list.add("abc");
        for(int i=0;i<list.size();i++)
            System.out.println(list.get(i));
    }
}
public class Ch5_11 {
    public static void main(String[] args) {
        A a=new A();
        a.f();
    }
}
```

程序中会有警告信息，如图 5.13 所示。

图 5.13 中的警告信息表示 List 类必须使用泛型才是安全的，如果想不显示这个警告信息有两种方法。一个将这个程序进行如下改写： List<String> list=new ArrayList<String>();即指明该泛型类的具体类型，此处是 String。

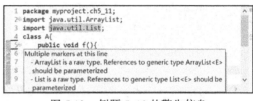

图 5.13　例题 5_10 的警告信息

另一种方法是使用@SuppressWarnings，将程序中类 A 改写为

```
class A{
    @SuppressWarnings("unchecked")
    public void f(){
    @SuppressWarnings("rawtypes")
    List list=new ArrayList();
    list.add("abc");
    for(int i=0;i<list.size();i++)
        System.out.println(list.get(i));
    }
}
```

或

```
class A{
    @SuppressWarnings({ "rawtypes", "unchecked" })
    public void f(){
    List list=new ArrayList();
    list.add("abc");
    for(int i=0;i<list.size();i++)
        System.out.println(list.get(i));
    }
}
```

分析：@SuppressWarnings 注解不同于@Override 和@Deprecated，它有一个 value 属性，可以通过这个 value 属性指定要抑制的警告类型名，如@SuppressWarnings({ "rawtypes", "unchecked" }) 表示分别要抑制没有用泛型的警告和未检查类型的警告。

小　　结

本章介绍了 Java abstract 关键字的用法，abstract 方法及 abstract 类的特点及其应用；Java 接口的定义格式及接口的特点；接口回调的用法；Java 内部类的定义及特点；匿名类的含义及实际应用方法；介绍了面向抽象（接口）编程的思想即实现方法；"工厂类"的含义及应用；简介了 Java 内置注释的含义及各自用法等。

习　　题　5

一、选择题

1. 关于 Java 的接口说法中不正确的是（　　）。
    A. 一个类可以同时实现多个接口
    B. 接口中的方法默认是 public abstract 方法
    C. 一个接口只能被某一个类来实现
    D. 若一个类声明实现某接口但没有实现该接口的所有方法，则该类一定是抽象类
2. 下面（　　）修饰符所定义的方法必须被子类所覆写。
    A. final   B. abstract   C. static   D. interface
3. 下面说法不正确的是（　　）。
    A. abstract 和 final 能同时修饰一个类
    B. 抽象类不仅可以做父类，也可以做子类
    C. 抽象方法不一定声明在抽象类中，也可以在接口中
    D. 声明为 final 的方法不能在子类中被覆写

二、上机实践题

1. 将 XiTi4_2_3 中图形类 Shape 的分别定义为抽象类，在其中定义方法 public abstract double getArea()，其余要求不变，测试程序；将 Shape 定义为接口，在其中包含方法 public double getArea()，让 Circle 类和 Rectangle 类分别实现接口 Shape，其余要求不变，测试程序。

2. 编写程序完成以下要求：

（1）定义一个抽象类 EntertainmentEquipment(娱乐设备)类，在其中定义三个抽象方法 abstract void playMusic()（听音乐），abstract void watchTV()（看电视），abstract void surfOnLine()（网上冲浪）。

（2）定义接口 Phone（电话），在其中定义方法 void call()（打电话）；定义接口 ChatTools（聊天工具），在其中定义方法 void chat()；定义接口 MessageTools（收发消息工具），在其中定义发送消息方法 void sendMessage()和接收消息方法 void recieveMessage()。

（3）定义计算机类 Computer 继承 EntertainmentEquipment 类，同时实 ChatTools,MessageTools

接口。定义智能手机类 SmartMobilePhone 继承 EntertainmentEquipment 类，同时实现 PhoneCall、ChatTools 以及 MessageTools 接口。

（4）定义主类 XiTi5_2_2，在 main 方法中分别创建 Computer 类以及 SmartMobilePhone 类的对象，并分别调用它们各自重写或实现过的方法进行测试。

3. 编写程序完成以下要求：

（1）定义名为 Software 的接口，在其中定义 4 个方法：String getSoftwareType() (得到软件类型)、String getSoftwareName()(得到软件名称)、String getSoftwareVersion()(得到软件版本)、String getSoftwareAuthor()(得到软件作者)。

（2）定义类 OS（操作系统）、OA（办公自动化系统），分别实现接口 Software。

（3）定义类 Computer，在其中定义方法 public void useComputer(Software s)，方法体中输出当前软件的类型、名称、版本以及作者等信息。

（4）在主类 XiTi5_2_3 的 main 方法中创建 Computer 类对象，调用 useComputer(Software s)，分别给参数 Software s 传递 OS 类对象和 OA 类对象，得到输出结果。

第 6 章　Java 数组与枚举

【本章内容提要】
- 一维数组；
- 数组间赋值；
- Arrays 类中处理数组的系统方法；
- Java 二维数组的定义及使用；
- Java 对象数组；
- Java 枚举类型。

6.1　一　维　数　组

数组是一组有序数据的集合，数组中的每个元素具有相同的数据类型。Java 中的数组是引用数据类型。

1．一维数组的声明

Java 一维数组的声明语法格式为：
数组元素基类型 数组名[]；
或：数组元素基类型[] 数组名；
如：int a[]; 或者：int[] a;

其中，数组元素基类型可以是 Java 中的任意数据类型，既可以是基本数据类型，也可以是引用数据类型如类或字符串等。符号"[]"说明声明的是一个数组对象。

注意：数组声明时，不能包含数组的长度。如 int a[10];就是错误的。

2．一维数组的创建

数组被声明以后，系统不会给数组分配内存空间。只有创建了数组对象或对数组进行初始化后，数组才会拥有存储空间。

和创建其他 Java 对象一样，使用 new 关键字创建一维数组对象，格式为
数组名 = new 数组元素基类型[元素个数];

可以先声明数组，再创建数组对象如：int[] a; a=new int[10]; 也可以将声明和创建放在一起，如 int[] a=new int[10];

注意：与 C 语言不同，Java 允许使用 int 型变量指定数组的大小。

如：int n=10;double arr[]=new double[n];

3. 数组元素的引用

与 C 语言一样，Java 也是采用下标引用方式引用数组元素。其中下标指数组中每个元素的索引值，数组中的第一个元素索引值是 0，第二个是 1，依次类推。在引用数组元素时要注意防止下标越界。

4. 数组的长度

在 Java 中提供了 length 属性来返回数组的长度，其格式如下：
数组名.length。

5. 一维数组初始化

数组初始化就是为数组元素指定初始值。通常在创建数组时，Java 会使每个数组元素初始化为一个默认值如：int 型默认为 0，float 型默认为 0.0f，引用类型默认为 null 等。

若不希望数组的各元素初值为默认值，则可以在声明数组的同时就为数组元素赋值。Java 编译程序会自动根据赋值的个数算出整个数组的长度，并分配相应的空间。

如：int a[]={1,2,3,4,5}; //定义一个数组 a，给数组各元素分别赋值为 1,2,3,4,5

数组成员是引用类型的也可以进行初始化，如定义了一个 Point 类如下：

```
class Point{
  int x,y;
  public Point(int x,int y){
     this.x=x;
     this.y=y;
  }
}
```

则可有

Point point[]={new Point(10,20),new Point(30,50),new Point(50,80)};
//定义一个 Point 对象数组，并以 Point 对象对其进行初始化。

【例题 6_1】输出数组中的各元素的值。

```
public class Ch6_1 {
  public static void main(String[] args) {
     int[] arr=new int[10];
     char[] ch={'A','B','C','D'};
     for(int i=0;i<arr.length;i++)
        arr[i]=i+1;                      //给数组 0-9 下标的对应的各元素分别赋值为 1~10
     System.out.println("第 1 种遍历并输出数组元素的方法: ");
     System.out.println("数组 arr 中各元素值为: ");
     for(int i=0;i<arr.length;i++)   //遍历数组 arr
        System.out.print(arr[i]+" ");//输出数组 arr 的各个元素值
     System.out.println();            //换行
     System.out.println("数组 ch 中各元素值为: ");
     for(int i=0;i<ch.length;i++)    //遍历数组 ch
        System.out.print(ch[i]+" ");//输出数组 ch 的各个元素值
     System.out.println();            //换行
     System.out.println("第 2 中遍历并输出数组元素的方法: ");
     System.out.println("数组 arr 中各元素值为: ");
     for(int a:arr)
```

```
            System.out.print(a+" ");
        System.out.println();                //换行
        System.out.println("数组 ch 中各元素值为: ");
        for(char c:ch)
            System.out.print(c+" ");
    }
}
```

程序运行结果如图 6.1 所示。

图 6.1　遍历输出数组元素

说明：for(int a:arr) System.out.print(a+" ");这种数组遍历方法是 JDK1.5 以后 Java 提供的遍历数组的方法。for(int a:arr)中的 int 表明要遍历数组的类型，a 代表各个数组元素，arr 是要遍历的数组名。

for(int a:arr) System.out.print(a+" ");与 for(int i=0;i<arr.length;i++)System.out.print(arr[i]+" ");是等价的。同理，for(char c:ch)System.out.print(c+" ");等价于 for(int i=0;i<ch.length;i++)System.out.print(ch[i]+" ");。

【例题 6_2】 运用"冒泡法"对 10 个整数按由小到大的次序排序。

```
public class Ch6_2 {
    public static void main(String[] args) {
        int a[]={324,34,222,79,67,444,668,89,3,1};
        int i,j,temp;
        int n=a.length;                  //数组元素个数
        System.out.println("排序前数组中的各元素为: ");
        for(i=0;i<a.length;i++)
            System.out.print(a[i]+" ");
        System.out.println();
        for(j=0;j<n-1;j++)               // 进行 n-1 趟比较
            for(i=0;i<n-j-1;i++)         // 每一趟中进行 n-1-j 次比较
                if (a[i]>a[i+1]){        // 数组中相邻两个数比较
                    temp=a[i];
                    a[i]=a[i+1];
                    a[i+1]=temp;
                }                        //交换
        System.out.println("排序后数组中的各元素为: ");
        for(i=0;i<a.length;i++)
            System.out.print(a[i]+" ");
    }
}
```

图 6.2　数组元素排序结果

程序运行结果如图 6.2 所示。

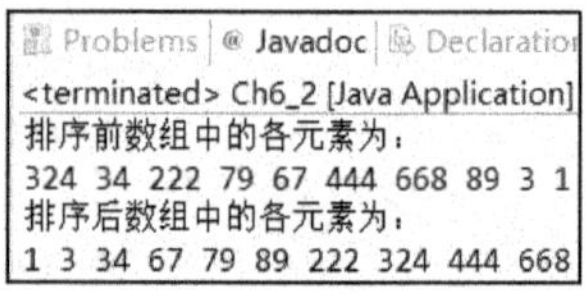

6.2　数组间赋值

两个同类型的数组引用（对象）间可以相互赋值，这时候，两个数组引用同时指向同一个数组空间。

如有：int[] arr1={1,2,3}; int[] arr2={4,5,6};arr1=arr2;

【例题 6_3】 整型数组之间赋值，观察赋值前后数组元素引用值、数组元素的变化。

```
public class Ch6_3 {
    public static void main(String[] args) {
```

```
        int[] arr1={1,2,3};
        int[] arr2={4,5,6};
        System.out.println("赋值前 arr1 的引用值: "+arr1);    //输出 arr1 的引用值
        System.out.println("赋值前 arr2 的引用值: "+arr2);    //输出 arr2 的引用值
        System.out.println("赋值前 arr1 的各元素值为: ");
        for(int a:arr1)
            System.out.printf("%3d",a);                //遍历输出 arr1 中的元素
        System.out.println();                          //换行
        System.out.println("赋值前 arr2 的各元素值为: ");
        for(int a:arr2)
            System.out.printf("%3d",a);                //遍历输出 arr2 中的元素
        arr1=arr2;                                     //将 arr2 赋值给 arr1 或称 arr1 指向 arr2
        System.out.println();                          //换行
        System.out.println("赋值后 arr1 的引用值: "+arr1);    //输出 arr1 的引用值
        System.out.println("赋值后 arr2 的引用值: "+arr2);    //输出 arr2 的引用值
        System.out.println("赋值后 arr1 的各元素值为: ");
        for(int a:arr1)
            System.out.printf("%3d",a);
        System.out.println();                          //换行
        System.out.println("赋值后 arr2 的各元素值为: ");
        for(int a:arr2)
            System.out.printf("%3d",a);
    }
}
```

程序运行结果如图 6.3 所示。

分析：在数组赋值前，arr1 和 arr2 各自拥有不同的引用值。当执行赋值语句 arr1=arr2;后，arr1 和 arr2 同时指向了 arr2 数组的地址空间，拥有了相同的引用值，此时，再次遍历输出 arr1 和 arr2 时，输出的数组元素都是它们所指向的 arr2 数组地址空间的数组元素值 4,5,6。

图 6.3　例题 6_3 运行结果

6.3　Arrays 类中处理数组的系统方法

Java.util 包中的 Arrays 类提供了方便进行数组处理的常用方法：
- public static void sort(int[] a)：用于将整型数组 a 中的元素按有效到达次序排序。
- public static int binarySearch(int[] a,int key)：使用二分法在已经排序好的 a 数组中来查找指定的 key 值在 a 中的下标位置，否则返回-1。如果数组包含多个与 key 相同的元素，则无法保证找到的是哪一个。
- public static int[] copyOfRange(int[] original,int from,int to)：将 original 数组中由 from 到 to 指定的下标范围的元素复制到另一个数组中。

当然，Arrays 类中提供的重载方法和其他方法还有很多，上面仅列出了其中的几个。

【例题 6_4】将整型数组 a 中的元素复制到整形数组 b 中，并在数组 b 中运用二分法查找指定的值。
```
import java.util.Arrays;
import java.util.Scanner;
public class Ch6_4 {
    public static void main(String[] args) {
```

```java
int a[]={56,43,232,4,666,788,89,60,3,45};
int b[]=Arrays.copyOfRange(a, 0, a.length);
//将数组 a 中所有元素复制到数组 b 中
System.out.println("数组 b 中的元素值为:");
for(int c:b)
    System.out.printf("%4d",c);
System.out.println();              //换行
Arrays.sort(b);                    //数组 b 中元素排序
System.out.println("排序后的数组 b 中元素值为:");
for(int c:b)
    System.out.printf("%4d",c);
System.out.println();              //换行
Scanner in=new Scanner(System.in);
System.out.println("请输入要在有序数组 b 中查找的元素值:");
int key=in.nextInt();              //读入待查找的 key 值
int index=Arrays.binarySearch(b,key);
//返回 key 在 b 中的下标值
if(index!=-1)                      //在 b 中找到了 key
    System.out.println(key+"在 b 中的位置为: "+index);
else{
    System.out.println("在 b 中未找到"+ key);
    System.out.println("index="+ index);
    }
  }
}
```

程序运行结果如图 6.4 所示。

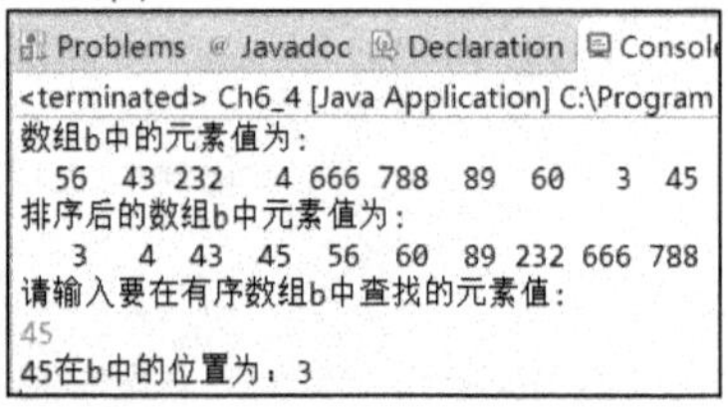

图 6.4　Arrays 类方法测试结果

6.4　Java 二维数组的定义及使用

1．二维数组的声明

二维数组的声明语法格式为：

数组元素基类型 数组名[ ][ ]；或数组元素基类型 [ ][ ] 数组名；

注意：二维数组声明时，也不能包含数组的长度，如 int a[2][3];就是错误的。

2．二维数组的创建

二维数组被声明以后，系统不会给数组分配内存空间。只有创建了数组对象或对数组进行初始化后，数组才会拥有存储空间。

创建二维数组对象的格式为：

数组名 = new 数组元素基类型[第一维大小] [第二维大小]；

可以先声明二维数组，再创建二维数组对象，如 "int[][] a; a=new int[2][3];"，也可以将声明和创建放在一起，如 "int[][] a =new int[2][3];"。

Java 也允许使用 int 型变量指定二维数组的第一、第二维的大小，如 "int a=2,b=3;double arr[][]=new double[a][b];"。

3．数组元素的引用

二维数组也是采用下标引用方式引用数组元素。每一维的下标值都是从 0 开始，在引用数组元素时要注意防止下标越界。

4．数组的长度

对于二维数组，用数组名.length 来返回该数组第一维的大小。

5．二维数组初始化

可以在声明一个二维数组的同时为其各个数组元素赋值，如
int arr1[][]={{1,2},{2,3},{3,4}};

也允许将数组初始化为第二维大小不尽相同（各行包含不同个数元素的不规则二维数组），如
"int arr2[][]={{1,2,7},{2,3},{3}};"。

也可先创建数组对象，在分别给各个数组元素一一赋，值如
double arr[][]=new double[2][3];
arr[0][0]=3.45;arr[0][1]=45.6;arr[0][2]=456.2;
arr[1][0]=98.0;arr[1][1]=232.2;arr[1][2]=33.1;

【例题 6_5】二维数组测试。

```
public class Ch6_5 {
    public static void main(String[] args) {
        int a=2,b=3;
        double arr[][]=new double[a][b];          //创建double型二维数组
        arr[0][0]=3.45;arr[0][1]=45.6;arr[0][2]=456.2;
        arr[1][0]=98.0;arr[1][1]=232.2;arr[1][2]=33.1;//给arr的各元素赋值
        int arr1[][]={{1,2},{2,3},{3,4}};          //声明二维数组arr1并初始化
        int arr2[][]={{1,2,7},{2,3},{3}};
//声明二维数组arr3,并将其初始化为不规则二维数组
        System.out.println("arr.length="+arr.length);
        System.out.println("arr1.length="+arr1.length);
        System.out.println("arr2.length="+arr2.length);   //测试length属性
        System.out.println("数组arr中的个元素为");
        for(int i=0;i<a;i++){                             //遍历输出arr的各元素值
            for(int j=0;j<b;j++)
                System.out.print(arr[i][j]+"  ");
        }
        System.out.println("\n数组arr1中的个元素为");
        for(int i=0;i<arr1.length;i++){                   //遍历输出arr1的各元素值
            for(int j=0;j<2;j++)
                System.out.print(arr1[i][j]+"  ");
        }
        System.out.println("\n数组arr2中的个元素为");
        /*遍历输出arr2的各元素值*/
        for(int j=0;j<3;j++)
            System.out.print(arr2[0][j]+"  ");
        for(int j=0;j<2;j++)
            System.out.print(arr2[1][j]+"  ");
        for(int j=0;j<1;j++)
            System.out.print(arr2[2][j]+"  ");
    }
}
```

程序运行结果如图 6.5 所示。

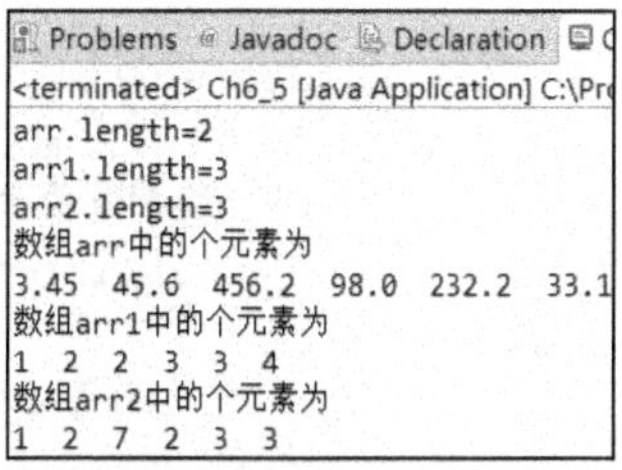

图 6.5　二维数组测试结果

【例题 6_6】将 3×4 的二维矩阵转置并输出结果。
```java
public class Ch6_6 {
    public static void main(String[] args) {
      int a[][]={{1,2,3,4},{5,6,7,8},{9,10,11,12}};
      //初始化二维数组，给矩阵各元素赋值
      int b[][]=new int[4][3];          //创建保存转置矩阵的二维数组
      System.out.println("将a按3×4矩阵方式输出:");
      for (int i=0;i<=2;i++){           //遍历数组a并输出各个元素
        for (int j=0;j<=3;j++){
          System.out.printf("%4d",a[i][j]);
          b[j][i]=a[i][j];              //转置
        }
       System.out.println();            //换行
      }
      System.out.println("将b按4×3矩阵方式输出:");
      for (int i=0;i<=3;i++){           //遍历数组b并输出各个元素
        for(int j=0;j<=2;j++)
          System.out.printf("%4d",b[i][j]);
          System.out.println();         //换行
      }
   }
}
```
程序运行结果如图 6.6 所示。

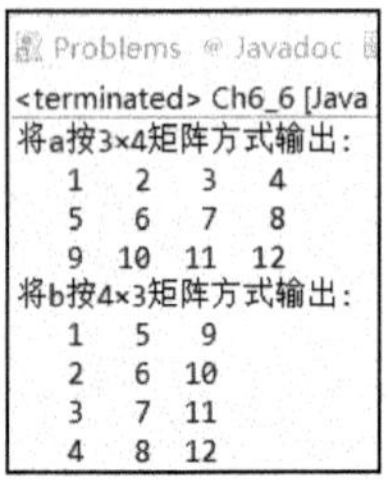

【例题 6_7】编写程序，寻找 $M \times N$ 的矩阵中值最大与最小的元素并输出，同时输出该最大值、最小值对应的行号与列号。

图 6.6 矩阵转置结果

```java
import java.util.Scanner;
public class Ch6_7 {
    public static void main(String[] args) {
        System.out.println("请输入矩阵的行数M以及列数N(M与N之间用空格分开): ");
        Scanner in=new Scanner(System.in);
        int M=in.nextInt();
        int N=in.nextInt();                  //读入M,N
        int arr[][]=new int[M][N];           //创建二维数组
        System.out.println("请输入矩阵的各个元素(各个元素之间用空格分开): ");
        for(int i=0;i<M;i++){
           for(int j=0;j<N;j++){
              arr[i][j]=in.nextInt();        //读入各个数组元素值并存放在 arr 中
           }
        }
        int max_row=0, max_col=0, max=arr[0][0]; //以 arr[0][0]为当前最大者
        int min_row=0, min_col=0,min=arr[0][0];  ///以 arr[0][0]为当前最小者
        System.out.println("该"+M+"*"+N+"阶矩阵是: ");
        for(int i=0;i<M;i++){
        //求解矩阵最大、最小值及其各自下标值并以矩阵形式输出各元素值
            for(int j=0;j<N;j++){
               System.out.printf("%-5d",arr[i][j]);
               if(max<arr[i][j]){
                   max = arr[i][j];
                   max_row = i ;
```

```
                max_col = j;
            }
            if(min>arr[i][j]){
                min=arr[i][j];
                min_row=i;
                min_col=j;
            }
        }
        System.out.println();
    }
    System.out.println("该矩阵各元素中的最大值是:"+max+",
该最大值位于矩阵的第"+(max_row+1)+"行"+"第"+(max_col+1)+"列");
    System.out.println("该矩阵各元素中的最小值
是:"+min+",
        该最小值位于矩阵的第"+(min_row+1)+"行"+"第"+(min_col+1)+"列");
    }
}
```
程序运行结果如图 6.7 所示。

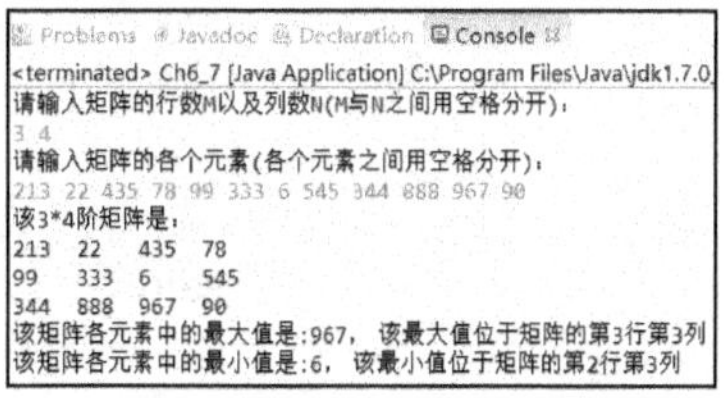

图 6.7　例题 6_7 运行结果

6.5　Java 对象数组

顾名思义，对象数组的元素应该是由某个类的对象组成。其声明和创建格式为
`类名 对象数组名[]=new 类名[数组长度];`

【例题 6_8】对象数组测试。
```
class Point{                        //定义"点"类
    int x,y;                        //坐标位置
    public Point(int x,int y){      //构造方法
        this.x=x;
        this.y=y;
    }
    void show(){                    //输出坐标位置
        System.out.println("x 坐标为位置为: "+x+" ,y 坐标为位置为:   "+y);
    }
}
public class Ch6_8{
    public static void main(String[] args) {
        Point point[]={new Point(10,20),new Point(30,50),new Point(50,80)};
        //声明对象数组并初始化
        for(int i=0;i<point.length;i++){//遍历对象数组并输出各个数组元素
            point[i].show();
        }
    }
}
```
程序运行结果如图 6.8 所示。

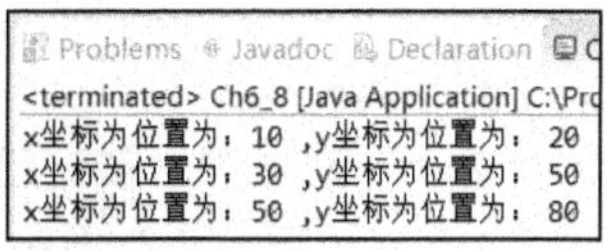

图 6.8　例题 6_8 运行结果

【例题 6_9】创建存储学生类对象的对象数组。
```
class Student{
    private String no;
```

```java
        private String name;
        private String sex;
        private int age;
        public Student(String no, String name, String sex, int age) {
            this.no = no;
            this.name = name;
            this.sex = sex;
            this.age = age;
        }
        public String getNo() {
            return no;
        }
        public String getName() {
            return name;
        }
        public String getSex() {
            return sex;
        }
        public int getAge() {
            return age;
        }
    }
    public class Ch6_9 {
        public static void main(String[] args) {
            Student stu[]=new Student[3];              //创建对象数组
            System.out.print("对象数组赋值前: ");
            for(Student s:stu){
                System.out.print(s+" ");
            //遍历数组并输出各个数组元素（学生对象）的引用值
            }
            /*给对象数组各元素赋值*/
            stu[0]=new Student("2015001","李明","男",17);
            stu[1]=new Student("2015002","张丽","女",17);
            stu[2]=new Student("2015003","王建设","男",18);
            System.out.println("\n对象数组赋值后: ");
            for(Student s:stu){                        //遍历对象数组
                System.out.println(s.getNo()+" "+s.getName()+" "+s.getSex()+" "+s.getAge());
                //得到各个学生对象的成员变量值并输出
            }
        }
    }
```

程序运行结果如图 6.9 所示。　　　　　　　　　　　　　　图 6.9　例题 6_9 运行结果

分析：对象数组被创建后，在未对其各个元素进行初始化时，各个元素的默认值为 null。本例是对 stu 对象数组各数组元素进行了逐一地赋值。也可以在声明对象数组时同时赋值如 Student stu[]={new Student("2015001"," 李 明 "," 男 ",17),new Student("2015002"," 张 丽 "," 女 ",17),new Student("2015003","王建设","男",18)};

6.6 Java 枚举类型

6.6.1 用 enum 定义枚举类型

在 JDK1.5 之后，引入了一个用于定义枚举类型的关键字 enum。定义一个枚举类型的格式如下：

```
[ public] enum 枚举类型名称{
    枚举成员1，枚举成员1，…，枚举成员n;
}
```

如定义一个颜色枚举类型如下：

```
enum Color{
    RED,GREEN,BLUE;         //每个枚举类型成员都是常量，它们都用大写字母表示
}
```

【例题 6_10】 枚举类型测试。

```
package myproject.ch6_10;
public enum WeekDay {
    MONDAY,TUESDAY,WEDNESDAY,THURSDAY,FRIDAY,SATURDAY,SUNDAY;
     //定义枚举成员
}
package myproject.ch6_10;
public class Ch6_10 {
    public static void main(String[] args) {
        WeekDay weekday=WeekDay.MONDAY;
        //用枚举类型名.枚举成员赋值给该枚举 WeekDay 的引用 weekday
        System.out.println(weekday);          //输出该枚举成员
    }
}
```

程序运行结果：MONDAY

说明：使用某个枚举成员时，需要将枚举类型名.枚举成员赋值给该枚举的引用，如"WeekDay weekday=WeekDay.MONDAY;"。

【例题 6_11】 输出所有的 WeekDay 枚举成员。

```
package myproject.ch6_11;
public enum WeekDay {
    MONDAY,TUESDAY,WEDNESDAY,THURSDAY,FRIDAY,SATURDAY,SUNDAY;
}
package myproject.ch6_11;
public class Ch6_11 {
    public static void main(String[] args) {
        System.out.println("所有枚举成员: ");
        for(WeekDay w:WeekDay.values())
           System.out.print(w+" ");
        System.out.println("\n所有枚举成员序号及名称: ");
        for(WeekDay w:WeekDay.values())
            System.out.println("序号: "+w.ordinal()+" 名称: "+w.name());
    }
}
```

程序运行结果如图 6.10 所示。

说明：枚举类型名可以调用 values()方法返回枚举成员组成的数组，遍历该数组就可以得到所有的枚举成员。此外，还可以用 java.lang.Enum 类中的 ordinal()方法返回枚举成员的序号（其中初始常量序号为 0）。用 java.lang.Enum 类中的 name()方法返回此枚举成员的名称。

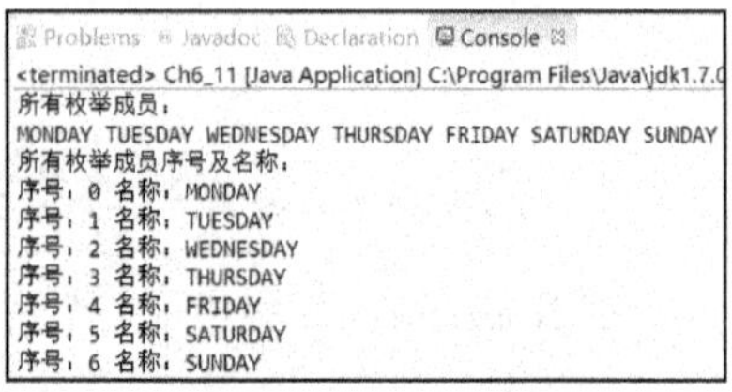

图 6.10　例题 6_11 运行结果

6.6.2　枚举类型的构造方法

有时，需要让各个枚举成员具有属性，如 MONDAY 具有属性"星期一"等。这时，需要给枚举类型定义构造方法，用于对枚举类型的成员进行初始化和属性赋值。但务必注意，构造方法必须被 private 修饰，否则出错，若构造方法前不加 private，编译系统会自动默认为 private。

【例题 6_12】 枚举类型构造方法测试。

```
package myproject.ch6_12;
public enum WeekDay {
    MONDAY("星期一"),TUESDAY("星期二"),WEDNESDAY("星期三"),THURSDAY("星期四"),FRIDAY("星期五"),SATURDAY("星期六"),SUNDAY("星期日");
//定义具有属性的枚举成员
    private String weekday;              //定义属性
    private WeekDay(String weekday) {    //定义构造方法
        this.weekday = weekday;
    }
    public String getWeekday() {         //定义getter()方法，用于获得枚举成员的属性值
        return weekday;
    }
}
package myproject.ch6_12;
public class Ch6_12 {
    public static void main(String[] args) {//遍历输出每个枚举成员的属性值
        for(WeekDay w:WeekDay.values())
            System.out.print(w.getWeekday()+" ");
    }
}
```

程序运行结果如图 6.11 所示。

说明：若将程序中的构造方法注释掉，则会报错（见图 6.12）。本例中通过 getWeekday()获得每个枚举成员的属性值。

图 6.11　例题 6_11 运行结果

图 6.12　出错信息

6.6.3　在 switch 结构中使用枚举类型

Java 允许在 switch 结构中根据枚举成员（常量）的取值来分情况执行不同的任务。

【例题 6_13】 在 switch 结构中应用枚举测试。

```
package myproject.ch6_13;
enum WeekDay {
```

```
        MONDAY,TUESDAY,WEDNESDAY,THURSDAY,FRIDAY,SATURDAY,SUNDAY;
//定义枚举成员
}
package myproject.ch6_13;
public class Ch6_13 {
    public void show(WeekDay w){            //以枚举类型实例作为形参
        switch(w){
          case MONDAY:System.out.println("星期一");break;
          case TUESDAY:System.out.println("星期二");break;
          case WEDNESDAY:System.out.println("星期三");break;
          case THURSDAY:System.out.println("星期四");break;
          case FRIDAY:System.out.println("星期五");break;
          case SATURDAY:System.out.println("星期六");break;
          case SUNDAY:System.out.println("星期日");break;
        }
    }
    public static void main(String[] args) {
        Ch6_13 test=new Ch6_13();
        test.show(WeekDay.WEDNESDAY);        //此处要写成"枚举类型.枚举成员"的形式
    }
}
```

程序运行结果如图 6.13 所示。

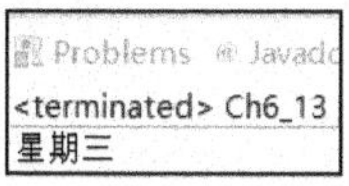

图 6.13 例题 6_13 运行结果

说明：在 switch 结构中的每个 case 后面只能写枚举成员名称如 case WEDNESDAY 等，不能写成 case WeekDay .WEDNESDAY 的形式。

小　　结

本章介绍了 Java 一维、二维数组的声明、创建、初始化方法及各自的典型应用方法；java.util 包中的 Arrays 类操作数组的常用方法的用法；Java 枚举类型的定义、枚举类型的构造方法及在 switch 结构中使用枚举类型的方法等。

习　题　6

上机实践题

1. 编程输出所有由 1，2，3，4 组成的互不相同且各位无重复数字的三位数，每输出 10 个满足要求的三位数就换行，并输出满足该要求的三位数的个数。

2. 编写程序求解一个 $M×N$ 的矩阵与一个 $N×K$ 的矩阵的乘积，以矩阵格式输出结果矩阵。要求 M、N、K 及各个矩阵元素的值由控制台读入。

3. 按以下要求设计各个类并输出结果：

（1）设计一个学生类 Student，包括成员变量学号、姓名、Java 程序设计、英语、数学等三门课程的成绩，给 Student 类定义一个含参构造方法。

（2）设计一个类表示多个学生的类 Students，以 Student 对象数组 Student[] stu 为其成员变量，定义一个成员方法 void printInfo(Student s)，用于输出学生信息。

（3）设计主类 XiTi6_1_3，在 main()方法中，在其中创建 Students 对象以及 Student 对象数组，创建多个 Student 对象给 students.stu[i]赋值，并用 Students 对象调用 printInfo(Student s)方法输出学生信息。

4. 编程完成以下要求：

（1）定义一个枚举类型 Seasons，其中包含枚举元素 SPRING,SUMMER,AUTUMN,WINTER。

（2）定义类 SeasonsAlternate，其中包含一个 public void alternate(Seasons s)方法，方法体中用 switch(s)的 s 的不同取值分别输出"春季""夏季""秋季""冬季"。

（3）定义主类 XiTi6_1_4，在主方法中创建 SeasonsAlternate 对象，测试 alternate(Seasons s)方法。遍历输出 Seasons 中的各个枚举元素值、对应的索引以及枚举元素的个数。

5. 改写第 4 小题，要求：

（1）其中的每个枚举元素含有参数，如 SPRING ("春季")。

（2）给 Seasons 定义私有成员变量 private String season，私有构造方法 private Seasons(String season)及一个用于返回 season 的方法 public String getSeason()。

（3）定义主类 XITi6_1_5，在主方法中用 getSeason()方法获取各枚举元素参数值；遍历 Seasons 并输出各枚举元素名称、参数值、索引值等。

第 7 章　Java 常用工具类

【本章内容提要】
- Object 类及其常用方法；
- 基本数据类型包装类；
- String 类；
- StringBuffer 类；
- 正则表达式；
- 字符串解析方法；
- 日期时间类；
- Math 类；
- BigInteger 类；
- Random 类；
- 其他常用类；
- Class 类与 Java 的反射机制简介。

7.1　Object 类及其常用方法

Object 类包含在 java.lang 包中，它是所有 Java 类的祖先类。在 Java 中如果定义了一个类并没有显式地继承任何类，那么它默认继承 Object 类，如

　　public class Person{ …//类体}　　// 当没有指定父类时，会默认 Object 类为其父类

上面的类定义等价于：public class Person extends Object { … //类体}

由于所有的类都是由 Object 类衍生出来的，所以 Oject 类中的方法适用于所有类。下面，通过实例，介绍 Object 类的几个常用方法的用法。

1. public String toString()

用于返回某对象的字符串描述。该字符串由类名标记符 "@" 和此对象哈希码的无符号十六进制表示组成，其形式如下：

　　getClass().getName() + " @ " +Integer.toHexString(对象名.hashCode());

【例题 7_1】toString()方法测试。
```
package myproject.ch7_1;
public class Person{
    String name;
```

```java
    int age;
    public Person(String name, int age) {
       this.name = name;
       this.age = age;
    }
}
package myproject.ch7_1;
public class Ch7_1{
  public static void main(String[] args) {
    Person zhangsan=new Person("张三",19);
    System.out.println("生成 zhangsan 对象的类名为: "+zhangsan.getClass().getName());
    System.out.println("zhangsan对象的哈希码: "+zhangsan.hashCode());
    System.out.println("zhangsan对象哈希码的无符号十六进制表示: "+Integer.toHexString(zhangsan.hashCode()));
    System.out.println(zhangsan);
    //会默认调用toString()方法，与下一行语句的输出结果相同
    System.out.println(zhangsan.toString());
  }
}
```

程序一种可能的运行结果如图 7.1 所示。

图 7.1 toString()方法测试结果

分析：由于程序的每一次运行对应的哈希码可能不同，所以说图 7.1 所示的结果是一种可能的运行结果。对象名.getClass().getName()可以获取生成该对象的类名（包括包名）。对象名.hashCode()可以获得该对象整型的哈希码值，Integer 类是基本数据类型 int 的包装类，其 toHexString()方法是以十六进制无符号整数形式返回一个整数参数的字符串表示形式。这些方法在本章后续内容中还会涉及。

【例题 7_2】改写例题 7_1 程序，重写 toString()方法如下：

```java
package myproject.ch7_2;
public class Person{
    String name;
    int age;
    public Person(String name, int age) {
       this.name = name;
       this.age = age;
    }
    @Override
    public String toString() {//重写Object类的toString()方法
       return "姓名: "+name+" 年龄: "+age;
    }
}
package myproject.ch7_2;
public class Ch7_2{
  public static void main(String[] args) {
      Person zhangsan=new Person("张三",19);
      System.out.println(zhangsan);
      System.out.println(zhangsan.toString());
  }
}
```

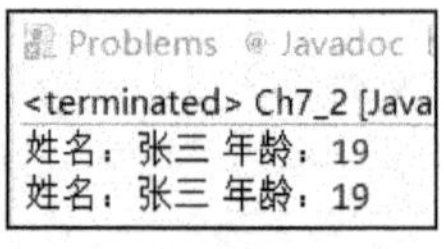

程序的运行结果如图7.2所示。

图 7.2 重写 toString()的结果

2. public boolean equals(Object obj)

用于判断两个对象内容是否相等，可以根据需要在用户自定义类中重写 equals 方法。

【例题 7_3】 equals 方法测试。

```
package myproject.ch7_3;
public class Person{
    String name;
    int age;
    public Person(String name, int age) {
        this.name = name;
        this.age = age;
    }
}
package myproject.ch7_3;
public class Ch7_3 {
    public static void main(String[] args) {
        Person p1=new Person("李四",18);
        Person p2=new Person("李四",18);
        System.out.println("p1==p2 的结果: "+(p1==p2));
        System.out.println("p1.equals(p2)的结果: "+p1.equals(p2));
    }
}
```

程序的运行结果如图 7.3 所示。

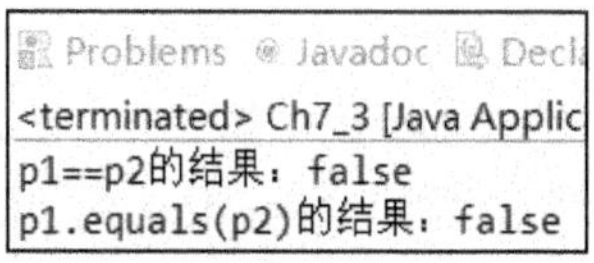

分析：运算符 "= ="用来比较两个对象是否具有相同的引用，只有两个对象具有相同的引用时，取值为 true。由于 p1、p2 是不同的对象，具有不同的引用值，因此 p1==p2 取值为 false。

图 7.3 equals 方法测试结果

p1.equals(p2)能否直接判断 p1 和 p2 对象是否具有相同的实体内容呢？显然，本例中 p1、p2 中的内容都是"李四",18，p1.equals(p2)应取值为 true，那么，为什么程序运行结果为 false 呢？这是因为对于自定义类 Person，应该重写 public boolean equals(Object obj)方法来实现当 p1 和 p2 的实体内容相等时，该方法返回 true。

改写例题 7_3 程序，重写 equals()方法如下：

```
package myproject.ch7_3;
public class Person{
    String name;
    int age;
    public Person(String name, int age) {
        this.name = name;
        this.age = age;
    }
    @Override
    /**重写 Object 类的 equals()方法,使其能够判定当前类的两个对象的内容是否相同*/
    public boolean equals(Object obj) {
        Person p=null;
        if(obj instanceof Person)
            p=(Person)obj;
        else
            return false;
        if(p.name==this.name&&p.age==this.age)//当前对象与 p 对象的内容一样
```

```
            return true;
        else
            return false;
    }
}
package myproject.ch7_3;
public class Ch7_3 {
    public static void main(String[] args) {
        Person p1=new Person("李四",18);
        Person p2=new Person("李四",18);
        System.out.println("p1==p2 的结果: "+(p1==p2));
        System.out.println("p1.equals(p2)的结果: "+p1.equals(p2));
    }
}
```

程序的运行结果如图 7.4 所示。

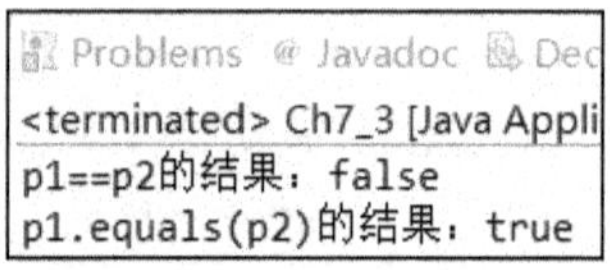

3. public int hashCode()

图 7.4 重写 equals()方法后的结果

该方法返回某对象的哈希码值。hashCode 是按照一定的算法得到的一个数值，是对象的散列码值。主要用来在集合中实现快速查找等操作，也可以用于对象的比较。通常需要子类对该方法进行重写。常见的 String 类及基本数据类型的包装类如 Integer 类等都已对 hashCode() 和 equals() 方法进行了重写，保证：若 obj1.equals(obj2)，则 obj1.hashCode()==obj2.hashCode()。其含义是：若对象 obj1 和对象 obj2 按对象的状态(或属性或内容)是相等的，则它们的 hashCode 值应相同。对于其他类对象，若没有重写 hashCode()方法，就不能保证以上结论成立。

【例题 7_4】 hashCode()方法测试。

```
package myproject.ch7_4;
public class Ch7_4 {
    public static void main(String[] args) {
        String s1=new String("Hello");
        String s2=new String("hello");
        String s3=new String("Hello");
        System.out.println("s1 的哈希码值:"+s1.hashCode());
        System.out.println("s2 的哈希码值:"+s2.hashCode());
        System.out.println("s3 的哈希码值:"+s3.hashCode());
        System.out.println("s1.equals(s2)的值: "+s1.equals(s2));
        System.out.println("s1.equals(s3)的值: "+s1.equals(s3));
    }
}
```

程序的运行结果如图 7.5 所示。

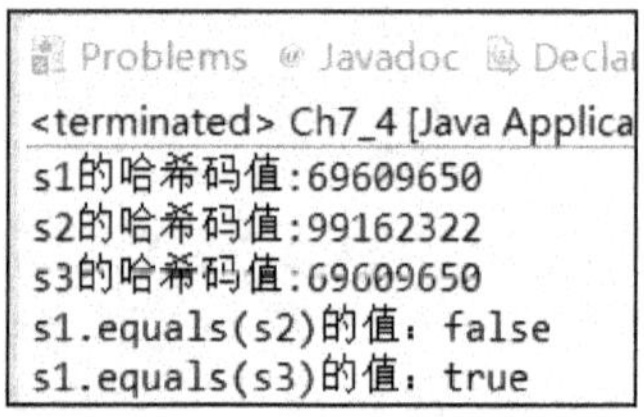

图 7.5 hashCode()方法测试结果

4. protected Object clone()throws CloneNotSupportedException

用于创建并返回此对象的一个克隆对象(副本)。要克隆某个类对象，要求该类必须实现 Cloneable 接口，并重写 Object 类的 clone()方法。

【例题 7_5】 clone()方法测试。

```
package myproject.ch7_5;
public class A implements Cloneable{    //实现 Cloneable 接口
```

```java
        private String no;
        private String name;
        public A(String no, String name) {
            this.no = no;
            this.name = name;
        }
        public String getNo() {
            return no;
        }
        public void setNo(String no) {
            this.no = no;
        }
        public String getName() {
            return name;
        }
        public void setName(String name) {
            this.name = name;
        }
        @Override
        protected Object clone() throws CloneNotSupportedException {
            return super.clone();     //调用父类的clone()方法完成克隆
        }
        @Override
        public String toString() {    //重写Object的toString()方法
           return "编号:"+no+" 名称:"+name;
        }
}
package myproject.ch7_5;
public class Ch7_5 {
    public static void main(String[] args) {
        A a1=new A("001","克隆前");
        System.out.println(a1);     //相当于System.out.println(a1.toString())
        A a2=null;
        try {
            a2=(A)a1.clone();        //克隆对象
        } catch (CloneNotSupportedException e) {
            e.printStackTrace();
        }
        a2.setNo("002");
        a2.setName("克隆后");
        System.out.println(a2);
    }
}
```
程序的运行结果如图7.6所示。

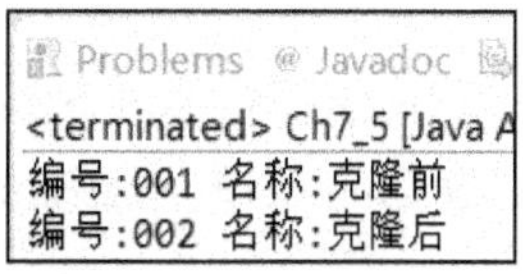

图7.6 clone()方法测试结果

7.2 基本数据类型包装类

对应于基本数据类型 boolean、byte、short、int、long、float、double、char，Java 提供了对应于这些基本数据类型的相关类，包含在 java.lang 包中，这些类分别是：Boolean、Byte、Short、

Integer、Long、Float、Double 和 Character。通过这些类实现了对基本数据类型的封装并为其提供了一系列操作方法。

如 Integer 类的常用方法：
- public int intValue()：以 int 类型返回该 Integer 的值。
- public static int parseInt(String s)throws NumberFormatException：将 String s 解析成 int 型数据并返回该数据。
- public static Integer valueOf(String s)throws NumberFormatException：返回 String s 指定的 Integer 对象。

JDK1.5 以后，基本类型数据和相应的包装类对象之间具备相互自动转换的功能，称作基本数据类型的自动装箱与拆箱（Autoboxing and Auto-Unboxing of Primitive Types）。其中，由基本数据类型到相应类对象的转换被称为"自动装箱"，反之，被称为"自动拆箱"。

【例题 7_6】Integer 类常用方法测试。

```java
package myproject.ch7_6;
public class Ch7_6{
    public static void main(String args[]){
        int i=100 ;
        Integer i1=new Integer(i);         // 将 i 包装为 Integer 类对象
        int x=i1.intValue();               // 转化为基本数据类型 int
        System.out.println("x="+x);
        /*自动装箱与拆箱**/
        Integer i2=100;                    //自动装箱
        int x1=i2;//自动拆箱
        System.out.println("x1="+x1);
        /*纯数字字符串转化为基本数据类型**/
        String s="100";
        try {
           int x2 = Integer.parseInt(s);
           System.out.println("x2="+x2);
              Integer i3=Integer.valueOf(s);
           System.out.println("i3="+i3); //自动拆箱为 int 型并输出
        } catch (NumberFormatException e) {
           e.printStackTrace();
        }
    }
}
```

程序的运行结果如图 7.7 所示。

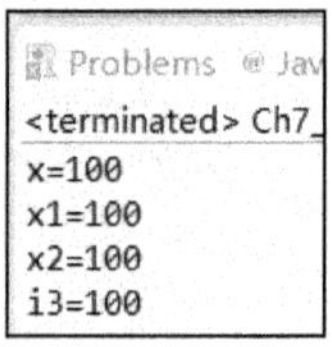

图 7.7 Integer 类常用方法测试结果

说明：在使用 parseInt 和 valueOf 方法时，都有可能引发 NumberFormatException 类型的异常，因此，要用 try，catch 语句块对其进行异常处理。

7.3 String 类

Java 语言中，把字符串作为对象来处理，类 String 和 StringBuffer 都可以用来表示一个字符串。在 Java 中，每个字符都是占用 16 bit 的 Unicode 字符。

1. 字符串常量

字符串常量对象是用双引号括起的若干字符序列，例如："Hello"、"A"、"学习"、"213.89"等。

2．字符串对象

声明字符串对象：String s;

创建字符串对象：s = new String("How are you");

在 Java 中，一个字符串常量也是一个 String 对象，可以将字符串常量直接赋值给字符串对象。

如：String s1="Hello";String s2=" Hello"; 这种情况下，Java 给 s1,s2 分配同一空间，因此，s1 与 s2 不仅具有相同的引用，也具有相同的实体内容。

而通过 String 类的构造方法创建 String 对象时，如：String s3=new String("Hello"); String s4=new String("Hello"); 这种情况下，Java 给 s3,s4 分配不同的空间，因此，s3 与 s4 具有不同的引用，但具有相同的实体内容。

【例题 7_7】equals()方法与"=="运算符用法比较。

```
package myproject.ch7_7;
public class Ch7_7 {
    public static void main(String[] args) {
        String s1 = "hello";
        String s2 = "hello";
        System.out.println(s1==s2);           //结果为 true
        System.out.println(s1.equals(s2));    //结果为 true
        String s3 = new String("hello");
        String s4 = new String("hello");
        System.out.println(s3==s4);           //结果为 false
        System.out.println(s3.equals(s4));    //结果为 true
    }
}
```

分析：由于 String 已对 equals()方法进行了重写，所以可以直接使用 equals 方法判断两个 String 对象的内容是否相等。

3．字符串的常用方法

- public boolean equals(String s)：是对 Object 类中的 equals 方法的重写，用于判定当前字符串是否与 s 的内容相同。
- public int length()：求字符串的长度（包含的字符个数）的方法。
- public String concat(String str)：将指定字符串连接到当前字符串的末尾并返回形成的新串。
- public boolean startsWith(String prefix)和 public boolean endsWith(String suffix)：这两个方法分别用来判断当前字符串的前缀和后缀是否为指定的子串，若是返回 true，否则返回 false。
- public int indexOf(int ch)：从前往后查找字符 ch 在当前字符串中第一次出现的位置，并返回字符 ch 出现的位置，若找不到返回-1。
- public int indexOf(String str,int fromIndex)：返回字符串中从 fromIndex 开始出现 str 的第一个位置。
- public int lastIndexOf(int ch)：返回 ch 在当前字符串中最后一次出现处的索引值。
- public char charAt(int index)：返回字符串中下标为 index 的字符。
- public boolean contains(CharSequence s)：若当前字符串中包含 s 时则返回 true，否则返回 false。
- public int compareTo(String anotherString)：用于按字典序比较两个字符串的大小，若当前字

串与参数字符串完全相同时返回 0；若当前字符串大于参数字符串时返回 1；否则返回-1。
- public boolean equalsIgnoreCase(String another)：忽略大小写地比较当前字符串与 another 是否一样。
- public String replace(char oldChar,char newChar)：在字符串中用 newChar 字符替换 oldChar 字符。
- public String toLowerCase( )：返回字符串的小写形式；public String toUpperCase( )：返回字符串的大写形式。
- public String substring(int beginIndex)：返回该字符串从 beginIndex 开始到结尾的子字符串。
- public String substring(int beginIndex,int endIndex)：返回该字符串从 beginIndex 开始到 endIndex-1 结尾的子字符串。
- public String trim()：返回将该字符串去掉开头和结尾空格后的字符串。
- public String[] split(String regex)可以将一个字符串按照 regex 指定的分隔符进行分隔，返回分隔后的字符串数组。

【例题 7_8】字符串常见方法测试。

```
package myproject.ch7_8;
public class Ch7_8 {
    public static void main(String[] args) {
        String s1="We are Students,";
        String s2="We study hard.";
        String stuno="2013000001";
        String stuname=" Rose ";
        String s3=s1.concat(s2);
        System.out.println(s3);
        System.out.println(s3.length());
        System.out.println(s3.startsWith("We"));
        System.out.println(s3.charAt(3));
        System.out.println(s3.lastIndexOf("We"));
        System.out.println("s3 中第 3 到第 5 个字符是: "+ s3.substring(3,6));
        System.out.println(s1.compareTo(s2));
        System.out.println(stuno.replace('3', '5'));
        System.out.println(stuname.equals("Rose"));
        String stuname1=stuname.trim();
        System.out.println(stuname1.equals("Rose"));
        System.out.println(stuname.equals("Rose"));
    }
}
```

程序的运行结果如图 7.8 所示。

4．字符串与简单类型的相互转化

（1）将简单数据类型转化为字符串

运用 String 的静态方法 public static String valueOf(参数)方法，可以将简单数据类型转化为对应的字符串表示。其中的"参数"可以是 byte、char、int、long、float、double 等简单数据类型的变量或常量值。

图 7.8　字符串常见方法测试结果

（2）将字符串转化为简单数据类型

用简单数据类型的包装类调用自己的 parseXXX(String s)方法将 String s 转化为对应的简单数

据类型如：
```
double x=Double.parseDouble("123.45");
int x=Integer.parseInt("234");
```
但应注意：parseXXX(String s)方法的原型是：public static double parseXXX(String s) throws NumberFormatException，当参数 String s 中包含非法字符时，会引发 NumberFormatException 异常。

5. 字符数组、字节数组与字符串

字符串与字符数组、字节数组之间有一些相互转化的方法。
String 类中有两个用字符数组创建字符串对象的构造方法：
- public String(char[] value)：该构造方法用指定的字符数组构造一个字符串对象。
- public String(char[] value,int offset,int count)：用指定的字符数组的一部分即从 offset 开始取 count 个字符构造一个字符串对象。

String 类的还提供了下列与字节数组、字符数组相关的方法：
- public byte[] getBytes()：使用平台的默认字符集将此 String 编码为 byte 序列，并将结果存储到一个新的 byte 数组中。
- public byte[] getBytes(Charset charset)：使用给定的 charset 将此 String 编码到 byte 序列，并将结果存储到新的 byte 数组。
- public void getChars(int srcBegin,int srcEnd,char[] dst,int dstBegin) ：将当前字符串中的一部分字符拷贝到参数 dst 指定的数组中，将字符串中从位置 srcBegin 到 srcEnd −1 位置上的字符拷贝的数组 dst 中，并从数组 dst 的 dstBegin 处开始存放这些字符。需要注意的是，必须保证数组 dst 能容纳下要被复制的字符。
- public char[] toCharArray()：将调用该方法的字符串转换为一个新的字符数组。该数组的长度与字符串的长度相等，并将字符串对象的全部字符拷贝到该数组中。

【例题 7_9】 统计一个字符串中数字字符、英文字符、空格字符以及其他字符的个数。
```
package myproject.ch7_9;
import java.util.*;
public class Ch7_9 {
 static int digitals=0;
 static int characters=0;
 static int blanks=0;
 static int others=0;
 public static void main(String[] args) {
   System.out.println("请输入一个字符串: ");
   Scanner scan=new Scanner(System.in);
   String s=scan.nextLine();            //读入字符串
   char[] ch=s.toCharArray();           //将字符串转化成字符数组
   for(int i=0;i<ch.length;i++){        //遍历数组
    if (ch[i]>='0'&& ch[i]<='9'){       //若是数字
      digitals++;
    }
    else if((ch[i]>='a'&&ch[i]<='z')||ch[i]>='A'&& ch[i]<='Z'){//若是字母
      characters++;
    }
    else if(ch[i]==' ') {               //若是空格
      blanks++;
    }
```

```
        else {
          others++;
        }
      }
      System.out.println("数字字符的个数: "+digitals);
      System.out.println("英文字符的个数: "+characters);
      System.out.println("空格字符的个数: "+blanks);
      System.out.println("其他字符的个数:"+others);
    }
  }
```

程序的运行结果如图 7.9 所示。

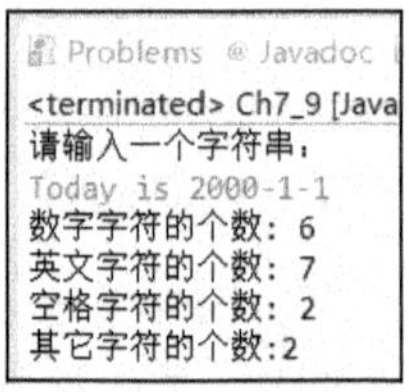

图 7.9 例题 7_9 运行结果

【例题 7_10】对字符串进行 MD5 加密。

MD5(Message Digest 5)是一种加密算法，能够对字节数组进行加密，有如下特点：不能根据加密后的信息（密文）得到加密前的信息（明文）；对于不同的明文，加密后的密文也是不同的。

在 Java 类库中，java.security.MessageDigest 类提供了对 MD5 算法实现的支持，因此我们不用了解 MD5 算法的细节也能实现它。基本过程为：首先把要加密的字符串转换成字节数组；获取 MessageDigest 对象，利用该对象的 digest 方法完成加密，返回字节数组。

```
package myproject.ch7_10;
import java.security.MessageDigest;
import java.security.NoSuchAlgorithmException;
import javax.swing.JOptionPane;
public class Ch7_10 {
    public static String generateMD5(String str){      //自定义 MD5 加密的方法
        MessageDigest md5=null;                         //声明 MessageDigest 对象
        try {
           md5 = MessageDigest.getInstance("MD5");     //得到 MessageDigest 对象
        }
        catch (NoSuchAlgorithmException e) {}
        byte[] srcBytes=str.getBytes();                 //将待加密的字符串 str 转成字节数组
        byte[] resultBytes= md5.digest(srcBytes);
        //使用 srcBytes 数组更新摘要，然后完成摘要计算,得到 resultBytes。
        String s=new String(resultBytes);               //将 resultBytes 转成字符串
        return s;                                       //返回加密串
    }
    public static void main(String[] args) {
        String str = JOptionPane.showInputDialog("请输入待加密的串: ");
        //输入对话框
        JOptionPane.showMessageDialog(null,"MD5 加密后的结果是: "
        + Ch7_10.generateMD5(str));                    //显示加密结果
    }
}
```

程序的运行结果如图 7.10（a）、(b) 所示。

（a）输入明文对话框　　　　　　（b）显示 MD5 加密结果的消息框

图 7.10 MD5 加密测试结果

7.4 StringBuffer 类

java.lang.StringBuffer 对象表示可变长的字符串。StringBuffer 类的常见构造方法：
- public StringBuffer()：构造一个空的字符串缓冲区，其初始容量为 16 个字符。
- public StringBuffer(int capacity)：构造一个不带字符，但具有 capacity 所指定初始容量的字符串缓冲区。
- public StringBuffer(String str)：构造一个字符串缓冲区，并将其内容初始化为指定的字符串内容，该字符串的初始容量为 16 加上字符串参数的长度。

StringBuffer 类的常用方法：
- 一组重载的 public StringBuffer append(...)方法，如 public StringBuffer append(Object obj)表示将参数 obj 的内容追加到当前 StringBuffer 对象后形成新的 StringBuffer 对象。
- public int capacity()：返回当前 StringBuffer 对象的容量。
- 一组重载的 public StringBuffer insert(...)方法，如 public StringBuffer insert(int offset,Object obj)表示在当前 StringBuffer 对象地 offset 下标处插入 obj，obj 将被转换成字符串。
- public StringBuffer delete(int start,int end)：用于删除从 start 开始到 end-1 为止的一段字符序列并返回修改后的该 StringBuffer 对象引用。
- public void setCharAt(int index,char ch)：用于将 index 处的字符设置为 ch。
- public StringBuffer reverse()：用于将字符序列逆序排列并返回修改后的该 StringBuffer 对象引用。

此外，StringBuffer 类还拥有 indexOf、lastIndexOf、length、replace 等方法，其各自含义及用法与 String 类的对应方法相同。

【例题 7_11】 StringBuffer 类常用方法测试。

```
package myproject.ch7_11;
public class Ch7_11 {
    public static void main(String[] args) {
        StringBuffer s1=new StringBuffer("ABCGHI");
        System.out.println("s1 的容量为"+s1.capacity());
        System.out.println("s1 的长度为"+s1.length());
        s1.insert(3,"DEF");
        System.out.println("在 s1 中插入'DEF'后的结果为: "+s1);
        s1.append("KKK");
        System.out.println("在 s1 后追加'KKK'的结果为: "+s1);
        s1.setCharAt(9,'J');
        System.out.println("将 s1 中 9 下标的字符替换为'J'后结果为: "+s1);
        s1.deleteCharAt(10);
        System.out.println("将 s1 中 10 下标的字符删除后结果为: "+s1);
        System.out.println("将 s1 中从 0 到 4 下标的字符替换为
         abcde 后结果为: "+s1.replace(0,5,"abcde"));
        System.out.println("将 s1 置逆后的结果为: "+s1.reverse());
    }
}
```

程序的运行结果如图 7.11 所示。

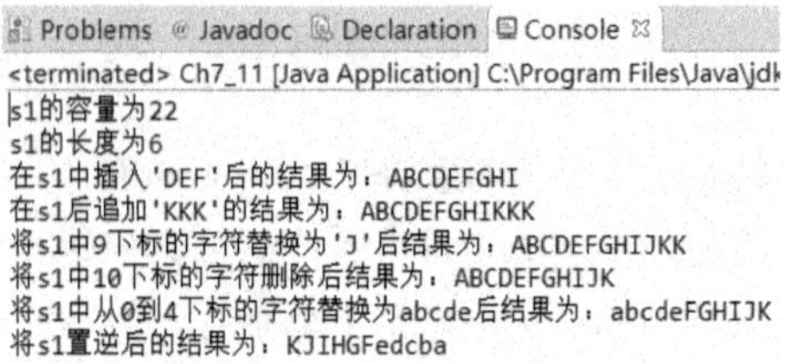
Problems @ Javadoc Declaration Console
<terminated> Ch7_11 [Java Application] C:\Program Files\Java\jdk
s1的容量为22
s1的长度为6
在s1中插入'DEF'后的结果为：ABCDEFGHI
在s1后追加'KKK'的结果为：ABCDEFGHIKKK
将s1中9下标的字符替换为'J'后结果为：ABCDEFGHIJKK
将s1中10下标的字符删除后结果为：ABCDEFGHIJK
将s1中从0到4下标的字符替换为abcde后结果为：abcdeFGHIJK
将s1置逆后的结果为：KJIHGFedcba

图 7.11　StringBuffer 类常用方法测试结果

7.5　正则表达式

7.5.1　正则表达式简介

正则表达式是按照一定规范构造的特殊字符串，用来描述特定的字符串模式，经常用于字符串的匹配与合法性校验等。

在正则表达式中，有些字符具有特殊含义，如表 7.1、表 7.2 所示。其中表 7.1 是常用的元字符表，运用其中的字符可以构造具有特定含义的正则表达式；表 7.2 是常用量词表，用于限定正则表达式中字符出现的次数。

表 7.1　常用元字符

<table>
<tr><th>元　字　符</th><th>在正则表达式中的写法</th><th>含　义</th></tr>
<tr><td>.</td><td>.</td><td>代表任何一个字符</td></tr>
<tr><td>\d</td><td>\\d</td><td>代表 0～9 的任何一个数字：[0～9]</td></tr>
<tr><td>\D</td><td>\\D</td><td>代表任何一个非数字字符：[^0～9]</td></tr>
<tr><td>\s</td><td>\\s</td><td>代表空白字符[\t\n\x0B\f\r]</td></tr>
<tr><td>\S</td><td>\\S</td><td>代表非空白符([^\s])</td></tr>
<tr><td>\w</td><td>\\w</td><td>代表单词字符：[a～z，A～Z，0～9]</td></tr>
<tr><td>\W</td><td>\\W</td><td>代表非单词字符：[^\w]</td></tr>
<tr><td>\uhhhh</td><td>\\ uhhhh</td><td>代表十六进制表示的 unicode 值为 hhhh 的字符</td></tr>
<tr><td>\p{Lower}</td><td>\\p{Lower}</td><td>小写字母[a～z]</td></tr>
<tr><td>\p{Upper}</td><td>\\p{Upper}</td><td>大写字母[A～Z]</td></tr>
<tr><td>\p{Alpha}</td><td>\\p{Alpha}</td><td>字母字符：[\p{Lower}\p{Upper}]</td></tr>
<tr><td>\p{Punct}</td><td>\\p{Punct}</td><td>标点符号：!"#$%&'()*+,-./:;<=>?@[\]^_`{|}~</td></tr>
<tr><td>\p{Blank}</td><td>\\p{Blank}</td><td>空格或制表符：[\t]</td></tr>
</table>

注意：由于“.”在正则表达式中是代表任何一个字符，若要在正则表达式中使用点字符本身，表示为[.]或\56。反斜线字符 '\'用于引用转义构造， \\ 与单个反斜线匹配，而 \{ 与左括号匹配。

在正则表达式中可以用“[]”将若干个字符括起来表示一个元字符。如：

- [abc]：代表 a、b 和 c 中的任意一个字符。
- [^abc]：代表除了 a、b 和 c 之外的任何字符。
- [a–zA–z]：代表从 a 到 z 或从 A 到 Z 的中的任何一个字符。
- [a–c[h–j]]：代表 a、b、c、h、i 和 j 中的任何字符（并集）。
- [a–z&&[hij]]：代表 h、i 或 j 中任何一个字符（交集）。

- [a-g&&[^b-d]]：代表a、e、f、g中的任意一个字符（差集）。

以上所述的各种元字符、量词等都只是具有特定含义的组成正则表达式的基本字符，实际应用中要将这些字符有机组合，并置于双引号中形成字符串才是正则表达式。

字符串对象或常量可以调用 public boolean matches (String regex)方法，当且仅当此字符串匹配 regex 给定的正则表达式时，返回 true。

表 7.2　常 用 量 词

量 词	含　义
X*	X 出现 0 次或多次
X+	X 出现 1 次或多次
X?	X 出现 0 次或 1 次
X{n}	X 恰好出现 n 次
X{n,}	X 至少出现 n 次
X{n,m}	X 出现 n 次到 m 次
XY	X 后跟 Y
X|Y	X 或 Y
(X)	定义捕获组

【例题 7_12】正则表达式测试。
```
public class Ch7_12 {
    public static void main(String[] args) {
        String s=new String("a");
        System.out.println(s.matches("[abc]"));              //true
        System.out.println(s.matches("\\p{Lower}"));         //true
        System.out.println("a".matches("[abc]"));            //true
        System.out.println("b".matches("[^abc]"));           //false
        System.out.println("k".matches("[a-zA-Z]"));         //true
        System.out.println("D".matches("[A-Z&&[F-P]]"));     //false
        System.out.println("\t\n\f\r".matches("\\s{4}"));    //true
        System.out.println("!.?".matches("\\p{Punct}+"));    //true
        System.out.println("abc_123".matches("\\w{7}"));     //true
    }
}
```

7.5.2　Pattern 与 Macther 类

java.util.regex 包中的 Pattern 与 Macther 类提供了一系列处理正则表达式的方法。Pattern 类的对象代表了一个以字符串形式指定的正则表达式。用字符串形式指定的正则表达式，必须先编译成 Pattern 类的实例。生成的模式用于创建 Matcher 对象，Matcher 对象可以根据正则表达式与任意字符序列进行匹配。使用步骤：

（1）使用正则表达式 regex 做参数得到一个 Pattern 对象 pattern。如：Pattern pattern = Pattern.compile(regex);

（2）对象 pattern 调用 matcher(CharSequence input)方法返回一个 Matcher 对象 matcher，如：Matcher matcher = pattern.matcher(input); 其中，input 表示待匹配的字符序列。

（3）Matcher 对象 matcher 可以使用下列方法寻找字符串 input 中是否有和正则表达式 regex 匹配的子序列。

- public boolean find():寻找 input 和 regex 匹配的下一子序列，如果成功该方法返回 true，否则返回 false。
- public String group()：返回和 regex 匹配的子序列。
- public boolean matches()：判断 input 是否完全和 regex 匹配。
- public boolean lookingAt()：判断从 input 的开始位置是否有和 regex 匹配的子序列。
- public String replaceAll(String replacement)：以参数 replacement 指定的字符串替换 input 中与 regex 匹配的所有子串，并返回替换后的字符串。

【例题 7_13】判断一个字符串是否是由纯数字字符组成的。

```java
package myproject.ch7_13;
import java.util.regex.Matcher;
import java.util.regex.Pattern;
import javax.swing.JOptionPane;
public class Ch7_13 {
    public static void main(String args[]){
    Pattern p;
    Matcher m;
    String regex="[0-9]+";           //正则表达式
    String input=JOptionPane.showInputDialog("请输入一个字符串: ");
    p=Pattern.compile(regex);        //编译正则表达式
    m=p.matcher(input);              //模式匹配
    if(m.matches()){                 // 若匹配
        JOptionPane.showMessageDialog(null,"该字符串是由数字字符组成的！");
    }
    else{
        JOptionPane.showMessageDialog(null,"该字符串包含非数字字符！");
    }
  }
}
```

程序的运行结果如图 7.12 所示。

图 7.12　例题 7_13 运行结果

思考："[0-9]+";可以替换为"\\d+";吗？

【例题 7_14】校验用户的手机号码和 QQ 号码的合法性。
```java
package myproject.ch7_14;
import java.util.Scanner;
import java.util.regex.Matcher;
import java.util.regex.Pattern;
public class  Ch7_14 {
    private static Scanner scan;
    public static void main(String args[]){
    Pattern p;
    Matcher m;
    String regex = "((13[0-9])|(14[5|7])|(15([0-3]|[5-9]))|(18[0,5-9]))\\d{8}";
    //验证手机号码的正则表达式
    String regex1="[1-9][0-9]{4,}";        //验证QQ号码的正则表达式
    scan = new Scanner(System.in);
    System.out.println("请输入您的手机号码: ");
    String phoneNumber=scan.nextLine();  //读入手机号
    p=Pattern.compile(regex);             //编译正则表达式
    m=p.matcher(phoneNumber);             //模式匹配
    if(m.matches())
        System.out.println("一个合法的手机号码！");
```

```
        else
            System.out.println("非法的手机号码! ");
        System.out.println("请输入您的QQ号码: ");
        String QQNumber=scan.nextLine();           //读入QQ号
        p=Pattern.compile(regex1);
        m=p.matcher(QQNumber);
        if(m.matches())
            System.out.println("一个合法的QQ号码! ");
        else
            System.out.println("非法的QQ号码! ");
    }
}
```

程序的运行结果如图 7.13 所示。

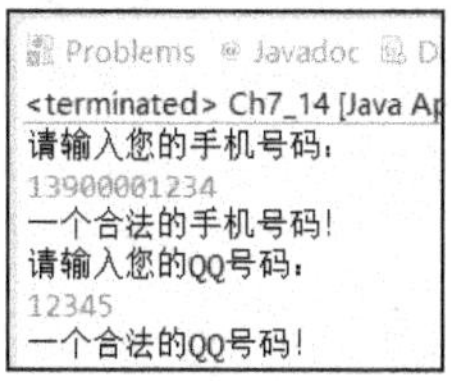

图 7.13 例题 7_14 运行结果

分析：程序中正则表达式"((13[0-9])|(14[5|7])|(15([0-3]|[5-9]))|(18[0,5-9]))\\d{8}"可以正确匹配目前常用的移动、电信、联通的移动电话号码，如目前移动电话号码为 11 位数字，移动号码的前 3 位以 139、138、137、136 等开头，电信号码的前 3 位以 133、189 等开头，联通号码的前 3 位以 130、131、136 等开头，后面的 8 位只要是数字即可。

程序中正则表达式"[1-9][0-9]{4,}"表示验证一个腾讯QQ号码的合法性，而腾讯QQ号是从 10000 开始，所以 QQ 号码的第一位是 1~9 之间的数字，而后面是连续的 4 位以上的数字组合即可。

【例题 7_15】文本查找与替换。

```java
package myproject.ch7_15;
import java.util.regex.Matcher;
import java.util.regex.Pattern;
public class Ch7_15 {
    public static void main(String[] args) {
        String str="apple 苹果 orange 橘子 banana 香蕉";   //待匹配字符串
        String regex = "[\\u4e00-\\u9fa5]+";              //匹配中文字符的正则表达式
        Pattern pattern = Pattern.compile(regex);
        Matcher matcher = pattern.matcher(str);
        System.out.print("提取出的内容为:");
        String s1=null;
        String s2=null;
        while(matcher.find()){
          s1=matcher.group();            //获取与 regex 匹配的子串
          System.out.print(s1+" ");      //输出每个与 regex 匹配的子串
        }
        System.out.println();            //输出空行
        s2=matcher.replaceAll(" ");      //以空格替换 str 中所有与 regex 匹配的子串
        System.out.println("替换后的字符串:"+s2);
    }
}
```

程序的运行结果如图 7.14 所示。

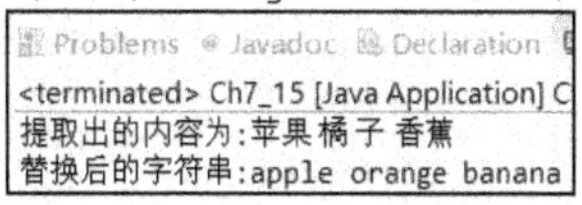

图 7.14 例题 7_15 运行结果

7.6 字符串解析方法

按照特定的分隔符、正则表达式可以将一个长字符串（文本）解析为单词、数字等。

1. 用 String 类的 split 方法

public String[] split(String regex)可以将一个字符串按照 regex 指定的分隔符进行分隔，返回分隔后的字符串数组。

2. 用 StringTokenizer 类

要用 java.util 包中的 StringTokenizer 类可以将字符串按照指定的分隔符解析为单词,首先需要创建 StringTokenizer 对象。

StringTokenizer 的常用构造方法有:

- public StringTokenizer(String str),其中 str 是待解析的字符串。使用默认的分隔符集即空格符、制表符、换行符、回车符和换页符等作为解析分隔符。
- public StringTokenizer(String str,String delim),其中 str 是待解析的字符串。delim 参数中的字符都是分隔标记的分隔符(可以有多个)。

StringTokenizer 对象可以使用下列方法获得解析出的各个单词以及统计出单词个数等。

- public int countTokens():得到 StringTokenizer 对象中计数变量的值,可以用作统计解析出的单词个数。
- public String nextToken():是一个"游标"方法,即每被调用一次,会依次返回解析出的一个单词。
- public boolean hasMoreTokens():通常与 nextToken()方法配合使用,当用 nextToken()方法取完解析出的所有单词后返回 false。

3. 用 Scanner 类

java.util 包中 Scanner 类可以将字符串按照指定的分隔符解析为单词,首先需要创建 Scanner 对象。

可以用构造方法 public Scanner(String source)创建一个 Scanner 对象,其中以 source 作为待解析字符串,用空格作为分隔符将字符串解析为单词。

Scanner 对象可以调用 public Scanner useDelimiter(String pattern)方法将一个正则表达式 pattern,待解析字符串 source 中和正则表达式 pattern 匹配的部分都作为解析字符串的分隔符。

Scanner 对象可以调用以下方法获得解析出的各个单词:

- public String next():是一个"游标"方法,即每被调用一次,会依次返回解析出的一个单词。
- public boolean hasNext():通常与 next()方法配合使用,当用 next ()方法取完解析出的所有单词后返回 false。

值得注意的是,Scanner 类功能强大,如可以用来包装输入流对象等。同时,Scanner 提供了用于读取各种基本数据类型的方法 nextInt()、nextLong()、nextDouble()等以及与之配合使用的 hasNextInt()、hasNextLong()、hasNextDouble()等方法。

【例题 7_16】 将字符串 "I am a student, I like to study Java." 解析为一个个单词并输出,同时统计解析出的单词个数并输出。

```
package myproject.ch7_16;
import java.util.Scanner;
import java.util.StringTokenizer;
public class Ch7_16{
    public static void main(String[] args) {
    String str="I am a student, I like to study Java.";//待解析的字符串
    String regex="[\\s\\p{Punct}]+";//构造由空格和常见标点符号组成的正则表达式
```

```java
        /**第 1 种解析方法*/
        String words[]=str.split(regex);              //以 regex 为分隔符解析 str
        System.out.println("###用 split 方法从解析单词###");
        System.out.print("str 中包含: ");
        for(int i=0;i<words.length;i++)
            System.out.print(" "+words[i]+" ");
        System.out.println("等"+words.length+"个单词");
        /**第 2 种解析方法*/
        System.out.println("***用 StringTokenizer 解析单词***");
        System.out.print("str 中包含: ");
        StringTokenizer tokenizer=new StringTokenizer(str," ,.");//指定空格、逗号和句点为分隔符
        int count=tokenizer.countTokens();              //解析出的单词个数
        while(tokenizer.hasMoreTokens()){              //若还有下一个单词
            String word=tokenizer.nextToken();         //取出该单词
            System.out.print(" "+word+" ");
        }
        System.out.println("等"+count+"个单词");
        /**第 3 种解析方法*/
        System.out.println("@@@用 Scanner 解析单词@@@");
        System.out.print("str 中包含: ");
        Scanner scan=new Scanner(str);
        scan.useDelimiter(regex);                      //指定分隔符
        count=0;
        while(scan.hasNext()){
            String word=scan.next();
            count++;
            System.out.print(" "+word+" ");
        }
        System.out.println("等"+count+"个单词");
    }
}
```

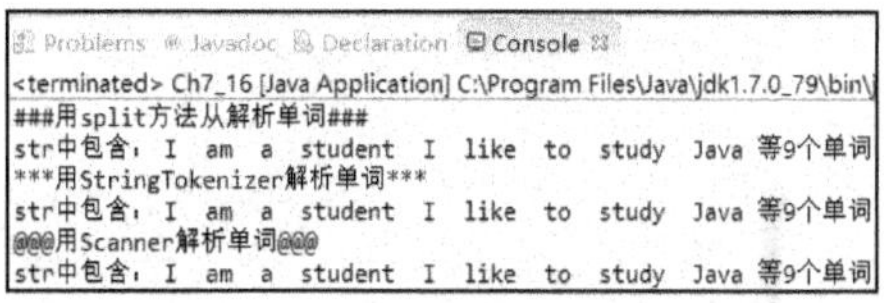

程序的运行结果如图 7.15 所示。

图 7.15 字符串解析结果

【例题 7_17】解析出字符串"语文:89.0 数学:87.5 英语:67.5"中的各个分数,求得各个分数之和并输出,解析出该字符串包含的中文单词以及单词个数并输出。

```java
package myproject.ch7_17;
import java.util.Scanner;
public class Ch7_17 {
    public static void main(String[] args) {
        String str="语文:89.0 数学:87.5 英语:67.5";        //待解析的字符串
        Scanner scan=new Scanner(str);
        String regex="[^0123456789.]+";                //除了 0-9 以及小数点以外的字符
        scan.useDelimiter(regex);                      //指定分隔符
        double sum=scan.nextDouble();                  //读入分数
        while(scan.hasNextDouble()){
            sum=sum+scan.nextDouble();
        }
        System.out.println("总成绩为: "+sum);
        String regex1="[^\\u4e00-\\u9fa5]+";            //非中文字符的正则表达式
        Scanner scan1=new Scanner(str);
        scan1.useDelimiter(regex1);
        int count=0;
```

```
            System.out.println("str 中包含的汉语单词为: ");
            while(scan1.hasNext()){
                String s=scan1.next();
                count++;
                System.out.print(s+" ");
            }
            System.out.println("\nstr 中有"+count+"个汉语单词");
        }
    }
```
程序的运行结果如图 7.16 所示。

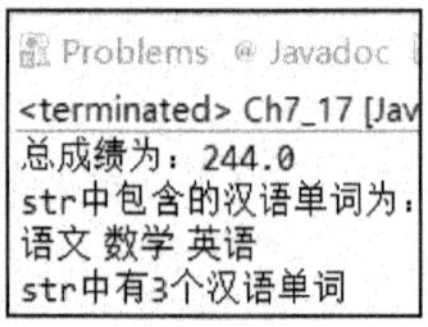

图 7.16 例题 7_17 运行结果

7.7 日期时间类

1．Date 类

运用 java.util.Date 类的无参构造方法 public Date()给 Date 类创建对象，可以得到本机系统当前时间。

2．日期格式化

java.text.DataFormat 类是抽象类，其实现类 java.text.SimpleDateFormat 用于定义日期的格式，该类主要用于将字符串解析成日期并将日期格式化成字符串。

字符串"yyyy-MM-dd-EEEE-hh-mm-ss"决定了日期的格式。"yyyy"表示长度为 4 的年份，"MM"表示月份，"dd"表示日期，"EEEE"表示星期，"hh""mm""ss"分别表示时、分、秒。

【例题 7_18】日期格式化测试。
```java
import java.util.Date;
import java.text.DateFormat;
import java.text.SimpleDateFormat;
public class Ch7_18 {
    public static void main(String[] args) {
        Date now=new Date();
        System.out.println("本机的当前系统时间为: "+now);
        DateFormat f=DateFormat.getDateTimeInstance(DateFormat.FULL,DateFormat.FULL);
        System.out.println(f.format(now));
        SimpleDateFormat sf = new SimpleDateFormat("yyyy年MM月dd日  hh时mm分ss秒");
        System.out.println(sf.format(now));
    }
}
```
程序的运行结果如图 7.17 所示。

图 7.17 日期格式化结果

说明：DateFormat 类中提供了若干个重载的 getDateTimeInstance 方法。如 public static final DateFormat getDateTimeInstance(int dateStyle, int timeStyle)用于得到一个 DateFormat 实例，该格式器具有默认语言环境的给定日期和时间格式化风格。其中，dateStyle 给定的日期格式化风格，timeStyle 给定的时间格式化风格。public final String format(Date date)方法是将一个 Date 对象格式化为设定的日期、时间字符串。public SimpleDateFormat(String pattern)方法是用 pattern 给定的模式和默认语言环境的日期格式创建 SimpleDateFormat 对象。

3. Calendar 类

通过 Date、DateFormat、SimpleDateFormat 等已经能够创建并格式化一个日期对象了。但如何才能设置和获取日期数据的特定部分年份、月份、日期、时、分、秒等信息呢？这就要通过 Calendar 类的相关方法才能完成。

java.util 包中的 Calendar 类是 Java 中提供的日历类。Calendar 类是一个抽象类，Calendar 类可以通过静态方法 public static Calendar getInstance()得到默认时区和语言环境的日历对象，返回的 Calendar 对象是基于本机系统当前时间的。例如："Calendar c = Calendar.getInstance();"。

【例题 7_19】 获取当前时间的特定部分。

```java
import java.util.Calendar;
public class Ch7_19 {
    public static void main(String[] args) {
        Calendar c=Calendar.getInstance();
        int year=c.get(Calendar.YEAR);
        int month=c.get(Calendar.MONTH)+1;              //0表示1月，1表示2月...
        int day = c.get(Calendar.DAY_OF_MONTH);
        int weekday=c.get(Calendar.DAY_OF_WEEK)-1; //1表示星期日，2表示星期一...
        int hour=c.get(Calendar.HOUR_OF_DAY);
        int second=c.get(Calendar.SECOND);
        int minute=c.get(Calendar.MINUTE);
        System.out.println("今天是"+year+"年"+month+"月"+day+"日"+",星期"+weekday);
        System.out.println("现在是"+hour+"时"+minute+"分"+second+"秒");
        c.set(2011,6,21);                               //将日历翻到指定年月日
        int year1=c.get(Calendar.YEAR);
        int month1=c.get(Calendar.MONTH)+1;             //0表示1月，1表示2月，...
        int day1 = c.get(Calendar.DAY_OF_MONTH);
        int weekday1=c.get(Calendar.DAY_OF_WEEK)-1;//1表示星期日，2表示星期一，...
        System.out.println("当前日历被翻到："+year1+"年"+month1+"月"+day1
            +"日"+",星期"+weekday1);
    }
}
```

程序的运行结果如图 7.18 所示。

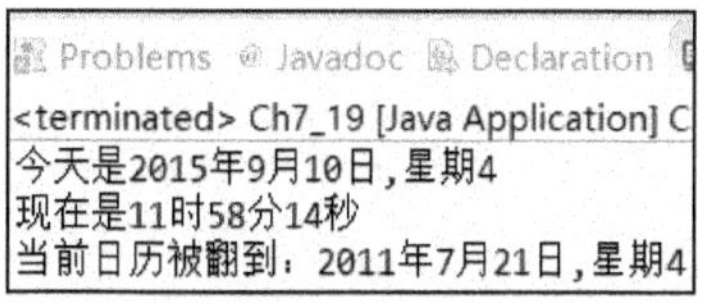

图 7.18 例题 7_19 运行结果

说明：Calendar 类中提供了若干个常量如 YEAR、MONTH、DAY_OF_MONTH、DAY_OF_WEEK、HOUR_OF_DAY、SECOND、MINUTE 等，用于表示当前日历对象的年、月、日、星期几、时、分、秒等特定部分，但要注意 MONTH 取值 0 表示 1 月，1 表示 2 月，……，DAY_OF_WEEK 取值 1 表示星期日，2 表示星期一，……get 方法用于获取这些信息。set 方法用于将日历翻到某一特定的日期和时间。

7.8 Math 类

Java.lang.Math 类提供了许多用于数学运算的常量和静态方法。

Math 类的定义：public final class Math extends Object，由于 Math 类是 final 类型的，因此不能派生子类；Math 类的构造方法是 private 类型的，因此 Math 类不能够被实例化。

Math 类的常量如下：

- public static final double E=2.7182818284590452354 //数学常数 e

- public static final doublePI=3.141592653589779323846 //圆周率常量 π

Math 类的常用方法如下：

- public static double abs(double a); //绝对值
- public static double floor(double a); //不大于 a 的最大整数
- public static double log(double a); //自然对数
- public static double log10(double a); //返回 a 的底数为 10 的对数
- public static double sqrt(double a); //开平方
- public static double random(); //产生 0 到 1 之间的随机数
- public static double pow(double a, double b); //乘方
- public static double rint(double a); //四舍五入
- public static double IEEEremainder(double a, double b); // 求余数(取模运算)
- public static double sin(double a);//返回角的三角正弦
- public static double tan(double a);//返回角的三角正切
- public static double cos(double a);//返回角的三角余弦
- public static double toDegrees(double angrad);//用弧度转换为角度
- public static double toRadians(double angdeg);//用角度转换为近似弧度表示的角

【例题 7_20】Math 常量及常见方法测试。

```
public class Ch7_20 {
    public static void main(String[] args) {
        System.out.println("e 的值为:"+Math.E);
        System.out.println("pi 的值为:"+Math.PI);
            System.out.println("abs(-100)="+Math.abs(-100));
        System.out.println("floor(34.56)=" + Math.floor(34.56));
        System.out.println("max(10,15)=" + Math.max(10,15));
        System.out.println("min(10,15)=" + Math.min(10,15));
        System.out.println("log(pow(Math.E,5.0))=" + Math.log(Math.pow(Math.E,5.0)));
        System.out.println("sqrt(256)=" + Math.sqrt(256));
        System.out.println("生成0到1之间的随机数: " + Math.random());
        System.out.println("34.56 四舍五入的结果值为: " + Math.rint(34.56));
        System.out.println("123/20 的余数为: " + Math.IEEEremainder(123, 20));
        System.out.println("sin(PI/2)=" + Math.sin(Math.PI/2));
        System.out.println("cos(PI)=" + Math.cos(Math.PI));
        System.out.println("tan(PI/2)=" + Math.tan(Math.PI/2));
        System.out.println("弧度 PI 对应的角度是: " + Math.toDegrees(Math.PI));
        System.out.println("角度 180° 对应的弧度: " + Math.toRadians(180));
    }
}
```

程序的运行结果如图 7.19 所示。

图 7.19 例题 7_20 运行结果

7.9 BigInteger 类

java.math 包中的 BigInteger 类提供了大整数类，可以进行任意精度的整数运算，解决了超出 long 型整数表示范围的大整数运算问题。可以 public BigInteger(String val) 构造一个十进制的 BigInteger 对象。

BigInteger 类的常用类方法：

- public BigInteger add(BigInteger val)：返回当前大整数对象与参数指定的大整数对象的和。
- public BigInteger subtract(BigInteger val)：返回当前大整数对象与参数指定的大整数对象的差。
- public BigInteger multiply(BigInteger val)：返回当前大整数对象与参数指定的大整数对象的积。
- public BigInteger divide(BigInteger val)：返回当前大整数对象与参数指定的大整数对象的商。
- public BigInteger remainder(BigInteger val)：返回当前大整数对象与参数指定的大整数对象的余。

【例题 7_21】 BigInteger 类方法测试。

```java
import java.math.BigInteger ;
public class Ch7_21{
    public static void main(String args[]){
        BigInteger b1 = new BigInteger("987654321") ;
        BigInteger b2 = new BigInteger("123456789") ;
        System.out.println("和: " + b1.add(b2)) ;
        System.out.println("差: " + b1.subtract(b2)) ;
        System.out.println("积: " + b1.multiply(b2)) ;
        System.out.println("商: " + b1.divide(b2)) ;
        System.out.println("余数: " + b1.remainder(b2)) ;
    }
}
```

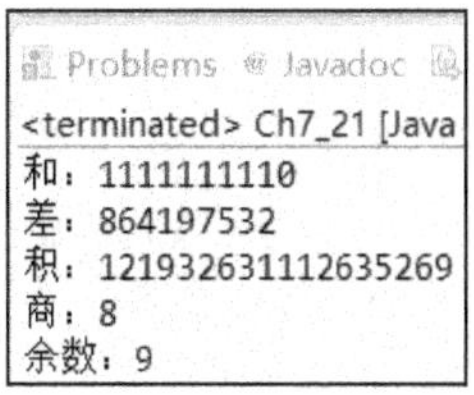

程序的运行结果如图 7.20 所示。

图 7.20 BigInteger 类方法测试结果

7.10 Random 类

java.util.Random 类提供了一系列用于生成伪随机数的方法。java.util.Random 类提供了一系列用于生成伪随机数的方法，之所以称之为伪随机数是因为它们是简单的均匀分布序列。

Random 类的如下构造方法：

- public Random()：创建一个 Random 对象。
- public Random(long seed)：使用参数 seek 指定的种子创建一个 Random 对象。

相同种子数的 Random 对象，相同次数生成的随机数字是完全相同的。也就是说，两个种子数相同的 Random 对象，第一次生成的随机数字完全相同，第二次生成的随机数字也完全相同。种子数只是随机算法的起源数字，和生成的随机数字的区间无关。

Random 类的常用方法：

- public int nextInt()：生成一个随机的介于 -2^{31} 和 $2^{31}-1$ 之间的 int 值。
- public int nextInt(int n)：生成一个随机的 int 值，该值在[0,n)区间内，如 random.nextInt(100); 返回一个 0~100 之间的随机整数(包括 0，但不包括 100)。
- public boolean nextBoolean()：生成一个随机的 boolean 值。
- public double nextDouble()：生成一个随机的 double 值，数值介于[0,1.0)之间。

【例题 7_22】带种子数的 Random 测试。
```java
import java.util.Random;
public class Ch7_22{
    public static void main(String args[]){
        Random r = new Random(10) ;
        Random r1 = new Random(10) ;         //创建两个种子数相同的 Random 对象
        System.out.println("r生成的随机序列: ") ;
        for(int i=0;i<5;i++){
          System.out.print(r.nextInt() + " ") ;
        }
        System.out.println("\n"+"r1生成的随机序列: ") ;
        for(int i=0;i<5;i++){
          System.out.print(r1.nextInt() + " ") ;
        }
    }
}
```
程序运行的结果如图 7.21 所示。

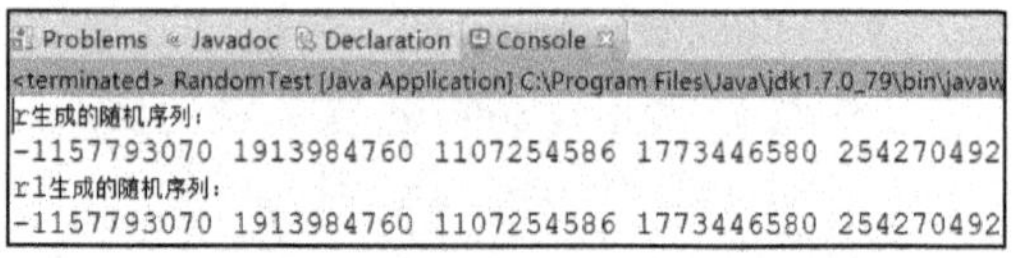

图 7.21 例题 7_22 的运行结果

【例题 7_23】Random 常用方法测试。
```java
import java.util.Random;
public class Ch7_23{
    public static void main(String args[]){
        Random r = new Random() ;
        System.out.println("[0,100)之间的随机序列: ") ;
        for(int i=0;i<5;i++){
          System.out.print(r.nextInt(100) + " ") ;
        }
        System.out.println("\n"+"随机boolean序列: ") ;
        for(int i=0;i<5;i++){
          System.out.print(r.nextBoolean() + " ") ;
        }
        System.out.println("\n"+"[0,1.0)之间随机 double 序列: ") ;
        for(int i=0;i<5;i++){
          System.out.println(r.nextDouble()) ;
        }
    }
}
```
程序运行的可能结果如图 7.22 所示。

图 7.22 例题 7_22 的运行结果

7.11 其他常用类

1. Runtime 类

每一个 Java 应用程序在运行时都会创建一个 java.lang.Runtime 类的实例。通过这个实例，应用程序可以和运行环境进行交互操作。Runtime 类没有构造方法，所以只能通过它提供的 getRuntime()方法来获取一个 Runtime 对象。一旦获得 Runtime 对象，就可以调用几个控制 Java 虚拟机的状态和行为的方法。

Runtime 定义的常用方法有：
- public static Runtime getRuntime()：返回与当前 Java 应用程序相关的运行时对象。
- public Process exec(String command)throws IOException：在单独的进程中执行指定的字符串命令。
- public long freeMemory()：返回 Java 虚拟机中的空闲内存量。

【例题 7_24】打开记事本程序并获取 JVM 空闲内存量。
```
import java.io.IOException;
public class Ch7_24{
    public static void main(String args[]){
        Runtime run = Runtime.getRuntime();        // 获得 Runtime 类对象
        System.out.println("当前JVM空闲内存量: "+run.freeMemory());
//获得Java虚拟机中的空闲内存量
        try {
            run.exec("notepad.exe");               //打开本机的记事本程序
        }
        catch (IOException e) {
            e.printStackTrace();
        }
        System.out.println("打开记事本后JVM空闲内存量: "+run.freeMemory());
    }
}
```
程序运行后首先会打开记事本程序，控制台的输出结果如图 7.23 所示。

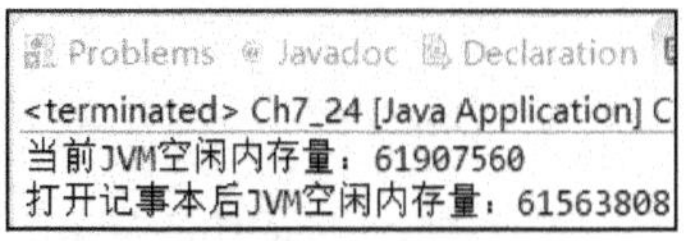

图 7.23 例题 7_24 运行结果

2. System 类

java.lang.System 类是系统中最常用的类，它定义了 3 个很有用的静态成员：out、in 和 err，分别表示标准的输出流、输入流和错误输出流。关于标准输入、输出流的用法，将在第 11 章中详细介绍。

System 类中还定义了一系列的静态方法，供程序与系统交互。例如：
- public static void exit(int status)：终止当前正在运行的 Java 虚拟机。
- public static long currentTimeMillis()：返回以毫秒为单位的当前时间，其值是本机系统当前时间与 1970 年 1 月 1 日午夜之间的时间差的毫秒值。该方法可以用于计量一个算法的运行时间。
- public static String getProperty(String key)：获取指定键指示的系统属性。Key 值可以取 java.version(Java 运行时环境版本)、java.home(Java 安装目录)、os.name(操作系统的名称、user.dir(用户的当前工作目录)等。

【例题 7_25】测试 for 循环的运行时间以及获取本机操作系统信息。
```
public class Ch7_25{
    public static void main(String args[]){
        long t1 = System.currentTimeMillis();
        for(int i=1;i<=80000000;i++){ }
        long t2 = System.currentTimeMillis();
        System.out.println("for 循环运行了:"+(t2-t1)+"毫秒。");
        //测试本 for 循环所用时间
```

```
        System.out.println("本机操作系统信息: "+
        System.getProperty("os.name")+ System.getProperty("os.version")) ;
        //获取本机操作系统信息
    }
}
```
程序运行的可能结果如图 7.24 所示。

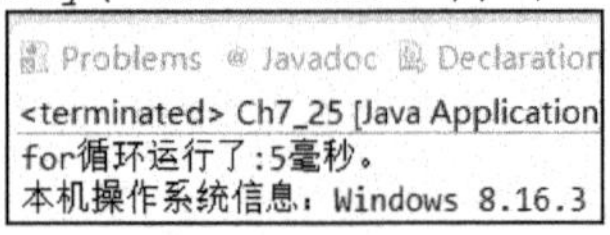

图 7.24　例题 7_25 运行结果

7.12　Class 类与 Java 的反射机制简介

反射（Reflection）是 Java 语言的特征之一。运行中的 Java 程序对自身进行检查，并能直接操作程序的内部属性，在运行时动态加载类、获取类信息、生成对象、操作对象的属性或方法等。

Java 中的反射机制主要用到 java.lang.Class 类以及 java.lang.reflect 包中的相关类：

- java.lang.Class 类：Class 类的对象表示正在运行的 Java 应用程序中的类和接口等。
- java.lang.reflect.Field 类：提供有关类或接口的属性（成员变量）的信息。
- java.lang.reflect.Constructor 类：提供关于类的构造方法信息。
- java.lang.reflect.Method 类：提供关于类或接口上某个成员方法的信息。
- java.lang.reflect.Array 类：提供了动态创建数组和访问数组的静态方法。该类中的所有方法都是静态方法。

1. Class 类

Java 中，运行中的类或接口在 JVM 中都会有一个对应的 Class 对象存在，它保存了对应类或接口的类型信息。JVM 为每种类型管理着一个独一无二的 Class 对象。

根据反射机制，运用 Class 类，可以方便地获得它所代表的实体（类、接口、数组、枚举、注解、基本类型或 void）的信息。

要想获取类和接口的相应信息，需要先获取这个 Class 对象。

2. 获得 Class 类对象

可以用以下方式获得 Class 类对象：

- 用 Object 类的 getClass()方法来得到 Class 对象。例如：Class c=对象名.getClass();
- 用 Class 类的 forName()静态方法获得与字符串对应的 Class 对象。

forName()方法定义如下：

public static Class<?> forName(String className) throws ClassNotFoundException：用于返回与参数字符串指定的类或接口相关的 Class 对象。注意：若 className 包含在某个包中，则必须以完整的 "包名.className" 形式传递。

- 使用类型名.class 获取该类型对应的 Class 对象。例如：Class c=某类名.class;

【例题 7_26】获得某类对应的 Class 对象。
```
package myproject.ch7_26;
class A{ }
public class Ch7_26{
    public static void main(String[] args) {
        /*第 1 种方式**/
        A a=new A();
```

```
        Class<?> c11=a.getClass();              //与 a 对应的 Class 对象
        Class<?> c12=new java.util.Date().getClass();//与 Date()对象对应的 Class 对象
        System.out.println(c11.getName());      //输出类名
        System.out.println(c12.getName());      //输出类名
        /*第 2 种方式**/
        try {
          Class<?> c21=Class.forName("myproject.ch7_26.A");
          Class<?> c22=Class.forName("java.util.Date");//必须有完整的包名,不能只写 Date
          System.out.println(c21.getName());
          System.out.println(c22.getName());
        }
        catch (ClassNotFoundException e) {
            e.printStackTrace();
        }
        /*第 3 种方式**/
        Class<?> c31=A.class;
        Class<?> c32=java.util.Date.class;
        System.out.println(c31.getName());
        System.out.println(c32.getName());
    }
}
```

程序运行的结果如图 7.25 所示。

图 7.25　例题 7_26 运行结果

说明：程序中运用 3 种不同的方式得到了自定义类 A 对应的 Class 对象 c11，c21，c31，并分别调用 getName()方法输出了 A 类完整的包名及类名。同时，也得到了系统类 java.util.Date 类所对应的 Class 对象 c12，c22，c32，并输出了 Date 类的包名及类名。

Class 是一个泛型类，所以 Class<?>来指定泛型类型。

在以上 3 种获得 Class 类对象的方式中，最常用的是用 Class 类的 forName()静态方法获得与字符串对应的 Class 对象的方式。

3．通过 Class 类获取类的信息

Class 类提供的用于获取此 Class 对象所表示的实体的信息的常用方法：

- public String getName()：返回此 Class 对象所表示的实体（类、接口、数组类、基本类型或 void）名称，若该类包含在某个包中，则返回"包名.类名"。
- public Package getPackage()：获取此 Class 对象所表示的实体的所在包。
- public Class<? super T> getSuperclass()：返回表示此 Class 所表示的实体（类、接口、基本类型或 void）的超类的 Class。
- public Constructor<?>[] getConstructors():获取此 Class 对象所表示的实体的所有 public 构造方法。
- public Constructor<?>[] getDeclaredConstructors()throws SecurityException：返回 Constructor 对象的一个数组，获取此 Class 对象所表示的实体的所有构造方法。
- public Field getField(String name) throws NoSuchFieldException,SecurityException：获取此 Class 对象所表示的实体的所有 public 成员变量。
- public Field[] getDeclaredFields() throws SecurityException：获取此 Class 对象所表示的实体的所有字段。

- public Method[] getMethods() throws SecurityException：返回一个包含某些 Method 对象的数组，获取此 Class 对象所表示的实体的所有 public 方法。
- public Method[] getDeclaredMethods() throws SecurityException：返回 Method 对象的一个数组，获取此 Class 对象所表示的实体的所有方法。

【例题 7_27】Class 类常用方法测试。

```
package myproject.ch7_27;
import java.lang.reflect.Constructor;
import java.lang.reflect.Field;
import java.lang.reflect.Method;
class A{
    int a;
    String s;
    public A(){}
    public A(int a, String s) {
        this.a = a;
        this.s = s;
    }
    public void f1() {}
    public void f2() {}
}
public class Ch7_27 {
    public static void main(String[] args) {
    try {
        Class<?> c=Class.forName("myproject.ch7_27.A");
        System.out.println("A 的类名"+c.getName());
        System.out.println("A 的父类"+c.getSuperclass());
        System.out.println("A 所处的包"+c.getPackage());
        Constructor<?>[] constructor=c.getDeclaredConstructors();
        //获取 c 所代表的类的所有构造方法
        System.out.println("A 中定义的所有构造方法");
        for(Constructor<?> con:constructor){         //遍历输出数组 constructor
            System.out.println(con);
        }
        Field[] field=c.getDeclaredFields();
        System.out.println("A 中定义的所有成员变量");
        for(Field f:field){                          //遍历输出数组 field
            System.out.println(f);
        }
        Method[] method=c.getDeclaredMethods();
        System.out.println("A 中定义的所有成员方法");
        for(Method m:method){                        //遍历输出数组 field
            System.out.println(m);
        }
    } catch (ClassNotFoundException e) {
        e.printStackTrace();
      }
    }
}
```

程序运行的可能结果如图 7.26 所示。

本例对 Class 类提供的常用方法进行了测试，可将程序中语句 Class<?> c=Class.forName e("myproject.ch7_27.A"); 中的 "myproject.ch7_27.A" 替换为某个 Java 系统类如 "java.util.Date"等进行测试。

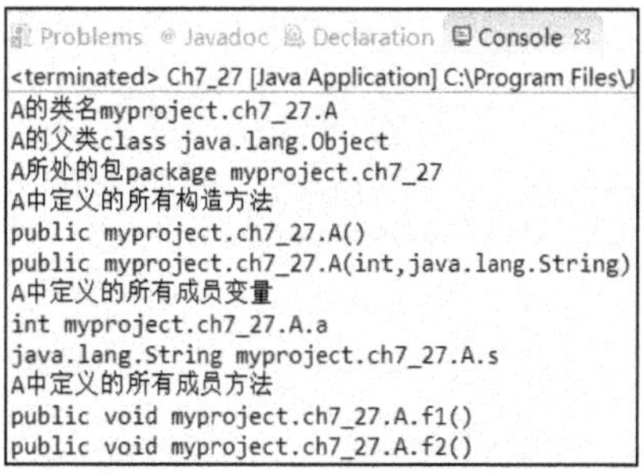

4. 运用 Class 实例化一个对象

可以通过 Java 反射 API，根据需要选择对应类的适当构造方法在运行时动态创建出对象。下面分两种情况来讨论利用反射创建对象的方式。

图 7.26 Class 类常用方法测试结果

（1）第 1 种方式步骤如下：

Step1：使用 Class 的 public static Class forName(String className) throws ClassNotFound Exception 方法得到一个和参数 className 指定的类相关的 Class 对象。

Step2：使用 Step1 中获得的 Class 对象调用 public Object newInstance() throws Instantiation Exceptio n,IllegalAccessException 方法就可以得到一个 className 类的对象。

需要注意的是：使用 Class 对象调用 newInstance()实例化一个 className 类的对象时，className 指定的类必须有无参的构造方法。

【例题 7_28】 运用 Class 实例化一个对象测试 1。

```
package myproject.ch7_28;
class Circle{
  private double radius;
  public Circle(){}                    //无参构造方法
  public Circle(double radius) {       //含参构造方法
    this.radius = radius;
  }
public double getRadius() {
    return radius;
  }
  public void setRadius(double radius) {
    this.radius = radius;
  }
  public double getArea(){
      return 3.14*Math.pow(radius, 2);
  }
}
public class Ch7_28 {
    public static void main(String[] args) {
        Class<?> cs=null;
        try {
            cs=Class.forName("myproject.ch7_28.Circle");//获取与 Circle 类对应的 Class 类对象
        } catch (ClassNotFoundException e) {
            e.printStackTrace();
        }
        try {
```

```
            Circle c=(Circle)cs.newInstance();        //得到Circle类对象
            c.setRadius(10.0);
            System.out.println("圆面积为: "+c.getArea());
        } catch (InstantiationException | IllegalAccessException e) {//捕获两
种类型的异常
            e.printStackTrace();
        }
    }
}
```

程序运行的可能结果如图 7.27 所示。

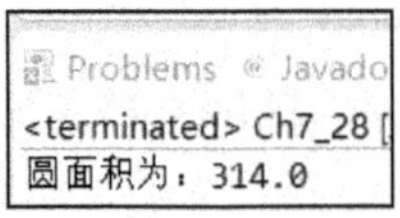

图 7.27　例题 7_28 运行结果

程序的运行结果表明可以用与 Circle 类对应的 Class 类对象调用 newInstance()方法给 Circle 类创建对象，可以替代传统用 new 运算符创建类对象的方式。

讨论： ①将程序中 Circle 类的无参构造方法、含参构造方法都去掉或注释起来，发现程序运行结果不变。

② 将程序中 Circle 类的无参构造方法去掉或注释起来，执行程序会发生如图 7.28 所示的异常：

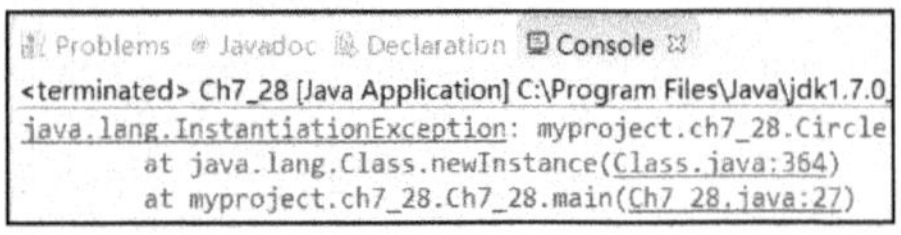

这是因为当定义了含参构造方法后，系统不再提供默认的无参构造方法，而用 newInstance()方法创建对象时一定需要无参构造方法造成的。

图 7.28　出错信息

③ 将程序中 Circle 类的含参构造方法去掉或注释起来，程序运行正常。

鉴于以上测试结果，一般地在用 Class 类给一个类创建对象时，在给该类给出含参构造方法的同时，务必同时定义无参构造方法。

（2）第 2 种方式步骤如下：

Step1：获取要创建对象的类对应的 Class 对象。

Step2：通过 Class 的 getConstrutors()得到该类的全部构造方法组成的 Constructor 数组。

Step3：调用指定 Constructor 对象的 newInstance 方法，传入对应的参数值，创建出对象。

【例题 7_29】 运用 Class 实例化一个对象测试 2。

```
package myproject.ch7_29;
import java.lang.reflect.Constructor;
import java.lang.reflect.InvocationTargetException;
class Circle{
  private double radius;
  public Circle(){}                        //无参构造方法
  public Circle(double radius) {           //含参构造方法
    this.radius = radius;
  }
  public double getRadius() {
    return radius;
  }
  public void setRadius(double radius) {
    this.radius = radius;
  }
  public double getArea(){
      return 3.14*Math.pow(radius, 2);
```

```
    }
}
public class Ch7_29 {
    public static void main(String[] args) {
        Class<?> cs=null;
        try {
            cs=Class.forName("myproject.ch7_29.Circle");//获取与Person类对应的Class类对象
        } catch (ClassNotFoundException e) {
            e.printStackTrace();
        }
        Constructor<?>[] con=cs.getConstructors();//得到Constructor对象数组
        Circle c=null;
        try {
            c = (Circle)con[1].newInstance(10.0);  //获得Circle类对象
        } catch (InstantiationException | IllegalAccessException | IllegalArgumentException
                | InvocationTargetException e) {    //捕获多种类型异常
            e.printStackTrace();
        }
        System.out.println("圆面积为: "+c.getArea());
    }
}
```

程序运行的可能结果如图7.27所示。

讨论：在 Constructor<?>[] con=cs.getConstructors();得到了 Circle 类的所有构造方法数组，由于程序中给 Circle 类定义了一个无参构造方法和一个含参构造方法，而通过 c = (Circle)con[1].new Instance(10.0);语句产生 Circle 类对象时传入了参数 10.0，说明使用的是 Circle 类的含参构造方法，它对应了 con[1]（下标 0 对应第一个构造方法即无参构造方法，下标 1 对应了第二个构造方法即含参构造方法，……）。

5. 反射的一个应用——简单工厂模式的改进

在第 5 章 5.7 中讨论了简单工厂模式，下面例题是利用反射机制对例题 5_8 的一个改进。

【**例题 7_30**】简单工厂模式测试。

```
package myproject.ch7_30;
/*存储器接口**/
interface Storage{
    void read();
    void write();
}
/*UsbDisk 类，被调用方**/
class UsbDisk implements Storage{
    @Override
    public void read() {           //实现 read()方法
        System.out.println("正在读取 UsbDisk...");
    }
    @Override
    public void write() {          //实现 write()方法
```

```java
            System.out.println("正在将数据写入UsbDisk...");
        }
    }
    /*HardDisk类,被调用方**/
    class HardDisk implements Storage{
        @Override
        public void read() {
            System.out.println("正在读取HardDisk...");
        }
        @Override
        public void write() {
            System.out.println("正在将数据写入HardDisk...");
        }
    }
    /*工厂类**/
    class Factory{
        public static Object getObject(String className) throws Exception{
            return Class.forName(className).newInstance();    //通过反射机制创建对象
        }
    }
    /*主类,调用方**/
    public class Ch7_30{
        public static void main(String[] args) throws Exception {
            Storage storage=null;                              //声明接口变量
            storage=(Storage)Factory.getObject("myproject.ch7_30.UsbDisk");
//需要包名.类名
            storage.read();
            storage.write();
            storage=(Storage)Factory.getObject("myproject.ch7_30.HardDisk");
//需要包名.类名
            storage.read();
            storage.write();
        }
    }
```

程序运行结果如图7.29所示。

图7.29 例题7_30运行结果

小　　结

本章介绍了Java Object类的及其常用方法的用法；基本数据类型的包装类、"自动装箱"与"自动拆箱"的概念与用法；String类常用方法的应用；字符数组、字节数组与字符串的转化；StringBuffer类的应用；Java正则表达式的含义及典型应用方法；日期时间工具类的应用；数学类的应用；Runtime类、System类的应用；Class类及Java反射机制的典型应用等。

习　题　7

上机实践题

1. 编写程序，运用StringBuffer类的reverse()方法判断一个字符串是否"回文"，所谓"回文"

是指将某数逆序排列后得到的字符串与原串相同，如数字 12321，普通字符串如"abcba""哈哈哈"等。

2. 分别用 String 类的 split()方法、StringTokenizer 类、Scanner 类将"Java 技术包括 JavaSE，JavaEE 以及 JavaME 三个分支。"中的英文部分提取出来并输出。

3. 从控制台读入一个字符串，运用正则表达式找出其中包含的所有合法电子邮箱地址并输出。

4. 编写类 CalendarAndDate，在其中定义几个静态方法并完成相应功能：

（1）public static Date stringToDate(String datestr)，功能是将方法参数中字符串形式的日期转换为相应的 Date 对象。

（2）public static String getWeekofYear(String datestr)，功能是判断方法参数中字符串形式的日期是当年的第几周。

（3）public static String getWeek(String datestr)，功能是判断方法参数中字符串形式的日期是星期几。

（4）在 main 方法中测试以上几个方法。

5. 编写程序，分别运用 java.lang.Integer 类的 toBinaryString(int i)、toOctalString(int i)、toHexString(int i)以下方法，输出一个十进制整数的二、八、十六进制形式字符串。

6. "猜数字"程序：运用 Random 类的 nextInt(int n)方法产生一个[0,100)之间的随机整数 randomnum，通过键盘输入你猜测的整数，若猜测的数字小于 randomnum，则输出"错误!太小，请重猜"，猜测的数字大于 randomnum 输出"错误!太大，请重猜"，直到猜对，输出"恭喜你，答对了！该数是:"+randomnum"，同时统计猜测的次数并输出。

第 8 章　异常处理机制

【本章内容提要】
- Java 异常处理概述；
- Java 异常类；
- Java 异常处理语法；
- 强制检查异常和非强制检查异常；
- 用户自定义异常。

程序中的错误有编译错误和运行错误两种。对于编译错误，是靠编译系统发现，并及时提示编程者修改的。而对于运行错误，则较难发现和处理，本章所说的异常就是用于处理这种运行错误的机制和方法。Java 的异常处理机制为保障程序的健壮性提供了有力支持。

8.1　Java 异常处理概述

程序在运行的过程中产生异常，就会中断程序的正常执行，为了保证程序在出现异常时依然能继续执行，就需要对异常进行处理。不同的程序设计语言提供了不同的异常处理机制。Java 异常处理机制为程序运行提供了健壮性保障，通过给可能产生异常的代码编写相应的处理代码，使得程序不会因为异常的发生而产生不可预见的结果。

异常是在程序运行过程中发生的、会打断程序正常执行的事件，如下面几种常见的异常：被 0 除等算术异常（ArithmeticException）、数组下标越界异常（ArrayIndexOutOfBoundsException）使用未经初始化的对象或者是不存在的对象时会出现空指针异常（NullPointerException）、文件未找到异常（FileNotFoundException）等。

Java 作为一个完全面向对象的程序设计语言，对异常的处理也是采用面向对象方法。Java 首先针对各种常见的异常定义了相应的异常类，每个异常类代表了一种运行错误，类中包含了该运行错误的信息和处理错误的方法等内容。

Java 程序的执行过程中若有某异常发生，就生成一个异常类对象，该异常对象封装了异常事件的信息并将被提交给 Java 运行时系统，这个过程称为抛出（throw）异常。

当 Java 运行时系统接收到异常对象时，会寻找能处理这一异常的代码并把当前异常对象交给其处理，这一过程称为捕获（catch）异常。

8.2　Java 异常类

Java 中，所有的异常类都是 java.lang.Throwable 类的子类，Throwable 有两个直接子类一个是

Exception，另一个是 Error（如图 8.1）。两个类各自派生出很多的子类，其中，主要讨论 Exception 的常见子类。

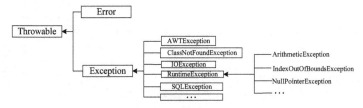

图 8.1 Java 异常类结构

Error 类定义了 Java 程序运行时出现的内部错误。通常用 Error 类来指明与 Java 运行环境相关的错误，如系统崩溃、虚拟机故障等，这类错误将导致应用程序中断，通常不由程序处理，用户也无法捕获。

Exception 类的子类对应了由程序和外部环境引起的常见异常，它是可以被捕获且可能恢复的异常情况。

8.3 Java 异常处理语法

Java 的异常处理是通过 try、catch、finally、throw、throws 这五个关键字完成的。

Java 中使用 try-catch-finally 语句来捕获并处理异常，try-catch-finally 语句的语法格式如下：

```
try
{
    // 可能会产生异常的程序代码
}
catch(Exception_1 e1)
{ // 处理异常 Exception_1 的代码 }
catch(Exception_2 e2)
{// 处理异常 Exception_2 的代码 }
…
catch(Exception_n en)
{ // 处理异常 Exception_n 的代码 }
finally
{
// 通常是释放资源的程序代码
}
```

整个语句由 try 语句块、catch 语句块和 finally 语句块三部分组成。catch 语句块和 finally 语句块都是可以省略的，但 Java 规范不允许两者同时省略。

说明：

- try 代码段包含可能产生异常的代码。
- try 代码段后跟有一个或多个 catch 代码段。
- 每个 catch 代码段声明其能处理的一种特定类型的异常并提供处理的方法。
- 当异常发生时，程序会中止当前发生异常的程序语句后面程序段的执行，根据获取异常的类型转去执行相应的 catch 代码段。

- finally 段的代码无论是否发生异常都会执行。

【例题 8_1】异常处理测试。
```java
public class Ch8_1 {
    public static void main(String[] args) {
      int a=2,b=0,result;
      result=a/b;
      System.out.println("result="+result);
      System.out.println("main()方法执行结束! ");
    }
}
```
程序运行结果如图 8.2 所示。

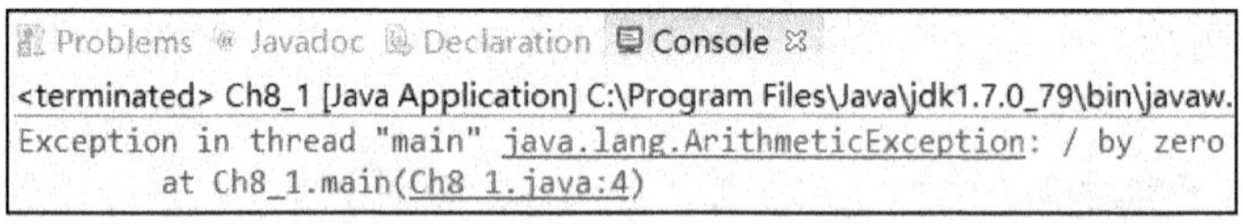

```
Problems  Javadoc  Declaration  Console
<terminated> Ch8_1 [Java Application] C:\Program Files\Java\jdk1.7.0_79\bin\javaw.
Exception in thread "main" java.lang.ArithmeticException: / by zero
        at Ch8_1.main(Ch8_1.java:4)
```

图 8.2　例题 8_1 的出错信息

分析：当程序执行至 result=a/b;时，发生了被 0 除的算数异常，因此终止程序的执行并由 JVM 自动抛出该异常对象，在控制台输出异常信息。result=a/b;后面的语句得不到执行。

Throwable 类提供了一些用于输出异常信息的方法：
- public String getMessage()：返回一个异常的描述的字符串。
- public void printStackTrace()：输出错误堆栈信息。
- public String toString()：重写了 Object 类的 toString()方法，用于返回此异常的简单描述信息。

【例题 8_2】输出异常信息的方法测试。
```java
public class Ch8_2 {
    public static void main(String[] args) {
        int a=2,b=0,result;
        try{                                          //可能发生异常的语句块
            result=a/b;
            System.out.println("result="+result);
        }
        catch(ArithmeticException e){                 //处理异常的语句块
            System.out.println("getMessage()方法的输出:"+e.getMessage());
            System.out.println("toString()方法的输出:"+e.toString());
            System.out.println("printStackTrace()方法的输出:");
            e.printStackTrace();
            System.out.println("被 0 除异常! ");      //自定义输出
        }
        finally{
            System.out.println("异常已经被成功捕获并处理! ");
        }
        System.out.println("main()方法执行结束! ");
    }
}
```
程序运行结果如图 8.3 所示。

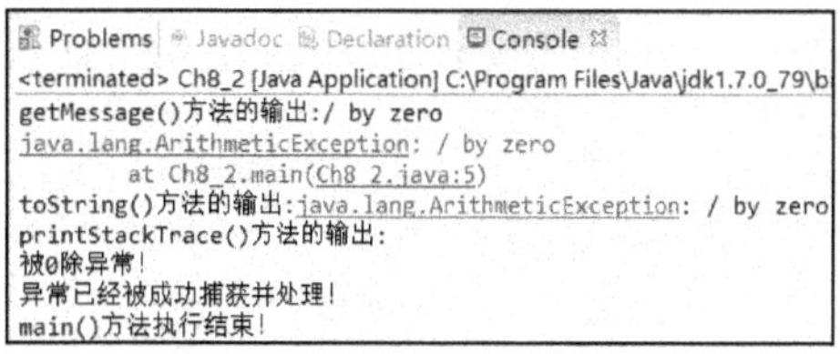

分析：被 try 关键字包围的语句块是程序中可能发生异常的若干语句；catch(ArithmeticException e)语句块用于捕获相应的异常并做处理，本例测试了 Throwable 类提供的几个用于输出异常信息的方法，其中，getMessage()方法只输出简单的异常原因字符串/by zero，toString()方法输出了异常类名以及异常原因信息，而 printStackTrace()不仅输出了异常类名、异常原因，而且输出了引发异常的具体语句行号等详细信息。除了用这些方法进行异常信息输出外，也可以自定义输出如 System.out.println("被 0 除异常！")。

图 8.3 例题 8_2 运行结果

当执行到 result=a/b;并引发 ArithmeticException 异常时，程序流程转向与 try 对应的 catch 语句块执行。不管有无异常发生，finally 语句块中的语句一定会得到执行，当然，finally 语句块也可以缺省。当整个 try-catch-finally 语句块执行完毕时，程序转向其后的语句继续执行。

【例题 8_3】多异常处理测试。

```
public class Ch8_3 {
    public static void main(String[] args) {
        int a,b,result=0;
        try{
            a=Integer.parseInt(args[0]);    //将第一个命令行参数转为整型作为被除数
            b=Integer.parseInt(args[1]);    //将第二个命令行参数转为整型作为被除数
            result=a/b;
        }
        catch(ArithmeticException e1){
            e1.printStackTrace();
            System.out.println("发生了被 0 除异常！");
        }
        catch(ArrayIndexOutOfBoundsException e2){
            e2.printStackTrace();
            System.out.println("发生了数组下标越界异常！");
        }
        catch(NumberFormatException e3){
            e3.printStackTrace();
            System.out.println("发生了数据转换格式异常！");
        }
        System.out.println("result="+result);
    }
}
```

本例可能引发不同类型的异常，分情况测试它们：

① 给出命令行参数 20 10，得到如如图 8.4（a）、（b）所示的结果。

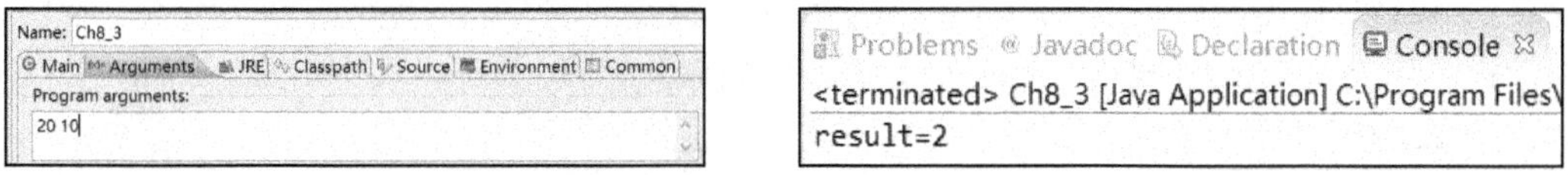

（a）命令行参数　　　　　　　　　　　　　　（b）执行结果

图 8.4　正确的命令行参数下的结果

② 给出命令行参数 20 0，得到如图 8.5（a）、（b）所示的结果。

（a）命令行参数　　　　　　　　　　　　（b）执行结果

图 8.5　命令行参数除数为 0 的结果

③ 给出命令行参数 20，得到如图 8.6（a）、（b）所示的结果。

（a）命令行参数　　　　　　　　　　　　（b）执行结果

图 8.6　命令行未给全的结果

这是因为只有一个命令行参数 20，将被存储在 args[0]中，程序中试图引用 args[1]，从而引发数组下标越界异常。

④ 给出命令行参数 2o 10，得到如图 8.7（a）、（b）所示的结果。

（a）命令行参数　　　　　　　　　　　　（b）执行结果

图 8.7　数值格式异常的结果

本例是通过将来自命令行的参数作为运算数，由于命令行参数是被保存在 String[] args 数组中，因此，在参与算数运算时要先进行数据类型转换，而 Integer 的 parseInt()方法定义如下：

int java.lang.Integer.parseInt(String s) throws NumberFormatException，表示在调用该方法时可能引发 NumberFormatException，当给 String s 传递由纯数字组成的字符串时，不会引发该异常，否则引发异常，如本例中故意将 20 写成 2o，就引发了异常。

8.4　强制检查异常和非强制检查异常

Java 异常可分为强制检查异常（checked exceptions)和非强制检查异常(unchecked exceptions)。Error 和 RuntimeException 类及其子类都属于 unchecked exceptions，表示编译器在编译程序时对这类异常不做检查。Exception 类及其子类(RuntimeException 除外)属于 checked exceptions，编译器在编译程序时，要检查程序是否对这类异常做了处理(如 try 捕获或 throws 声明抛出异常)，若没有处理，则编译器会报错，要求对此必须进行处理。

java.lang.RuntimeException 及它的子类都是非强制检查异常：如：错误的类型转换异常 ClassCastException，数组下标越界异常 ArrayIndexOutOfBoundsException，空指针访问：NullPointerException，算术异常(除 0 溢出) ArithmeticException 等。

常见的强制检查(checked)异常如：没有找到指定名称的类异常 ClassNotFoundException，访问不存在的文件异常 FileNotFoundException，操作文件时发生的异常 IOException，操作数据库时发生的异常 SQLException 等。

【例题 8_4】 强制检查的异常测试。

```
import java.io.File;
import java.io.FileInputStream;
import java.io.FileNotFoundException;
import java.io.IOException;
public class Ch8_4{
    public static void main(String[] args) {
        File f=new File("E:\\a.txt");              //File对象f代表E:\\a.txt
        byte[] b=new byte[8192];
        FileInputStream fin=null;
        try {
            fin=new FileInputStream(f);            //创建文件字节输入流对象
            System.out.println("文件E:\\a.txt 中的内容是: ");
            try {
                while(fin.read(b)!=-1){            //读文件
                    System.out.println(new String(b));//输出文件内容
                }
            } catch (IOException e) {
                e.printStackTrace();
            }
        } catch (FileNotFoundException e) {
            e.printStackTrace();
        }
        finally{                                   //关闭输入流对象
            try {
                fin.close();                       //关闭文件字节输入流对象
            } catch (IOException e) {
                e.printStackTrace();
            }
        }
    }
}
```

程序运行结果如图 8.8（a）、(b) 所示。

（a）a.txt 内容 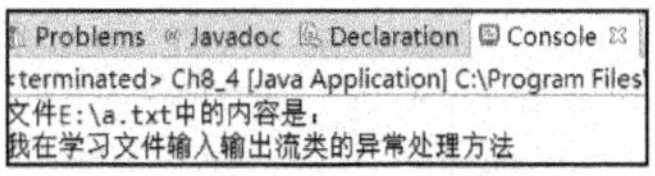　　　　　　（b）读出 a.txt 内容并输出

图 8.8　例题 8_4 运行结果

说明：本例是一个简单的用于读取 E 盘下文件 a.txt 并将文件内容输出在控制台的程序，程序中包含了 FileNotFoundException 以及 IOException 两个强制检查异常。FileNotFoundException 发生在创建文件字节输入流类 FileInputStream 对象创建的时候，IOException 是 fin.read(b)的 read()方法以及 fin.close()方法可能引发的异常。若不使用 try-catch 对可能引发异常的语句进行处理，则编译报错。

将例题 8_4 的程序的改写如下：
```java
import java.io.File;
import java.io.FileInputStream;
import java.io.FileNotFoundException;
import java.io.IOException;
public class Ch8_4 {
    public static void main(String[] args) {
        File f=new File("E:\\a.txt");
        byte[] b=new byte[8192];
        FileInputStream fin=null;
        try {
            fin=new FileInputStream(f);
            System.out.println("文件 E:\\a.txt 中的内容是: ");
            while(fin.read(b)!=-1){
                System.out.println(new String(b));
            }
            fin.close();
        }//end of try
        catch (FileNotFoundException e) {
            e.printStackTrace();
        }
        catch (IOException e1) {
            e1.printStackTrace();
        }
    }
}
```

说明：程序经过改写后并不影响运行结果，但是避免了嵌套的 try-catch 语句块的复杂性。

继续改写例题 8_4 的程序如下：
```java
import java.io.File;
import java.io.FileInputStream;
public class Ch8_4 {
    public static void main(String[] args) {
        File f=new File("E:\\a.txt");
        byte[] b=new byte[8192];
        FileInputStream fin=null;
        try {
            System.out.println("文件 E:\\a.txt 中的内容是: ");
            fin=new FileInputStream(f);
            while(fin.read(b)!=-1){
                System.out.println(new String(b));
            }
            fin.close();
        }//end of try
        catch (Exception e) {
            e.printStackTrace();
        }
    }
}
```

说明：程序经过改写后并不影响运行结果，只是一种笼而统之地捕获可能发生异常类的父类 Exception 异常。

再次改写例题 8_4 的程序如下：
```java
import java.io.File;
import java.io.FileInputStream;
import java.io.FileNotFoundException;
import java.io.IOException;
public class Ch8_4 {
    public static void main(String[] args) throws FileNotFoundException, IOException {
        File f=new File("E:\\a.txt");
        byte[] b=new byte[8192];
        FileInputStream fin=null;
        fin=new FileInputStream(f);
        System.out.println("文件 E:\\a.txt 中的内容是: ");
        while(fin.read(b)!=-1){
            System.out.println(new String(b));
        }
        fin.close();
    }
}
```

说明：程序经过改写后并不影响运行结果，只是"偷懒"地将可能引发的异常用 throws 关键字在 main()方法头部进行声明，一旦程序中引发某种异常，则由调用 main()方法的 Java 虚拟机来处理异常。

当然，public static void main(String[] args) throws FileNotFoundException,IOException 也可以写成 public static void main(String[] args) throws Exception。

8.5 用户自定义异常

用户自定义异常及其处理的步骤如下：
- 通过继承 java.lang.Exception 类来自定义异常类。
- 在可能发生异常的方法头部用 throws 关键字声明该方法可能抛出的异常。
- 在可能发生异常的方法体中创建自定义异常类的对象，并用 throw 关键字将其抛出。
- 在调用可能发生异常的方法时，用 try、catch、finally 语句块进行异常处理。

声明抛出异常的语法格式如下：
throws <异常列表>
例如，public int read() throws IOException { ... }
声明抛出异常是一个说明性的子句，只是表明一个方法可能会抛出异常，而真正抛出异常的动作是由抛出异常语句来完成的，语法格式如下：
throw <异常对象>;其中，<异常对象>必须是 Throwable 类或其子类的对象。

【例题 8_5】编写求解两个正整数的最大公约数的程序，通过自定义异常类来处理如下异常：当给求最大公约数的方法传递的参数不是正整数或除数大于被除数时,引发异常并做相应的处理。

```java
package myproject.ch8_5;
public class MyException extends Exception {        //自定义异常类
    int m,n;
    public MyException(int m, int n) {              //构造方法
        this.m = m;
        this.n = n;
    }
    public String warningMess() {                   //显示出错信息的方法
        return "该方法的两个参数都必须为正整数！";
    }
}
package myproject.ch8_5;
public class Ch8_5 {
    public void commonDivisor(int m,int n) throws MyException{
        int temp,r;
        if(m<=0||n<=0||m<n)
            throw new MyException(m,n);             //抛出自定义异常类对象
        if(m<n){                                    //交换
            temp=m;
            m=n;
            n=temp;
        }                                           //m是被除数，n是除数
        System.out.println(m+"与 "+n+"的最大公约数为: ");
        while(n!=0) {                               //"辗转相除法"求最大公约数
            r=m%n;
            m=n;
            n=r;
        }
        System.out.println(m);
    }
    public static void main(String args[]) {
        Ch8_5 test=new Ch8_5();
        try {
            test.commonDivisor(81,54);
            /**以下3条语句都会引发异常*/
            test.commonDivisor(25, 225);
            //test.commonDivisor(125, 0);
            //test.commonDivisor(-125, 25);
        }
        catch (MyException e) {                     //捕获并处理异常
            System.out.println("异常信息: "+e.warningMess());
        }
    }
}
```

程序运行结果如图8.9所示。

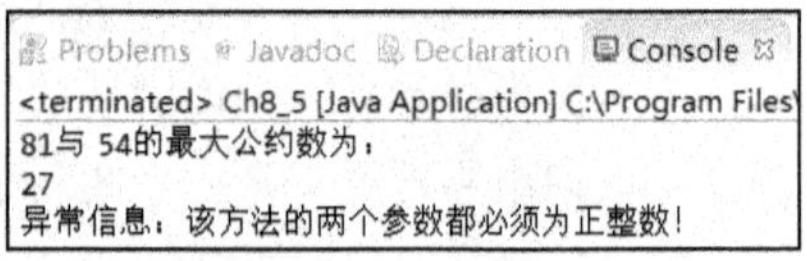

图8.9 自定义异常测试结果

小　　结

本章介绍了Java异常处理机制的作用；Java系统异常类及其分类；Java异常处理的语法、流程及相关关键字的用法；讨论了强制检查异常和非强制检查异常各自的含义及应用；自定义异常类及其使用方法等。

习　题　8

上机实践题

按以下要求自定义异常类并做测试：

（1）自定义异常类OverLoadingException，包含成员变量String warnmess以及构造方法public OverLoadingException(String warnmess) {super(warnmess);}。

（2）定义大巴车 Bus 类，在其中定义方法 public void loading(int x)throws OverLoading Exception。设该大巴车的核准载客人数为50人，在 loading 方法体中判断，若x>50，则抛出OverLoadingException异常并输出异常信息"超载！大巴车载客人数不能超过50人"；否则输出"目前乘车人数为："+x+",可以发车"。

（3）定义主类XiTi8_3_1，在main方法中创建Bus类对象并调用loading方法进行测试。

第 9 章　Java 组件与事件编程

【本章内容提要】
- Java AWT 与 Swing 简介；
- 容器和组件；
- Java 布局管理器；
- Java Swing 常用的中间容器；
- Java 事件编程机制；
- 事件监听器对象的几种实现；
- Swing 的常用组件及其事件编程；
- 其他几个应用；
- 字体与颜色；
- GUI 图形绘制。

GUI（Graphics User Interface）编程是 Java 编程的重要组成部分。运用 Java 提供的组件类以及各种事件编程机制，可以构建出友好的 GUI 程序。

9.1　Java AWT 与 Swing 简介

Java 提供了 AWT（Abstract Window Toolkit）组件与 Swing 组件，分别包含在 java.awt.*包中及 javax.swing.* 包中。其中，AWT 组件被称为重量组件，用它们编写的可视界面在不同的平台下的显示风格略有变化； Swing 组件是用 Java 实现的轻量级（ light-weight）组件，没有本地代码，不依赖操作系统的支持，这是它与 AWT 组件的最大区别。因此，目前主要运用 Swing 组件编写桌面 GUI 应用程序。

Swing 是建立在 AWT 基础之上的，它利用了 AWT 的下层组件，如图形、颜色、字体、布局管理器等，同时，也使用了 AWT 的事件处理机制。正因为如此，在实际的 Java GUI 应用程序中，往往需要导入 javax.swing.*包，也需要导入 java.awt.*包、java.awt.event.* 等包。

9.2　容器和组件

Java 把 Container 类的子类或间接子类创建的对象称为一个容器，如窗口、面板等。Java 把 Component 类的子类或间接子类创建的对象称为一个组件，如按钮、文本框、单选按钮、下拉列表等。容器用于容纳组件，组件只有添加到容器中才能显示。容器本身也是一种特殊的组件。

大部分 AWT 组件在 Swing 中都有等价的组件，它们在类名上差一个"J"，如在 AWT 中，窗体类名为 Frame，按钮、文本框、菜单组件类名分别为 Button、Text、Menu，而在 Swing 中，窗体类名为 JFrame，按钮、文本框、菜单组件类名分别为 JButton、JText、JMenu 等。

下面，首先通过编写一个 AWT 窗体和一个 Swing 窗体来了解一下两者的异同（见图 9.1 和图 9.2）。

【例题 9_1】 Swing 窗体与 AWT 窗体。

 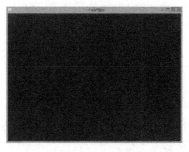

图 9.1　Swing 窗体　　　　　　　　图 9.2　AWT 窗体

```java
import java.awt.Color;
import java.awt.Container;
import javax.swing.JFrame;
public class Ch9_1_1{
    public static void main(String[] args) {
        JFrame frame=new JFrame("一个 Swing 窗体");
        frame.setBounds(0, 0, 1024, 768);         //设置窗体显示的坐标位置及窗体的大小
        Container con=frame.getContentPane();     //获取 Container 对象
        con.setBackground(Color.GREEN);           //给窗体设置背景色
        frame.setResizable(true);                 //设置窗体可调整大小
        frame.setVisible(true);                   //设置窗体为可见
        frame.setDefaultCloseOperation(JFrame.EXIT_ON_CLOSE);//设置窗体关闭事件
    }
}
```

一个 AWT 窗体的程序如下：

```java
import java.awt.Color;
import java.awt.Frame;
import java.awt.event.WindowAdapter;
import java.awt.event.WindowEvent;
public class Ch9_1_2 {
    public static void main(String[] args) {
        Frame frame=new Frame("一个 AWT 窗体");
        frame.setBounds(0, 0, 1024, 768);         //设置窗体显示的坐标位置及窗体的大小
        frame.setBackground(Color.BLUE);          //给窗体设置背景色
        frame.setResizable(true);                 //设置窗体可调整大小
        frame.setVisible(true);                   //设置窗体为可见
        frame.addWindowListener(new WindowAdapter()
        {
            public void windowClosing(WindowEvent e)
            {
                System.exit(0);
            }
        }
        );
    }
}
```

分析：上面两段程序分别创建了一个空的（没有在其上添加组件）Swing 窗体和 AWT 窗体。通过查阅 JDK 开发文档发现，Swing 的 JFrame 和 AWT 的 Frame 类的类继承关系如图 9.3 所示。

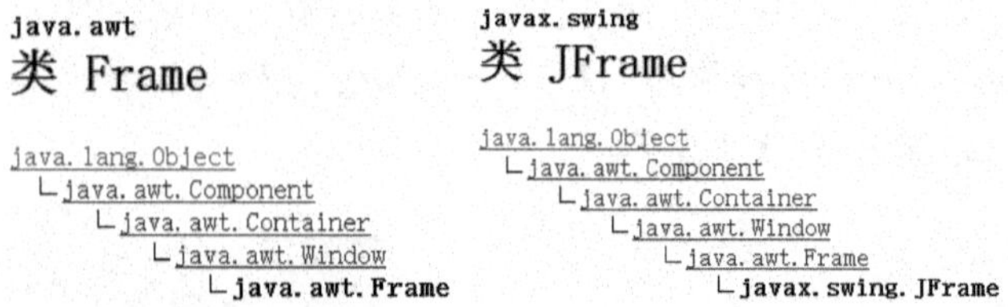

```
java.awt                          javax.swing
类 Frame                          类 JFrame

java.lang.Object                  java.lang.Object
  └java.awt.Component               └java.awt.Component
     └java.awt.Container               └java.awt.Container
        └java.awt.Window                  └java.awt.Window
           └java.awt.Frame                   └java.awt.Frame
                                                └javax.swing.JFrame
```

图 9.3　Frame 类和 JFrame 类各自的继承关系

两者是通过创建 JFrame 类对象或 Frame 类对象生成窗体对象，继而调用 public void setBounds(int x,int y,int width,int height)设置窗体左上角相对于以屏幕左上角为坐标原点的坐标系中的水平偏移量 x 和垂直偏移量 y，窗体本身的宽度 width 和高度 height。事实上，Java 中屏幕坐标系是以屏幕左上角为坐标原点，水平向右的方向为屏幕坐标系的 x 轴正方向，垂直向下的方向为 y 轴正方向；窗体本身也是以自己的左上角为窗体坐标系的原点，水平向右的方向为屏幕坐标系的 x 轴正方向，垂直向下的方向为 y 轴正方向，若在窗体中添加了组件，则对组件设置坐标(x, y)，是对窗体坐标系而言的，如图 9.4 所示。

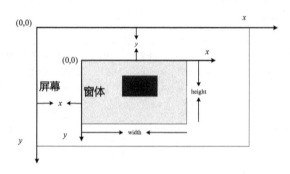

图 9.4　Java 中屏幕、窗体坐标系示意图

程序中也可以用 frame.setSize(1024, 768);（设置窗体或组件的大小）和 frame.setLocation(0, 0);（设置窗体或组件在各自坐标系中的位置）两条语句代替 frame.setBounds(0, 0, 1024, 768)。

对于 Swing 窗体，要对其设置背景色等操作时，要通过窗体对象调用 getContentPane()方法获取 Container 对象，再用 Container 对象调用相应方法；而 AWT 窗体对象可直接调用设置背景色等方法。

Swing 窗体的关闭操作较简单，AWT 窗体的关闭则要通过给窗体对象添加窗口事件监听器来实现。

9.3　Java 布局管理器

Java 通过布局管理器来实现对容器内的组件进行布局管理。布局管理器决定容器的布局风格及容器内组件的排列顺序、组件大小与位置，以及当窗口移动或调整大小后组件如何变化等。

Java 提供了 FlowLayout、BorderLayout、GridLayout、CardLayout、BoxLayout、GridBagLayout 等布局管理器类。

每一种布局管理器类对应一种布局策略，每个容器一般都有默认的布局管理器。当创建一个容器时，Java 自动地为之创建一个默认的布局管理器。例如：JFrame、Window 和 Dialog 默认的布局管理器是 BorderLayout，而 JPanel 和 Applet 默认的布局管理器是 FlowLayout。若想改变某容器的默认布局方式，可以通过用该容器对象调用 setLayout( )方法进行改变。

除 BoxLayout 外（包含在 javax.swing 包中），其他布局管理器都包含在 java.awt 包中，并实现了 LayoutManager 接口。

9.3.1 FlowLayout

其布局策略是按组件添加到容器的先后次序，从左到右、自上向下，一个接一个地摆放到容器中。

FlowLayout 的构造方法：

- public FlowLayout()：创建一个居中对齐的，组件间默认的水平和垂直间隙是 5 个像素的流式布局管理器。
- public FlowLayout(int align)：创建一个以 align 指定的对齐方式，组件间默认的水平和垂直间隙是 5 个像素的流式布局管理器。align 的值必须是以下值之一：FlowLayout.LEFT、FlowLayout.RIGHT、FlowLayout.CENTER、FlowLayout.LEADING 或 FlowLayout.TRAILING。
- public FlowLayout(int align,int hgap,int vgap)：创建一个具有指定的对齐方式以及指定的水平和垂直间隙的流式布局管理器。

【例题 9_2】创建 10 个按钮对象并将它们按 FlowLayout 布局方式添加到窗体中，如图 9.5 所示。

```
import java.awt.Container;
import java.awt.FlowLayout;
import javax.swing.JButton;
import javax.swing.JFrame;
public class Ch9_2 {
    public static void main(String[] args) {
        JFrame frame=new JFrame();
        JButton[] jb=new JButton[10];           //按钮对象数组
        frame.setTitle("FlowLayout 布局测试");
        frame.setBounds(0, 0, 240, 180);
        FlowLayout flow=new FlowLayout();
        Container con=frame.getContentPane();
        con.setLayout(flow);                    //设置窗体的布局方式为流式布局
        for(int i=0;i<10;i++){
            jb[i]=new JButton("按钮"+i);        //创建按钮对象并给按钮数组赋值
            con.add(jb[i]);                     //将按钮对象添加到窗体
        }
        frame.setVisible(true);
        frame.setDefaultCloseOperation(JFrame.EXIT_ON_CLOSE);
    }
}
```

图 9.5　FlowLayout 布局

注意：在给 JFrame 窗体中添加组件或设定布局管理方式时，假定 JFrame 对象为 frame，则要通过 Container con=frame.getContentPane();获取与该 JFarmer 关联的 Container 对象 con，再用 con. setLayout(布局管理器类对象)设置布局管理方式。

9.3.2 BorderLayout

BorderLayout 是 Window、Frame、Dialog 类的默认布局管理器，它将容器划分成东、南、西、北、中 5 个区域来摆放组件，这 5 个区域分别由 BorderLayout 中的常量 EAST、SOUTH、、WEST、NORTH、 CENTER 来表示。

BorderLayout 的构造方法：

- public BorderLayout()：创建一个各个组件之间没有间距的 BorderLayout 布局管理器。
- public BorderLayout(int hgap,int vgap)：创建一个具有指定组件间距的 BorderLayout 布局管理器。水平间距由 hgap 指定，垂直间距由 vgap 指定。

采用 BorderLayout 的容器对象，可以调用 public void add(Component comp,Object constraints) 方法在其上添加一个组件 comp，同时可以用 constraints 指定组件在容器中的区域。

constraints 可选值为 BorderLayout 提供的 5 个常量 EAST、SOUTH、WEST、NORTH 和 CENTER。

【例题 9_3】 用 BorderLayout 将 5 个按钮对象布局到窗体中，如图 9.6 所示。

```
import java.awt.BorderLayout;
import java.awt.Container;
import javax.swing.JButton;
import javax.swing.JFrame;
public class Ch9_3 {
public static void main(String[] args) {
    JFrame frame=new JFrame();
    frame.setTitle("BorderLayout 布局测试");
    frame.setBounds(0, 0, 600, 240);
    Container con=frame.getContentPane();
    con.add(new JButton("我在窗体的东部区域"),BorderLayout.EAST);
    con.add(new JButton("我在窗体的南部区域"),BorderLayout.SOUTH);
    con.add(new JButton("我在窗体的西部区域"),BorderLayout.WEST);
    con.add(new JButton("我在窗体的北部区域"),BorderLayout.NORTH);
    con.add(new JButton("我在窗体的中部区域"),BorderLayout.CENTER);
    frame.setVisible(true);
frame.setDefaultCloseOperation(JFrame.EXIT_ON_CLOSE);
    }
}
```

图 9.6　BorderLayout 布局

9.3.3 GridLayout

网格布局管理器 GridLayout 是流布局管理器的扩展，以矩形网格形式对容器的组件进行布置。容器被分成若干大小相等的矩形，一个矩形网格中放置一个组件，放置的顺序是从上到下，从左到右。

其构造方法有：

- public GridLayout()：创建具有默认值的网格布局，即每个组件占据一行一列。
- public GridLayout(int rows,int cols)：创建具有指定行数和列数的网格布局。给布局中的所有组件分配相等的大小。
- public GridLayout(int rows,int cols,int hgap,int vgap)：创建具有指定行数和列数的网格布局。给布局中的所有组件分配相等的大小。此外，将水平和垂直间距设置为指定值。rows 和 cols 中的一个可以为零，但不能两者同时为零。

当网格布局管理器对应的窗口发生变化时，内部组件的相对位置并不变化，只有大小发生变化。

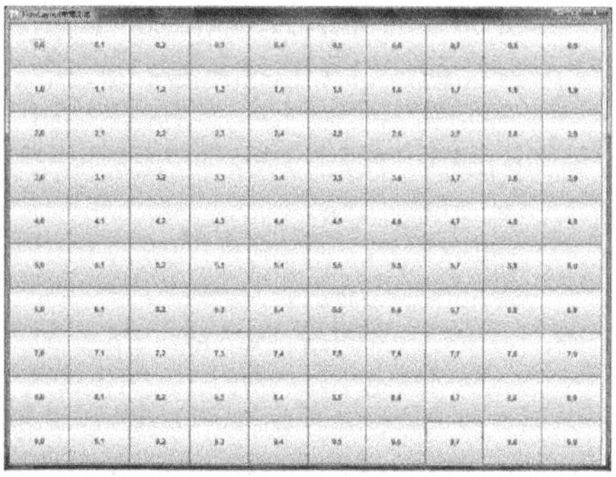

【例题 9_4】将 100 个按钮以 GridLayout 方式排列在窗体中，每个按钮的名称为数字 00～99，如图 9.7 所示。

```java
import java.awt.Container;
import java.awt.GridLayout;
import javax.swing.JButton;
import javax.swing.JFrame;
public class Ch9_4{
    public static void main(String[] args) {
        JFrame frame=new JFrame();
        JButton[][] jb=new JButton[10][10];
        frame.setTitle("FlowLayout布局测试");
        frame.setBounds(0, 0, 1024, 768);
        GridLayout grid=new GridLayout(10,10);
        Container con=frame.getContentPane();
        con.setLayout(grid);          //设置窗体的布局方式为网格布局
        for(int i=0;i<10;i++)
        {
            for(int j=0;j<10;j++)
            {
                jb[i][j]=new JButton(i+","+j);
                con.add(jb[i][j]);
            }
        }
        frame.setVisible(true);
        frame.setDefaultCloseOperation(JFrame.EXIT_ON_CLOSE);
    }
}
```

图 9.7　GridLayout 布局

9.3.4　CardLayout

CardLayout 布局方式是将添加到窗体中的组件对象看成"一叠卡片"，在任何时候只有其中一个卡片（组件对象）是可见的，该卡片（组件对象）占据容器的整个区域。

CardLayout 的构造方法：public CardLayout(int hgap,int vgap)。

参数 hgap 表示卡片和容器的左右边界之间的间隙，参数 vgap 表示卡片和容器的上下边界的间隙。

显示卡片的方法有：

- public void first(Container parent)：翻转到容器的第一张卡片。
- public void last(Container parent)：翻转到容器的最后一张卡片。
- public void next(Container parent)：翻转到指定容器的下一张卡片。
- public void previous(Container parent)：翻转到指定容器的前一张卡片。
- public void show(Container parent, String name)：翻转到使用 addLayoutComponent 添加到此布局的具有指定 name 的组件。

【例题 9_5】在窗体中以 CardLayout 方式放置 3 个名为"按钮 0""按钮 1""按钮 2"的按钮，当单击"上一张"或"下一张"按钮时分别显示它们，如图 9.8 所示。

图 9.8　CardLayout 布局

```java
import java.awt.BorderLayout;
import java.awt.CardLayout;
import java.awt.Container;
import java.awt.GridLayout;
import java.awt.event.ActionEvent;
import java.awt.event.ActionListener;
import javax.swing.JButton;
import javax.swing.JFrame;
import javax.swing.JPanel;
class CardLayoutWin extends JFrame implements ActionListener {
    JButton[] jb=new JButton[3];
    JButton previous,next;
    JPanel p1,p2;
    CardLayout card=null;
    GridLayout grid=null;
    public CardLayoutWin(){
        card=new CardLayout();
        grid=new GridLayout(1,2);
        p1=new JPanel();                 //创建面板对象
        p2=new JPanel();
        p1.setLayout(card);              //设置p1的布局方式为卡片布局
        for(int i=0;i<3;i++){
            jb[i]=new JButton("按钮"+i);
            p1.add(jb[i],"按钮"+i);//将按钮添加到面板p1上
        }
        p2.setLayout(grid);              //设置p2的布局方式为网格布局
        previous=new JButton("上一张");
        next=new JButton("下一张");
        /**给previus,next按钮添加事件监听对象*/
        previous.addActionListener(this);
        next.addActionListener(this);
        p2.add(previous);
        p2.add(next);
        /**将面板p1,p2添加到窗体*/
        this.add(p1, BorderLayout.CENTER);
        this.add(p2, BorderLayout.SOUTH);
    }
    /**actionPerformed是ActionListener的抽象方法,
    对其的实现就是"上一张"、"下一张"按钮的事件处理代码*/
    public void actionPerformed(ActionEvent e) {
        if(e.getSource()==previous)   //若事件源为previous
            card.previous(p1);        //调用previous方法
        if(e.getSource()==next)       //若事件源为next
            card.next(p1);            //调用next方法
    }
}
public class Ch9_5 {
    public static void main(String[] args) {
        CardLayoutWin win=new CardLayoutWin();
        win.setTitle("CardLayout布局测试");
        win.setBounds(0, 0, 512, 384);
        win.setVisible(true);
        win.setDefaultCloseOperation(JFrame.EXIT_ON_CLOSE);
```

        }
    }

分析：本例在窗体中添加了两个面板对象 p1 和 p2，分别以 CardLayout 和 GridLayout 方式进行布局，分别用于容纳"按钮 0""按钮 1""按钮 2"，以及"上一张""下一张"按钮，然后又将 p1 和 p2 添加到窗体的 CENTER 和 SOUTH 区域。

本例中的"上一张"和"下一张"按钮是事件源，需要给它们添加事件监听器并进行相应的事件处理，关于事件处理机制在下一小节详细述及。

当要在一个用 CardLayout 布局的容器中添加组件时，需要调用 void add(Component comp,Object constraints)方法将，其中，comp 表示容器对象，constraints 表示为组件指定的名字如 p1.add(jb[i],"按钮"+i);

9.3.5 GridBagLayout

GridBagLayout 布局管理器是最灵活、复杂的布局管理器，它是在 GridLayout 的基础上发展而来，但它不需要组件的尺寸大小一致，每个组件可以占有一个或多个网格单元，组件也可以按任意顺序添加到容器的任意位置。

为了使用 GridBagLayout 布局管理器，必须构造一个 GridBagConstraints 对象，这个对象指定了组件显示的区域在网格中的位置，以及应该如何摆放组件，它通过设置下列 GridBagConstraints 的变量来实现。

- gridx、gridy：指定组件左上角在网格中的行与列。容器中最左边列的 gridx=0，最上边行的 gridy=0。这两个变量的默认值为 GridBagConstraints.RELATIVE，表示对应的组件将放在前面放置组件的右边或下面。
- gridwidth、gridheight：指定组件显示区域所占的列数与行数，以网格单元而不是像素为单位，默认值为 1。GridBagConstraints.REMAINDER 指定组件是所在行或列的最后一个组件，GridBagConstraints.RELATIVE 指定组件是所在行或列的倒数第二个组件。
- weightx、weighty：用来指定在容器大小改变时，增加或减少的空间如何在组件间分配。默认值是 0，即所有的组件将聚拢在容器的中心，多余的空间将放在容器边缘与网格单元之间。每一列组件的 weightx 值指定为该列组件的 weightx 的最大值；每一行组件的 weighty 值指定为该行组件的 weighty 的最大值。weightx 和 weighty 的取值一般在 0.0~1.0 之间，数值大表明组件所在的行或列将获得更多的空间。
- ipadx、ipady：指定组件的内部填充宽度，即为组件的最小宽度、最小高度添加多大的空间，默认值为 0。
- fill：指定单元大于组件的情况下，组件如何填充此单元，默认为组件大小不变。fill 的取值：GridBagConstraints.NONE（组件大小不变）、GridBagConstraints.HORIZONTAL（水平填充）、GridBagConstraints.VERTICAL（垂直填充）、GridBagConstraints.BOTH（填充全部区域）。
- anchor：指定组件在显示区域中的摆放位置。其值可以为 GridBagConstraints.CENTER (默认值)、GridBagConstraints.NORTH、GridBagConstraints. NORTHEAST、GridBag Constraints. EAST、GridBagConstraints. SOUTHEAST、GridBagConstraints. SOUTH、GridBagConstraints. SOUTHWEST、GridBagConstraints. WEST、GridBagConstraints. NORTH WEST。

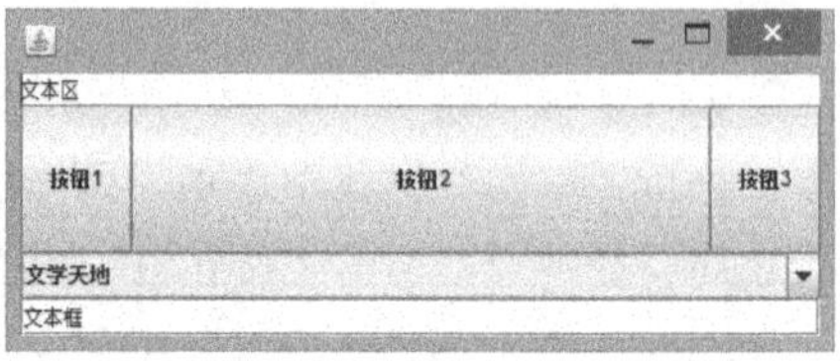

【例题 9_6】用 GridBagLayout 布局方式将多个组件对象布局到窗体，如图 9.9 所示。

图 9.9　GridBagLayout 布局

```java
import java.awt.*;
import javax.swing.*;
public class Ch9_6 extends JFrame{
    JComboBox comboBox;
    JTextArea area;
    JTextField text;
    JButton b1, b2, b3;
    GridBagLayout gbLayout;
    GridBagConstraints gbConstraints;
    public Ch9_6() {
        gbLayout = new GridBagLayout();
        this.setLayout(gbLayout);                       //将窗体布局为GridBagLayout
        gbConstraints = new GridBagConstraints();//创建GridBagConstraints对象
        /*创建所需的各种组件对象*/
        area = new JTextArea("文本区");
        comboBox= new JComboBox();
        comboBox.addItem("文学天地");
        comboBox.addItem("体育天地");
        comboBox.addItem("音乐天地");
        text = new JTextField("文本框");
        b1 = new JButton("按钮1");
        b2 = new JButton("按钮2");
        b3 = new JButton("按钮3");
        gbConstraints.weightx = 0;
        gbConstraints.weighty = 0;
        gbConstraints.fill = GridBagConstraints.BOTH;
        /**调用addComponent()方法将各个组件对
            象按设定的GridBagConstraints值添加到窗体*/
        this.addComponent(area, gbLayout, gbConstraints, 0,0,3,1);
        this.addComponent(b1, gbLayout, gbConstraints, 1,0,1,1);
        gbConstraints.weightx = 1;
        gbConstraints.weighty = 1;
        this.addComponent(b2, gbLayout, gbConstraints, 1,1,1,1);
        gbConstraints.weightx = 0;
        gbConstraints.weighty = 0;
        this. addComponent(b3, gbLayout, gbConstraints, 1,2,1,1);
        this.addComponent(comboBox, gbLayout, gbConstraints, 2,0,3,1);
        this.addComponent(text, gbLayout, gbConstraints, 3,0,3,1);
        this.setSize(500,200);
        this.setVisible(true);
    }
    /*自定义的addComponent()方法*/
    private void addComponent(Component c, GridBagLayout g,GridBagConstraints gc, int row, int column, int width, int height) {
        gc.gridx = column;           //设置组件显示区域的开始边单元格
        gc.gridy = row;              //设置组件显示区域的顶端单元格
        gc.gridwidth = width;        //设置组件显示区域一列的单元格数
        gc.gridheight = height;      //设置组件显示区域一行的单元格数
        g.setConstraints(c, gc);     //给组件对象c设置布局的约束条件
        this.add(c);                 //将组件对象c添加到窗体
    }
    public static void main(String args[]) {
```

```
        new Ch9_6();
    }
}
```
分析：GridBagConstraints 的 fill、weightx、weighty、gridx、gridy、gridwidth、gridheight 的值来确定某个组件对象在窗体中的布局约束条件。本例中自定义了方法 addComponent()，方法体中完成了给组件对象 c 设置布局约束条件并将其按约束条件添加到窗体的任务。

一般地，使用网格包布局管理器要涉及到一个辅助类 GridBagContraints，该类包含 GridBagLayout 类用来保存组件布局大小和位置等约束条件的全部信息，其使用步骤如下：

（1）创建一个 GridBagLayout 对象，并将其设置为当前容器的布局管理器。
（2）创建一个 GridBagContraints 对象。
（3）通过 GridBagContraints 为组件对象设置布局信息（约束条件）。
（4）将组件对象添加到容器中。

9.3.6 BoxLayout

使用盒式布局的容器将组件排列在一行或一列。可以使用 Box 类的静态方法 createHorizontalBox()获得一个具有行型盒式布局的盒式容器；使用 Box 类的类（静态）方法 createVerticalBox()获得一个具有列型盒式布局的盒式容器。

运用 Box 类调用其静态方法 createHorizontalStrut(int width)可以得到一个不可见的水平 Struct 对象，称作水平支撑。Box 类调用静态方法 createVerticalStrut(int height)可以得到一个不可见的垂直 Struct 对象，称作垂直支撑。水平支撑和垂直支撑决定了组件对象间的间隔大小。

【例题 9_7】 使用 BoxLayout 布局方式制作一个用户登录窗体，如图 9.10 所示。

图 9.10 BoxLayout 布局

```
import java.awt.BorderLayout;
import java.awt.Container;
import javax.swing.*;
class LoginFrame extends JFrame  {
    JLabel welcome;
    Box vbox,hbox1,hbox2,hbox3,hbox4;
    JPanel panel;
    Container con;
    public LoginFrame(){
        panel=new JPanel();
        vbox=Box.createVerticalBox();
        //产生一个具有列型盒式布局的盒式容器
            welcome=new JLabel("欢迎登录学生信息管理系统");
        welcome.setFont(new java.awt.Font("宋体",1,20));   //给标签对象设置字体
        hbox1=Box.createHorizontalBox();
        //产生一个具有行型盒式布局的盒式容器
        hbox1.add(welcome);
        vbox.add(hbox1);
        vbox.add(Box.createVerticalStrut(20));            //垂直支撑
        hbox2=Box.createHorizontalBox();
        hbox2.add(new JLabel("姓 名:"));
        hbox2.add(Box.createHorizontalStrut(30));          //水平支撑
        hbox2.add(new JTextField(20));
        vbox.add(hbox2);
        vbox.add(Box.createVerticalStrut(20));
```

```
            hbox3=Box.createHorizontalBox();
            hbox3.add(new JLabel("密 码:"));
            hbox3.add(Box.createHorizontalStrut(30));
            hbox3.add(new JPasswordField (20) );
            vbox.add(hbox3);
            vbox.add(Box.createVerticalStrut(20));
            hbox4=Box.createHorizontalBox();
            hbox4.add(new JButton("确定"));
            hbox4.add(Box.createHorizontalStrut(30));
            hbox4.add(new JButton("取消"));
            vbox.add(hbox4);
            panel.add(vbox);
            con=this.getContentPane();
            con.add(panel,BorderLayout.CENTER);
        }
}
public class Ch9_7 {
    public static void main(String[] args) {
    LoginFrame win=new LoginFrame();
    win.setTitle("登录窗体");
    win.setBounds(0, 0, 360, 240);
    win.setVisible(true);
    }
}
```

9.3.7 空布局

在需要人工排列或布局组件时,也可以使用 setLayout(null)方法取消所有的布局方式,由用户自行安排组件的合适位置。各个组件调用 setBounds(int a,int b,int width,int height)方法可以设置本身的大小和在容器中的位置。

【例题 9_8】将 3 个特定大小的按钮对象首尾相接地放置在窗体中,如图 9.11 所示。

```
import javax.swing.*;
public class Ch9_8 extends JFrame{
    JButton bt1,bt2,bt3;
    JPanel panel ;
    public Ch9_8(){
        panel=new JPanel();
        panel.setLayout(null);//panel 设置为空布局
        bt1=new JButton("按钮 1");
        bt1.setBounds(20,20, 100, 100);
        panel.add(bt1);
        bt2=new JButton("按钮 2");
        bt2.setBounds(120,120, 100, 50);
        panel.add(bt2);
        bt3=new JButton("按钮 3");
        bt3.setBounds(220,170, 150, 150);
        panel.add(bt3);
        this.add(panel);
        this.setSize(500,400);
        this.setVisible(true);
    }
    public static void main(String[] args) {
        new Ch9_8();
    }
}
```

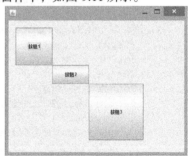

图 9.11 空布局示例

9.4 Java Swing 常用的中间容器

JFrame 是用于容纳组件的底层容器，相对于 JFrame，在进行窗口组件的布局时，还需要借助于一些中间容器，这些中间容器必须被添加到底层容器上才能起作用。Java 中常见的中间容器有：

（1）JPanel（面板）。使用 JPanel 创建面板对象，然后在该面板对象上添加组件，最后把这个面板对象添加到其他容器（如窗体或其他面板）中。JPanel 面板的默认布局是 FlowLayout 布局。

（2）JScrollPane（滚动面板）。滚动面板对象常常用于给添加在其上的组件提供滚动条。例如，可将一个文本区对象放到一个滚动窗格中：

JScorollPane scroll=new JScorollPane(new JTextArea());

（3）JSplitPane（拆分面板）。可以对其进行水平拆分和垂直拆分。JSplitPane 的常用构造方法：

public JSplitPane(int newOrientation,Component newLeftComponent,Component new RightComponent)

其中，newOrientation 的取值为 JSplitPane.HORIZONTAL_SPLIT 或 JSplitPane.VERTICAL_SPLIT，决定是水平还是垂直拆分；newLeftComponent 是将出现在被水平拆分面板的左边或被垂直拆分面板顶部的组件对象；newRightComponent 是将出现在被水平拆分面板的右边或被垂直拆分面板底部的组件对象。

publicJSplitPane(int newOrientation,boolean newContinuousLayout,Component new LeftComponent,Component newRightComponent)

其中，boolean newContinuousLayout 为 true 时表示当拆分线移动时，组件是否连续变化。

（4）JLayeredPane（分层面板）

其常用方法有：

- public add(Jcomponent com, int layer)：其中，com 是要添加在其上的组件对象；layer 可以取值为 DEFAULT_LAYER、PALETTE_LAYER、MODAL_LAYER、POPUP_LAYER 以及 DRAG_LAYER 之一，用户指定 com 所在的层。
- public void setLayer(Component c,int layer)：可以重新设置组件 c 所在的层。
- public int getLayer(Component c)：可以获取组件 c 所在的层数。

【例题 9_9】拆分面板以及面板嵌套，如图 9.12 所示。

```
import java.awt.GridLayout;
import javax.swing.*;
class SplitPaneWin extends JFrame {
    JPanel panel;
    JSplitPane hsplitepane,vsplitepane;
    JScrollPane jspanel;
    public SplitPaneWin(){
        this.setTitle("JSplitPane 示例");
        panel=new JPanel();
        panel.setLayout(new GridLayout(1,3));
        hsplitepane=new JSplitPane(JSplitPane.HORIZONTAL_SPLIT,new JLabel
        ("水平拆分面板左半部"),new JLabel("水平拆分面板右半部"));
        //水平拆分面板
        vsplitepane=new JSplitPane(JSplitPane.VERTICAL_SPLIT,new JLabel("垂直
        拆分面板上半部"),new JLabel("垂直拆分面板下半部"));
        //垂直拆分面板
        jspanel=new JScrollPane(new JTextArea(10,20));
```

图 9.12 拆分面板示例

```
            //滚动面板上容纳多行文本编辑区
            panel.add(hsplitepane);
            panel.add(vsplitepane);
            panel.add(jspanel);
            //将一个水平拆分面板、一个垂直拆分面板以及一个滚动面板添加到panel
            this.getContentPane().add(panel);
            //将panel添加到窗口
            this.setSize(640, 360);
            this.setVisible(true);
            this.setDefaultCloseOperation(JFrame.EXIT_ON_CLOSE);
        }
    }
    public class Ch9_9 {
        public static void main(String[] args) {
            new SplitPaneWin();
        }
    }
```

分析：程序中将一个 JPanel 对象 panel 按照 1 行 3 列进行网格布局，在其中嵌套添加一个水平拆分面板、一个垂直拆分面板以及一个滚动面板对象，再将 panel 添加到窗口上。为了看到滚动面板上的滚动条，特意将本例源码内容粘贴到了文本区中。

【例题 9_10】用分层面板在窗体中放置若干颜色不同的标签对象，如图 9.13 所示。

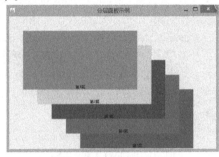

图 9.13　分层面板示例

```
    import java.awt.Color;
    import java.awt.Point;
    import java.util.Random;
    import javax.swing.*;
    public class Ch9_10 {
        public static void main(String[] args) {
            new WinWithLayeredPane();
        }
    }
    class WinWithLayeredPane extends JFrame{
        JLayeredPane layeredpane;
        JLabel[] label=new JLabel[5];
        Random random;
        int[]
layerconstraints={JLayeredPane.DEFAULT_LAYER,JLayeredPane.PALETTE_LAYER,
JLayeredPane.MODAL_LAYER,JLayeredPane.POPUP_LAYER,JLayeredPane.DRAG_LAYER};
        //初始化JLayeredPane的层，取值为JLayeredPane的各个常量
        public WinWithLayeredPane (){
            this.setTitle("分层面板示例");
            layeredpane=this.getLayeredPane();
            //获得JLayeredPane对象layeredpane
            random=new Random();
            for(int i=0;i<5;i++){
              label[i]=new JLabel("第"+(i+1)+"层",JLabel.CENTER);
              label[i].setVerticalAlignment(JLabel.BOTTOM);
              label[i].setBackground(new Color(random.nextInt(255),random.nextInt(255),
              random.nextInt(255)));
              //为各个标签对象随机设定颜色
              label[i].setSize(400, 200);
              label[i].setLocation(50+50*i, 50+50*i);
```

```
            //为各个标签对象设定位置
            label[i].setOpaque(true);
            //若为 true，该标签对象绘制其边界内的所有像素
            layeredpane.add(label[i], layerconstraints[i]);
            //将各个标签对象添加到 layeredpane 的不同层上
        }
        this.setSize(750,500);
        this.setVisible(true);
        this.setDefaultCloseOperation(JFrame.EXIT_ON_CLOSE);
    }
}
```

9.5 Java 事件编程机制

先来学习几个基本概念：

（1）事件：用户对组件对象施加的动作。

（2）事件源：能够引发某种事件的组件对象。例如，单击按钮对象、菜单项、工具栏工具按钮，在单行文本框中输入文本并回车，组件对象获得或失去焦点等时，这些组件对象就成为事件源。

（3）事件类：Java 针对各种事件，定义了相应的事件类，这些事件类包含在 java.awt.event 包以及 javax.swing.event 包中。例如：ActionEvent、ItemEvent、MouseEvent、KeyEvent 等。每个事件类都是对某种事件的抽象。

（4）事件监听器接口：对应于每个事件类，Java 定义了与其对应的事件处理接口，如 ActionListener、ItemListener、MouseListener、KeyListener 等。在每个事件处理接口中都包含了至少一个抽象方法，该抽象方法给出的方法体就是事件处理代码。

（5）事件监听器：事件源通过调用相应的方法将某个对象注册为自己的事件监听器。给某个事件源注册事件监听器的方法是：

```
事件源.addXXXListener(事件监听器);
```

如有

```
JButton bt; bt.addActionListener(事件监听器);
```

那么，什么样的对象才可以充当事件监听器呢？答案是：凡是实现过相应事件监听器接口的类对象就可以充当该事件源的事件监听器。

事件源注册事件监听器之后，当该事件源上发生某种事件时，事件监听器就会监听到该事件，调用重写过的相应事件处理接口的方法做事件处理。

为了清楚地理解 Java 事件处理机制，举一个简单的例子来说明。

【例题 9_11】文本框事件处理，如图 9.14 所示。

图 9.14　文本框事件处理结果

```
import java.awt.FlowLayout;
import java.awt.event.ActionEvent;
import java.awt.event.ActionListener;
import javax.swing.*;
class EventDeal extends JFrame implements ActionListener{
```

```java
    JLabel label ;
    JTextField text;
    public EventDeal(){
        this.setTitle("文本框事件处理");                    //窗体标题
        this.setLayout(new FlowLayout());                  //将窗体布局为FlowLayout
        label=new JLabel("请输入文本并回车确认: ");         //创建标签对象
        text=new JTextField(20);                           //创建文本框对象
        text.addActionListener(this);                      //给事件源text注册事件监听器
        this.add(label);                                   //将label添加到窗体中
        this.add(text);                                    //将text添加到窗体中
        this.setBounds(50, 100, 400, 100);                 //设置窗体位置及大小
        this.setVisible(true);                             //使得窗体可见
        this.setDefaultCloseOperation(JFrame.EXIT_ON_CLOSE);//关闭窗体
    }
    @Override
    //事件处理代码
    public void actionPerformed(ActionEvent e) {
        String str=text.getText();                         //从文本框中取出文本
        System.out.println("文本框中的内容为: "+str);      //将str输出到控制台
    }
}
public class Ch9_11 {
    public static void main(String[] args) {
        new EventDeal();                                   //创建窗体对象
    }
}
```

分析：本例中的事件源是 JTextField 对象 text，当在 text 上输入文本并回车时，text 会发生 ActionEvent 类型的事件。要能够处理该事件，就需要有某个类实现 ActionEvent 对应的事件监听器接口 ActionListener 并重写其抽象方法 public void actionPerformed(ActionEvent e)，本例中是窗体类 EventDeal 实现类 ActionListener 并重写了其 actionPerformed 方法。因此，事件源 text 通过调用 addActionListener(this)给其注册事件监听器，this 代表的正是当前类对象（EventDeal 类对象）。当在 text 发生事件后，this 监听到该事件并调用 actionPerformed 方法进行事件处理。

对于其他事件源，其各自的事件处理机制都一样，只不过是触发的事件类型、处理事件的监听器接口不一样而已。

Swing 包是在 AWT 包的基础上创建的，Swing 仍然使用基于监听器的事件处理机制。因此 Swing 中基本的事件处理仍然需要使用 java.awt.event 包中的事件类及其相应事件监听器接口（见表 9.1），另外 javax.swing.event 包中也增加了一些新的事件类及对应的事件监听器接口（见表 9.2）。

表 9.1 java.awt.event 包中常用的事件类、对应接口及事件处理方法

事件类名	事件监听器接口名	接口的抽象方法（事件处理方法）
ActionEvent	ActionListener	void actionPerformed(ActionEvent e)
AdjustmentEvent	AdjustmentListener	void adjustmentValueChanged(AdjustmentEvent e) 在可调整的值发生更改时调用该方法
ComponentEvent	ComponentListener	void componentResized(ComponentEvent e)：组件大小更改时调用。 void componentMoved(ComponentEvent e)：组件位置更改时调用。 void componentShown(ComponentEvent e)：组件变得可见时调用。 void componentHidden(ComponentEvent e)：组件变得不可见时调用

续表

事 件 类 名	事件监听器接口名	接口的抽象方法（事件处理方法）
ContainerEvent	ContainerListener	void componentAdded(ContainerEvent e)：已将组件添加到容器中时调用。 void componentRemoved(ContainerEvent e)：已从容器中移除组件时调用
FocusEvent	FocusListener	void focusGained(FocusEvent e)：组件获得键盘焦点时调用。 void focusLost(FocusEvent e)：组件失去键盘焦点时调用
ItemEvent	ItemListener	void itemStateChanged(ItemEvent e) 在用户已选定或取消选定某项时调用
KeyEvent	KeyListener	void keyTyped(KeyEvent e)：键入某个键时调用。 void keyPressed(KeyEvent e)：按下某个键时调用。 void keyReleased(KeyEvent e)：释放某个键时调用
MouseEvent	MouseListener	void mouseClicked(MouseEvent e)：鼠标在组件上单击时调用。 void mousePressed(MouseEvent e)：鼠标在组件上按下时调用。 void mouseReleased(MouseEvent e)：鼠标按钮在组件上释放时调用。 void mouseEntered(MouseEvent e)：鼠标进入到组件上时调用。 void mouseExited(MouseEvent e)：鼠标离开组件时调用
MouseMotionEvent	MouseMotionListener	void mouseDragged(MouseEvent e)：鼠标按键在组件上按下并拖动组件时调用。 void mouseMoved(MouseEvent e)：鼠标光标移动到组件上但无按键按下时调用
TextEvent	TextListener	void textValueChanged(TextEvent e) 文本的值已改变时调用
WindowEvent	WindowListener	void windowOpened(WindowEvent e)：窗口首次变为可见时调用。 void windowClosing(WindowEvent e)：关闭窗口时调用。 void windowClosed(WindowEvent e)：因对窗口调用 dispose 而将其关闭时调用。 void windowIconified(WindowEvent e)：窗口从正常状态变为最小化状态时调用。 void windowDeiconified(WindowEvent e)：窗口从最小化状态变为正常状态时调用。 void windowActivated(WindowEvent e)：将 Window 设置为活动 Window 时调用。 void windowDeactivated(WindowEvent e)：当 Window 不再是活动 Window 时调用

表 9.2 javax.swing.event 包中常用的事件类、对应接口及事件处理方法

事 件 类 名	事件监听器接口名	接口的抽象方法（事件处理方法）
CaretEvent	CaretListener	void caretUpdate(CaretEvent e)：当插入符的位置被更新时调用
ChangeEvent	ChangeListener	void stateChanged(ChangeEvent e)：当事件源已更改其状态时调用
DocumentEvent	DocumentListener	void insertUpdate(DocumentEvent e)：对文档执行了插入操作时调用。 void removeUpdate(DocumentEvent e)：移除文档时调用。 void changedUpdate(DocumentEvent e)：文档发生了更改时调用
HyperlinkEvent	HyperlinkListener	void hyperlinkUpdate(HyperlinkEvent e)：在更新超文本链接时调用
MenuEvent	MenuListener	void menuSelected(MenuEvent e)：选择某个菜单时调用。 void menuDeselected(MenuEvent e)：取消选择某个菜单时调用。 void menuCanceled(MenuEvent e)：取消菜单时调用

续表

事件类名	事件监听器接口名	接口的抽象方法（事件处理方法）
PopupMenuEvent	PopupMenuListener	void popupMenuWillBecomeVisible(PopupMenuEvent e)：弹出菜单变得可见之前调用 void popupMenuWillBecomeInvisible(PopupMenuEvent e)：在弹出菜单变得不可见之前调用。 void popupMenuCanceled(PopupMenuEvent e)：在弹出菜单被取消时调用
TableModeEvent	TableModeListener	void tableChanged(TableModelEvent e)：表格单元格、行或列的发生更改时调用
TreeModelEvent	TreeModelListener	void treeNodesChanged(TreeModelEvent e)在已经以某种方式更改结点（或同级结点集）后调用。 void treeNodesInserted(TreeModelEvent e)在已将结点插入树中以后调用。 void treeNodesRemoved(TreeModelEvent e)在已从树中移除结点后调用。 void treeStructureChanged(TreeModelEvent e)在树结构中从某个给定结点开始向下的地方发生彻底更改之后调用

【例题 9_12】 编写求解介于两个正整数 m 和 n 之间素数，界面要求如图 9.15 所示。

图 9.15 素数求解界面

```
import java.awt.BorderLayout;
import java.awt.Container;
import java.awt.event.ActionEvent;
import java.awt.event.ActionListener;
import javax.swing.*;
class PrimeFrame extends JFrame implements ActionListener{
    JPanel p1,p2;
    JScrollPane js;
    JTextField t1,t2,t3;
    JButton bt;
    JTextArea area;
    int m,n,p,i,j;
    boolean isPrime;
    public PrimeFrame(){
        p1=new JPanel();
        p2=new JPanel();                   //创建面板对象
        t1=new JTextField(10);
        t2=new JTextField(10);
        t3=new JTextField(10);             //创建文本框对象
        area=new JTextArea(10,40);         //创建多行文本编辑区对象
        js=new JScrollPane(area);          //创建滚动面板对象，使文本区带上滚动条
        bt=new JButton("输出素数");          //创建按钮对象
        bt.addActionListener(this);        //给按钮注册事件监听器
```

```java
            p1.add(new JLabel("请输入 m:"));              //面板 p1 上添加标签
            p1.add(t1);                                    //将文本框 t1 添加到 p1
            p1.add(new JLabel("请输入 n:"));
            p1.add(t2);
            p1.add(new JLabel("请输入每行显示素数的个数 p:"));
            p1.add(t3);
            p1.add(bt);
            p2.add(new JLabel("求解得到的素数如下:"));
            p2.add(js);
            Container con=this.getContentPane();
            con.add(p1,BorderLayout.NORTH);
            con.add(p2,BorderLayout.CENTER);              //在窗体上添加 p1,p2
        }
        /**事件处理代码*/
        public void actionPerformed(ActionEvent e) {
            int num=0;
            area.append(null);                             //清空文本区
            try{
                m=Integer.parseInt(t1.getText().trim());
                n=Integer.parseInt(t2.getText().trim());
                p=Integer.parseInt(t3.getText().trim());
                //从各个文本框中取出文本并将其转化为整数
                for(i=m;i<=n;i++){
                    isPrime=true;
                    int k=(int)Math.sqrt(i);
                    for(j=2;j<=k;j++){
                     if(i%j==0){
                         isPrime=false;
                         break;
                     }
                   }
                    if(isPrime)  {                         //是素数
                      area.append(String.valueOf(i)+"    ");//显示素数
                      System.out.println(String.valueOf(i)+"    ");
                      num++;
                      if(num%p==0)
                         area.append("\n");                //换行
                    }
                }
             area.append("\n"+m+"~"+n+"之间的素数个数为: "+num+"个");
            }
            catch(NumberFormatException e1){
                System.out.println(e1.getMessage()+"m,n,p 不能为空, 且应为整数");
            }
        }
    }
    public class Ch9_12{
        public static void main(String args[]) {
            PrimeFrame prime= new PrimeFrame();  //创建窗体对象
            prime.setBounds(0,0,800,300);
            prime.setVisible(true);
            prime.setDefaultCloseOperation(JFrame.EXIT_ON_CLOSE);
       }
    }
```

9.6 事件监听器对象的几种实现

既然实现过相应事件监听器接口的类对象可以充当该事件源的事件监听器，那么，事件监听器就有多种实现方式。下面就以实例来说明事件监听器对象的几种不同实现方式。

9.6.1 窗体类自身实现相应事件监听器接口的方式

【例题 9_13】编写程序，在窗体上放置一个文本框，给窗体添加鼠标事件，当鼠标移入窗口时，在文本框中显示"鼠标移入窗体"，同时显示当前鼠标在窗体中的坐标；当鼠标离开窗体时，在文本框中显示"鼠标移出了窗体"；当鼠标在窗口中进行了点击操作，则在文本框中显示"鼠标在窗体中进行了点击"。

```java
import java.awt.BorderLayout;
import java.awt.event.MouseEvent;
import java.awt.event.MouseListener;
import javax.swing.*;
class MyFrame extends JFrame implements MouseListener{
    JTextField text;
    public MyFrame(){
        super("窗体类自身实现事件监听器接口");        //设置窗体标题
        text=new JTextField("显示鼠标在窗体中的各种动作");
        //创建文本框对象并设置初始文本
        this.getContentPane().add(text,BorderLayout.NORTH);//将text添加到窗体
        this.addMouseListener(this);                //给窗体对象设置鼠标事件监听器
        this.setSize(500,200);
        this.setVisible(true);
    }
    public void mouseClicked(MouseEvent e) {
        text.setText("鼠标在窗体中进行了点击");
        //给text设置文本为"鼠标在窗体中进行了点击"
    }
    public void mousePressed(MouseEvent e) {}
    public void mouseReleased(MouseEvent e) {}
    public void mouseEntered(MouseEvent e) {
        text.setText("鼠标移入窗体,鼠标在窗体中的坐标为:"+e.getX()+","+e.getY());
    }
    public void mouseExited(MouseEvent e) {
        text.setText("鼠标移出了窗体");
    }
}
public class Ch9_13_1 {
    public static void main(String[] args) {
        new MyFrame();
    }
}
```

运行结果如图 9.16 所示。

程序分析：任何事件源都可以发生鼠标事件，鼠标事件对应的事件类是 MouseEvent，对应的事件监听器接口是 MouseListener，MouseListener 中提供了鼠标在事件源上进行各种不同操作如将鼠标移入事件源、在事件源

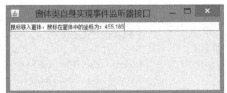

图 9.16 窗体实现鼠标事件

上单击鼠标等的抽象方法（见表 9.1），对这些方法进行重写就可以实现鼠标事件编程。

MouseEvent 类中的 getX() 方法可以获取鼠标指针在事件源坐标系中的 x 轴坐标；getY() 方法可以获取鼠标指针在事件源坐标系中的 y 轴坐标。

本例中的事件源是窗体对象，因此，用 this.addMouseListener(this);给窗体对象注册事件监听器，其中，addMouseListener 方法前面的 this 表示当前的事件源即窗体对象本身。而 addMouseListener 方法参数 this 表示事件监听器，这是由于窗体类自身实现了 MouseListener 接口，所以由窗体类对象本身 this 监听自己的鼠标事件。

9.6.2 自定义外部类实现相应事件监听接口的方式

改写例题 9_13 如下：
```
//监听器类 Listener
import java.awt.event.MouseEvent;
import java.awt.event.MouseListener;
import java.awt.BorderLayout;
import javax.swing.*;
class Listener implements MouseListener {
    JTextField t;
    public void setT(JTextField t) {
        this.t = t;
    }
    public void mouseClicked(MouseEvent e) {
        t.setText("鼠标在窗体中进行了点击");
    }
    public void mousePressed(MouseEvent e) {}
    public void mouseReleased(MouseEvent e) {}
    public void mouseEntered(MouseEvent e) {
        t.setText("鼠标移入窗体，鼠标在窗体中的坐标为: "+e.getX()+","+e.getY());
    }
    public void mouseExited(MouseEvent e) {
        t.setText("鼠标移出了窗体");
    }
}
//MyFrame 类
class MyFrame1 extends JFrame {
    JTextField text;
    public MyFrame1(){
        super("外部类实现事件监听器接口");
        text=new JTextField("显示鼠标在窗口中的各种动作");
        this.getContentPane().add(text,BorderLayout.NORTH);
        Listener listener=new Listener();
        //创建 Listener 对象
        listener.setT(text);
        //调用 set()方法给 Listener 的成员变量 t 赋值
        this.addMouseListener(listener);
        //给窗体对象注册鼠标事件监听器为 listener
        this.setSize(500,200);
        this.setVisible(true);
    }
}
public class Ch9_13_2 {
    public static void main(String[] args) {
```

```
        new MyFrame1();
    }
}
```
分析：改写后的程序与例题 9_13 执行结果一样，只是单另定义了一个专门实现 MouseListener 接口的 Listener 的类，因此，用 this.addMouseListener(listener);给窗体类对象注册事件监听器。

9.6.3 自定义外部类继承相应的事件适配器类的方式

Java 为每个包含两个或两个以上方法的监听器接口提供了事件适配器类。可以通过创建适配器类的子类对象的方式实现事件监听器。

常见的事件适配器类有：ComponentAdapter、ContainerAdapter、FocusAdapter、KeyAdapter、MouseAdapter、WindowAdapter 及 MouseMotionAdapter 等。这些适配器类已经实现了相应事件监听器接口中的多个方法，因此，只需要定义某个适配器类的子类并有选择地重写其中部分方法即可。

继续改写改写例题 9_13 如下：
```java
import java.awt.event.MouseAdapter;
import java.awt.event.MouseEvent;
import java.awt.event.MouseListener;
import java.awt.BorderLayout;
import javax.swing.*;
//Listener1 类
class Listener1 extends MouseAdapter {               //继承 MouseAdapter 类
    JTextField t;
    public void setT(JTextField t) {
        this.t = t;
    }
    /**重写需要的方法*/
    public void mouseClicked(MouseEvent e) {
        t.setText("鼠标在窗体中进行了点击");
    }
    public void mouseEntered(MouseEvent e) {
        t.setText("鼠标移入窗体，鼠标在窗体中的坐标为: "+e.getX()+","+e.getY());
    }
    public void mouseExited(MouseEvent e) {
        t.setText("鼠标移出了窗体");
    }
}
class MyFrame2 extends JFrame {
    JTextField text;
    public MyFrame2(){
        super("外部类继承事件适配器类");
        text=new JTextField("显示鼠标在窗口中的各种动作");
        this.getContentPane().add(text,BorderLayout.NORTH);
        Listener1 listener=new Listener1();
        listener.setT(text);
        this.addMouseListener(listener);
        this.setSize(500,200);
        this.setVisible(true);
    }
}
public class Ch9_13_3 {
    public static void main(String[] args) {
        new MyFrame2();
```

        }
    }
分析：一个类实现某接口，则一定要实现其所有的抽象方法；而一个类继承某个父类，则只需要选择需要的方法进行重写。与前两种方法相比，本例中无须实现 mousePressed 及 mouseReleased 方法。

9.6.4 匿名类实现事件监听器

匿名类分为两种情况，与相应的事件监听器接口相关的匿名类；与相应的事件适配器类相关的匿名类。
继续改写改写例题 9_13 如下：

```
import java.awt.BorderLayout;
import java.awt.event.MouseEvent;
import java.awt.event.MouseListener;
import javax.swing.*;
class MyFrame3 extends JFrame {
    JTextField text;
    public MyFrame3(){
        super("与MouseListener接口相关的匿名类方式实现鼠标监听");
        text=new JTextField("显示鼠标在窗口中的各种动作");
        this.getContentPane().add(text,BorderLayout.NORTH);
        this.addMouseListener(new MouseListener(){
            //用实现过MouseListener接口的匿名类对象作为事件监听器
            public void mouseClicked(MouseEvent e) {
                text.setText("鼠标在窗体中进行了点击");
            }
            public void mousePressed(MouseEvent e) {}
            public void mouseReleased(MouseEvent e) {}
            public void mouseEntered(MouseEvent e) {
                text.setText("鼠标移入窗体,鼠标在窗体中的坐标为: "+e.getX()+","+e.getY());
            }
            public void mouseExited(MouseEvent e) {
                text.setText("鼠标移出了窗体");
            }
        });
        this.setSize(500,200);
        this.setVisible(true);
    }
}
public class Ch9_13_4 {
    public static void main(String[] args) {
        new MyFrame3();
    }
}
```

继续改写改写例题 9_13 如下：

```
import java.awt.BorderLayout;
import java.awt.event.MouseAdapter;
import java.awt.event.MouseEvent;
import javax.swing.*;
class MyFrame4 extends JFrame {
    JTextField text;
    public MyFrame4(){
        super("与MouseAdapter类相关的匿名类方式进行鼠标事件监听");
        text=new JTextField("显示鼠标在窗口中的各种动作");
```

```java
            this.getContentPane().add(text,BorderLayout.NORTH);
            this.addMouseListener(new MouseAdapter(){
                //用 MouseAdapter 类的匿名子类对象作为事件监听器
                public void mouseClicked(MouseEvent e) {
                    text.setText("鼠标在窗口中进行了点击");
                }
                public void mouseEntered(MouseEvent e) {
                    text.setText("鼠标移入窗口,鼠标在窗口中的坐标为: "+e.getX()+","+e.getY());
                }
                public void mouseExited(MouseEvent e) {
                    text.setText("鼠标移出了窗口");
                }
            });
            this.setSize(500,200);
            this.setVisible(true);
        }
    }
    public class Ch9_13_5 {
        public static void main(String[] args) {
            new MyFrame4();
        }
    }
```

9.7 Swing 的常用组件及其事件编程

本小节中只对 Swing 中的常用组件类及其常用方法、事件编程方法进行讲解，通过对典型组件用法的掌握，达到举一反三、触类旁通。

Swing 类包几乎包含了开发 Java GUI 程序所需的所有组件，非常丰富。各个组件类都包含了多个重载的构造方法，提供了大量成员方法，同时继承了多级父类的各种成员方法。学习和实践中要善于利用 JDK 开发文档，在编程实践中逐步熟悉所有组件类的用法。也可以灵活运用 javap 这个 JDK 命令查看某个组件类的属性和方法。

9.7.1 JButton

JButton 是 Swing 组件用于创建按钮的类。其常用构造方法有：

- public JButton(Icon icon)：创建一个带图标的按钮，如工具栏按钮等。
- public JButton(String text)：创建一个带文本（按钮名称）的按钮。
- public JButton(String text,Icon icon)：创建一个带初始文本和图标的按钮。

9.7.2 JLabel

用于创建标签的组件类。其常用构造方法有：

- public JLabel(String text)：创建具有指定文本的标签对象。
- public JLabel(Icon image)：创建具有指定图像的标签对象。

其常用方法有：

- public void setText(String text)：设置标签文本。
- public String getText()：返回标签文本。

9.7.3　JTextField

JTextField 是具有单行文本编辑功能的文本框组件。其常用构造方法有：
- public JTextField(int columns)：创建一个具有指定列数的文本框对象。初始字符串设置为 null，如 JTextField text =new JTextField(10);。
- public JTextField(String text)：创建一个用指定文本初始化的文本框对象。例如，JTextField text =new JTextField("请输入聊天内容");。

其常用方法有：
- public void addActionListener(ActionListener l)：给文本框对象注册事件监听器。
- public void setText(String t)：给文本框设置文本。该方法是 JTextField 继承自其父类 JTextComponent 的方法。
- public String getText()：返回文本框中包含的文本。该方法也是 JTextField 继承自其父类 JTextComponent 的方法。

9.7.4　JTextArea

JTextArea 是一个具有多行文本编辑功能的组件。其常用构造方法有：
- public JTextArea(String text)：创建显示指定文本的的多行文本编辑区对象，行、列设置为 0。
- public JTextArea(int rows,int columns)：创建具有指定行数和列数的多行文本编辑区对象，初始字符串为 null。例如，JTextArea area=new JTextArea(5,40);。
- public JTextArea(String text,int rows,int columns)：创建具有指定文本、行数和列数多行文本编辑区对象。

9.7.5　JRadioButton

是用于创建单选按钮的组件。其常用构造方法有：
- public JRadioButton(String text)：创建一个具有指定文本，状态为未选择的单选按钮。
- public JRadioButton(String text,boolean selected)：创建一个具有指定文本和选择状态的单选按钮。

注意：JRadioButton 需要与与 ButtonGroup 配合使用。目的是将多个单选按钮对象设置为一个按钮组，从而实现一次只能选择其中的一个按钮。通过 add()方法将 JRadioButton 对象添加到该组中。例如：

```
JRadioButton male=new JRadioButton("男");
JRadioButton female=newJRadioButton("女");
ButtonGroup group=new ButtonGroup();
group.add(male);group.add (female);
```

9.7.6　JCheckBox

是用于创建复选框的组件。其常用构造方法有：
- public JCheckBox(String text)：创建一个带文本的、最初未被选定的复选框。
- public JCheckBox(String text,boolean selected)：创建一个带文本的复选框，并指定其最初是否处于选定状态。

9.7.7 JComboBox

用于创建下拉列表的组件类。其常用构造方法有：
- public JComboBox()：创建空的下拉列表对象，可使用 addItem 方法给其添加待选项。
- public JComboBox(Object[] items)：创建包含指定数组中的元素为待选项的下拉列表对象。

其常用方法有：
- public void addItem(Object anObject)：给其添加待选项。
- public void addItemListener(ItemListener aListener)：给其注册事件监听器。当下拉列表中的待选项发生变化时，会引发 ItemEvent 类型事件，处理该事件的监听器接口为 ItemListener。
- public Object getItemAt(int index)：返回指定索引处的列表项。如果 index 超出范围（小于零或者大于等于列表大小），则返回 null。
- public int getItemCount()：返回列表中的项数。
- public Object getSelectedItem()：返回当前所选项。

9.7.8 JList

允许用户在多个条目中做出选择的列表框组件，其常用构造方法有：
- public JList()：创建一个具有空的列表框对象。
- public JList(Object[] listData)：创建一个列表框对象，使其显示指定数组中的元素。

JList 的 public void setSelectionMode(int selectionMode)用来设置列表的选择模式，参数有以下可选值：ListSelectionModel.SINGLE_SELECTION——一次只能选择一项；ListSelectionModel.SINGLE_INTERVAL_ SELECTION——允许选择连续范围内的多个项，如果用户选中了某一项，接着按住【Shift】键，单击另一个项，那么这两项之间的所有项都会被选中；ListSelectionModel.MULTIPLE_INTERVAL_SELECTION——这是列表框的默认选择模式。用户既可以选择连续范围内的多个项，也可以选择不连续的多个项。只要按住【Ctrl】键，单击列表框的多个项，这些项都会被选中。

当在列表框中选择一些项时，将触发 ListSelectionEvent 事件，ListSelectionListener 监听器负责处理该事件。

下面，通过一个实例说明上述组件类的应用。

【例题 9_14】编写一个用于学生信息注册的窗体，在其上可以输入学生姓名、学号、出生日期、选择性别、专业、主修课程、兴趣爱好、输入备注信息等功能（见图 9.17）。

```
import java.awt.Container;
import java.awt.FlowLayout;
import javax.swing.*;
class StudentInfoInput extends JFrame {
    JTextField name,no,birthday;
    JRadioButton male,female;
    ButtonGroup group;
    JComboBox profession;
    JList course;
    JCheckBox fav1,fav2,fav3,fav4;
    JTextArea info;
    JButton commit,cancel;
    String[] pro={"计算机科学与技术","信息安全","信息管理与信息系统","软件工程"};
```

图 9.17 学生信息注册窗体

```
    String[] courses={"操作系统","组成原理","数据结构","Java SE程序设计"};
    public StudentInfoInput(){
       this.setLayout(new FlowLayout());         //将窗体设为FlowLayout布局
       /*创建所需的各个组件对象*/
       name=new JTextField(15);no=new JTextField(15);birthday=new JTextField(20);
       male=new JRadioButton("男");female=new JRadioButton("女");
       group=new ButtonGroup();group.add(male);group.add(female);
       //将两个单选按钮组合在一起
       profession=new JComboBox();
       profession=new JComboBox(pro);            //以pro数组中的元素作为待选项
       course=new JList(courses);                //以courses数组中的元素作为待选项
       fav1=new JCheckBox("体育");fav2=new JCheckBox("音乐");fav3=new
       JCheckBox("美术");fav4=new JCheckBox("书法");
       info=new JTextArea(5,20);
       commit=new JButton("提交"); cancel=new JButton("取消");
       Container con=this.getContentPane();      //得到Container对象
       /*将各个组件对象添加到窗体*/
       con.add(new JLabel("请输入学生姓名: "));con.add(name);
       con.add(new JLabel("请输入学生学号: "));con.add(no);
       con.add(new JLabel("请输入学生出生日期: "));con.add(birthday);
       con.add(new JLabel("请选择学生性别: "));con.add(male);con.add(female);
       con.add(new JLabel("请选择学生专业"));con.add(profession);
       con.add(new JLabel("请选择学生主修课程: "));con.add(course);
       con.add(new JLabel("请选择学生兴趣爱好: "));con.add(fav1);con.add(fav2);
       con.add(fav3);con.add(fav4);
       con.add(new JLabel("请添加学生备注: "));con.add(info);
       con.add(commit);con.add(cancel);
       this.setBounds(100, 100, 290, 530);
       this.setTitle("学生信息录入");
       this.setVisible(true);
    }
}
public class Ch9_14
{
    public static void main(String[] args){
      new StudentInfoInput();                    //创建窗体对象
    }
}
```

说明：为了凸显各个常用组件的创建方法，程序中没有涉及事件编程。同时，组件布局方式值简单地采用了流式布局。

思考：该注册窗体中没有密码设置项，读者试为其增加密码输入框。提示：JPasswordField组件是用于编辑密码的文本框组件，在其中输入时，用户输入的文本并不真正显示出来，而是显示回显字符如'*'。JPasswordField的常用方法有：

- public char[] getPassword()：返回JPasswordField的文本内容。
- public char getEchoChar()：获取密码的回显字符。
- public void setEchoChar(char c)：设置密码的回显字符。

【**例题9_15**】 编程实现一个简单的"计算器"，执行效果如图9.18所示。

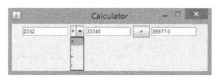

图9.18 简单计算器

```java
import java.awt.Container;
import java.awt.FlowLayout;
import java.awt.event.ActionEvent;
import java.awt.event.ActionListener;
import java.awt.event.ItemEvent;
import java.awt.event.ItemListener;
import javax.swing.*;
class Calculator extends JFrame implements ActionListener,ItemListener{
  JTextField t1,t2,t3;
  JComboBox<String> box;
  JButton bt;
  String operator=null;
  public Calculator(){
    this.setTitle("Calculator");
    this.setLayout(new FlowLayout());
    t1=new JTextField(10);
    t2=new JTextField(10);
    t3=new JTextField(10);
    box=new JComboBox<String>();
    box.addItem("+");box.addItem("-");
    box.addItem("*");box.addItem("/");
    bt=new JButton("=");
    Container con=this.getContentPane();            //得到 Container 对象
    /**将各个组件对象添加到窗体*/
    con.add(t1);
    con.add(box);
    con.add(t2);
    con.add(bt);
    con.add(t3);
    box.addItemListener(this);                      //注册事件监听器
    bt.addActionListener(this);                     //注册事件监听器
    this.setVisible(true);
    this.setBounds(100,100,500,100);
    this.setResizable(false);
  }
   /*下拉列表 box 的事件处理代码**/
   public void itemStateChanged(ItemEvent arg0) {
      operator=(String)box.getSelectedItem();   //得到下拉列表的当前选项
   }
    /*"="按钮的事件处理代码**/
   public void actionPerformed(ActionEvent arg0) {
     String s1=t1.getText();
     String s2=t2.getText();                        //获取两个运算数
     double d1=Double.parseDouble(s1);
     double d2=Double.parseDouble(s2);              //将两个运算数转成 double 型
     if(operator==null||operator.equals("+")){ //注意此条件的写法
        t3.setText(String.valueOf(d1+d2));
     }
     else if(operator.equals("-")){
        t3.setText(String.valueOf(d1-d2));
     }
     else if(operator.equals("*")){
        t3.setText(String.valueOf(d1*d2));
     }
     else if(operator.equals("/")){
```

```
            t3.setText(String.valueOf(d1/d2));
        }
    }
}
public class Ch9_15 {
    public static void main(String args[]){
        new Calculator();
    }
}
```

讨论：若将程序中 if(operator==null||operator.equals("+"))改写成 if(operator.equals("+"))，则当运行程序并默认选择"+"，即不对 JComboBox 做选项更改时会出错。这是因为只有对 JComboBox 的当前选项该选为其他选项时，才会触发 ItemEvent 类型的事件，operator 才有具体的"+""-""*""/"之一。不做更改时 operator==null。

9.7.9 JMenuBar、JMenu 与 JMenuItem

Java Swing 的菜单编程中涉及的类有菜单条类 JMenuBar、菜单类 JMenu 及菜单项类 JMenuItem 等。其关系是：一个菜单条 JMenuBar 中可以包含多个菜单 JMenu，一个菜单 JMenu 中可以容纳多个菜单项 JMenuItem。

（1）JMenuBar

其构造方法为 public JMenuBar()：创建新的菜单栏。

其常用方法有：

- public JMenu add(JMenu c)：将指定的菜单追加到菜单栏的末尾。
- public JMenu getMenu(int index)：返回菜单栏中指定位置的菜单。
- public boolean isSelected()：如果当前已选择了菜单栏的组件，则返回 true。

（2）JMenu

其常用构造方法为 public JMenu(String s)：创建一个新菜单对象，用 s 作为菜单名称。

其常用方法有：

- public JMenuItem add(JMenuItem menuItem)：将某个菜单项追加到此菜单的末尾。
- public JMenuItem add(String s)：创建具有指定文本的新菜单项，并将其追加到此菜单的末尾。
- public void addMenuListener(MenuListener l)：添加菜单事件的监听器。
- public void addSeparator()：将菜单分隔线添加到菜单的末尾。
- public JPopupMenu getPopupMenu()：返回与此菜单关联的弹出菜单。若不存在弹出菜单，则将创建一个。
- public boolean isSelected()：如果菜单是当前选择的（即高亮显示的）菜单，则返回 true。
- public void remove(JMenuItem item)：从此菜单移除指定的菜单项。

（3）JMenuItem

其常用构造方法：

- public JMenuItem(String text)：创建带有指定文本的 JMenuItem 对象。
- public JMenuItem(String text,Icon icon)：创建带有指定文本和图标的 JMenuItem 对象。

其常用方法有：

- public void addActionListener(ActionListener l)：给菜单项注册单击事件监听器，此方法是从

JMenuItem 的父类 javax.swing.AbstractButton 继承来的方法。
- public KeyStroke getAccelerator()：返回菜单项快捷键的 KeyStroke 对象。
- public void setAccelerator(KeyStroke keyStroke)：给菜单项设置快捷键。

9.7.10　JPopupMenu

其常用构造方法 public JPopupMenu(String label)可以创建一个具有指定标题的 JPopupMenu（弹出式菜单）对象。

其常用方法有：
- public JMenuItem add(JMenuItem menuItem)：将指定菜单项添加到此菜单的末尾。
- public JMenuItem add(String s)：创建具有指定文本的菜单项，并将其添加到此菜单的末尾。
- public void insert(Component component,int index)：将指定组件插入到菜单的给定位置。
- public void addPopupMenuListener(PopupMenuListener l)：注册事件监听器。
- public boolean isVisible()：如果弹出菜单可见，则返回 true。
- public String getLabel()：返回弹出菜单的标签。
- public void setVisible(boolean b)：设置弹出菜单的可见性。
- public void show(Component invoker,int x,int y)：在组件调用者的坐标空间中的位置（x、y）显示弹出菜单。

若鼠标事件是弹出式菜单触发事件，则调用弹出式菜单对象的 show 方法来显示弹出式菜单。

如：JPopupMenu popupMenu = new JPopupMenu("菜单");下面的 showJPopupMenu 方法在收到触发器事件就会显示弹出式菜单，代码如下：

```
public void showJPopupMenu(MouseEvent e) {
    if (e.isPopupTrigger()) {
        popupMenu.show(ivoker,e.getX(),e.getY());
    }
}
```

9.7.11　JToolBar

在 GUI 开发中，工具栏是不可或缺的一部分，它提供了对菜单项功能的快捷、直观地等价实现方式。JToolBar 类用于实现工具栏，在其上可以添加若干功能按钮。

下面，通过一个实例来说明以上几个常用组件的用法。

【例题 9_16】在窗体中添加"文件"及"编辑"下拉菜单，同时，添加工具栏，工具栏中包含"打开""保存""复制""剪切""粘贴"等 5 个工具按钮。当在文本区中编辑了文本，则当选中文本后，菜单、弹出式菜单以及工具栏中的"复制"、"剪切"、"粘贴"可以完成相应功能；当单击菜单或工具栏的"打开""保存"命令时，可以弹出"文件"或"保存"对话框（效果如图 9.19 所示）。

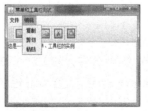

图 9.19　菜单与工具栏

图 9.19 菜单与工具栏（续）

```java
import java.awt.BorderLayout;
import java.awt.event.ActionEvent;
import java.awt.event.ActionListener;
import javax.swing.*;
import java.awt.event.MouseAdapter;
import java.awt.event.MouseEvent;
import javax.swing.JPopupMenu;
public class Ch9_16 extends JFrame implements ActionListener{
    JMenuBar menubar;
    JMenu file,edit;
    JMenuItem open,save,exit,copy,cut,paste;
    JPopupMenu popupmenu;
    JMenuItem copy1,cut1,paste1;
    JToolBar toolbar;
    JButton opentool,savetool,copytool,cuttool,pastetool;
    JTextArea area;
    JScrollPane pane;
    JFileChooser filechooser;             //声明文件对话框对象
    PopupListener popuplistener;          //声明弹出式菜单的监听器对象
    public Ch9_16(){
        menubar=new     JMenuBar();
        file=new JMenu("文件");
        edit=new JMenu("编辑");
        open=new JMenuItem("打开");
        save=new JMenuItem("保存");
        exit=new JMenuItem("退出");
        copy=new JMenuItem("复制");
        cut=new JMenuItem("剪切");
        /*以下代码是将菜单项添加到菜单中，菜单添加到
         * 菜单条中，最后将菜单条加入窗体*/
        paste=new JMenuItem("粘贴");
        file.add(open);file.add(save);file.add(exit);
        edit.add(copy);edit.add(cut);edit.add(paste);
        menubar.add(file);menubar.add(edit);
        this.setJMenuBar(menubar);
        /*菜单项添加事件监听器*/
        open.addActionListener(this);save.addActionListener(this);
        exit.addActionListener(this);copy.addActionListener(this);
        cut.addActionListener(this);paste.addActionListener(this);
        /*创建弹出式菜单，并给各个菜单项添加事件监听器*/
        popupmenu=new JPopupMenu();
        copy1=new JMenuItem("复制");
        cut1=new JMenuItem("剪切");
        paste1=new JMenuItem("粘贴");
```

```java
            popupmenu.add(copy1);popupmenu.add(cut1);
            popupmenu.add(paste1);
            copy1.addActionListener(this);cut1.addActionListener(this);
            paste1.addActionListener(this);
            /*创建工具栏,给工具栏按钮上添加图片,将工具栏按钮添加到工具栏并添加事件监听器*/
            toolbar=new JToolBar();
            opentool=new JButton(new ImageIcon(getClass().getResource("open.jpg")));
            savetool=new JButton(new ImageIcon(getClass().getResource("save.jpg")));
            copytool=new JButton(new ImageIcon(getClass().getResource("copy.jpg")));
            cuttool=new JButton(new ImageIcon(getClass().getResource("cut.jpg")));
            pastetool=new JButton(new ImageIcon(getClass().getResource("paste.jpg")));
            toolbar.add(opentool);toolbar.add(savetool);
            toolbar.add(copytool);toolbar.add(cuttool);toolbar.add(pastetool);
            opentool.addActionListener(this);
            savetool.addActionListener(this);
            copytool.addActionListener(this);
            cuttool.addActionListener(this);
            pastetool.addActionListener(this);
            area=new JTextArea();
            pane=new JScrollPane(area);
            this.getContentPane().add(toolbar,BorderLayout.NORTH);
            this.getContentPane().add(pane,BorderLayout.CENTER);
            /*给文本区添加事件监听器,让弹出式菜单显示在鼠标点击的位置*/
            popuplistener=new PopupListener();
            popuplistener.setPopupmenu(popupmenu);//向PopupListenr传递popupmenu对象
            area.addMouseListener(popuplistener);
            filechooser=new JFileChooser();//创建文件对话框
        }
    /*实现菜单项及工具栏工具按钮的事件代码*/
        public void actionPerformed(ActionEvent e) {
            if(e.getSource()==open||e.getSource()==opentool)
                filechooser.showOpenDialog(this);   //在当前窗体中弹出文件打开对话框
            if(e.getSource()==save||e.getSource()==savetool)
                filechooser.showSaveDialog(this);   //在当前窗体中弹出文件保存对话框
            if(e.getSource()==exit)
                System.exit(0);                     //关闭程序
            if(e.getSource()==copy||e.getSource()==copy1||e.getSource()==copytool)
                area.copy();                        //实现文本区文本的复制操作
            if(e.getSource()==cut||e.getSource()==cut1||e.getSource()==cuttool)
                area.cut();                         //实现文本区文本的剪切操作
            if(e.getSource()==paste||e.getSource()==paste1||e.getSource()==pastetool)
                area.paste();                       //实现文本区文本的粘贴操作
        }
        public static void main(String args[]){
            Ch9_16 frame=new Ch9_16();
            frame.setTitle("菜单和工具栏测试");
            frame.setBounds(0, 0, 512, 384);
            frame.setVisible(true);
        }
    }
    //处理弹出式菜单鼠标事件的类 PopupListener
    class PopupListener extends MouseAdapter {
        JPopupMenu popupmenu;
        public void setPopupmenu(JPopupMenu popupmenu) {
```

```
        this.popupmenu = popupmenu;
    }
    public void mousePressed(MouseEvent e) {
        this.showPopupMenu(e);
    }
    public void mouseReleased(MouseEvent e) {
        this.showPopupMenu(e);
    }
    //显示弹出式菜单的方法
    public void showPopupMenu(MouseEvent e){
        if(e.isPopupTrigger()){
            popupmenu.show(e.getComponent(),e.getX(),e.getY());
        }
    }
}
```

说明：为了使得工具栏上的各个功能按钮上显示图片，需要预先准备好 open.jpg、save.jpg、copy.jpg、cut.jpg 及 paste.jpg 等素材图片，并将它们保存在源程序同一路径下。

思考：试给各个菜单项增加分隔线与快捷键。如何给工具栏工具按钮添加提示文字？

提示：可以用 JMenu 对象调用 addSeparator()方法在分割各个菜单项。例如：file.add(open);file.addSeparator();file.add(save);表示给 open 菜单项和 save 菜单项之间增加分隔线。

可以用 open.setAccelerator(KeyStroke.getKeyStroke('O', KeyEvent.CTRL_MASK));给打开菜单项设置快捷键，即按下组合键【Ctrl+O】和单击"打开"菜单项的作用一样（见图9.20）。

可以用 opentool.setToolTipText("打开");给工具栏中"打开"按钮添加提示文字，当鼠标移动到该按钮时，出现提示文字（见图9.21）。

图 9.20　添加了分割线及快捷键的效果　　　　图 9.21　工具栏按钮的提示文字

本例只是实现了"打开"和"保存"对话框的弹出，那如何实现真正的文件代开与保存呢？

提示：要通过文件对话框实现真正的打开和保存文件，需要输入/输出流的支持，将在后面章节讲解。

9.7.12　JTable

表格组件 JTable 以行和列的形式显示数据，允许对表格中的数据进行编辑。JTable 主要数据的显示方式，表格数据处理主要由数据模型类负责。通常用数据模型类的对象来保存数据，数据模型类派生于抽象类 AbstractTableModel 类，并且必须重写抽象模型类的 getColumn Count、getRowCount 及 getValueAt 等几个方法。因为表格会从这个数据模型的对象中自动获取数据，数据模型类的对象负责表格大小、数据填写、表格单元更新等与表格有关的属性和操作。Java 在

javax.swing.table 包中提供了几个表格模型类，比较常用的是 DefaultTableModel。

JTable 的常用构造方法（JTable 的构造方法有 7 个，此处只列出 3 个）：
- public JTable()：创建一个 JTable 对象，使用默认的数据模型、默认的列模型和默认的选择模型对其进行初始化。
- public JTable(int numRows,int numColumns)：使用 DefaultTableModel 创建具有 numRows 行和 numColumns 列个空单元格的 JTable 对象。列名称采用 "A"、"B"、"C" 等形式。
- public JTable(Object[][] rowData,Object[] columnNames)：创建一个 JTable 来显示二维数组 rowData 中的值，其列名称为 columnNames 中的元素值。

JTable 的常用方法有：
- public int getSelectedColumn()：返回第一个选定列的索引，若没有选定的列，则返回-1。
- public int getSelectedRow()：返回第一个选定行的索引，若没有选定的行，则返回-1。
- public Object getValueAt(int row,int column)：返回 row 和 column 位置的单元格值。

通过对表格中的数据进行编辑，可以修改表格中二维数组 data 中对应的数据数据。在表格中输入或修改数据后，需按【Enter】键或用鼠标单击表格的单元格确定所输入或修改的结果。当表格需要刷新显示时，让表格对象调用 repaint()方法。

JTable 自身没有滚动条。因此，当表格中数据量较大时，应将表格添加到滚动面板 JScrollPane 中。

【例题 9_17】 在窗体中创建表格，用于输入学生的基本信息，当用鼠标点选表格的某个单元格时，在窗体的标签上显示当前点选单元格的行号、列号以及单元格值（见图 9.22）。

图 9.22 编辑表格并显示表格行、列、单元格值

```
import java.awt.BorderLayout;
import java.awt.event.MouseEvent;
import java.awt.event.MouseListener;
import javax.swing.*;
class TableWin extends JFrame implements MouseListener{
    String [] title={"班级","学号","姓名","性别","年龄","籍贯"};//定义表格的列标题;
    Object [][]data;                       //定义表格数据单元格
    JTable table;
    JScrollPane scroll;
    JLabel state;
    public TableWin(String s){
        this.setTitle(s);                  //设置窗体标题
        data=new Object[50][6];            //创建表格数据单元格
        for(int i=0;i<50;i++){
            for(int j=0;j<6;j++)
                data[i][j]=null;
        }//表格单元格数据初始化
        table=new JTable(data,title);      //创建表格对象
        scroll=new JScrollPane(table);     //创建滚动面板对象
        state=new JLabel();
        this.getContentPane().add(scroll,BorderLayout.CENTER);
        this.getContentPane().add(state,BorderLayout.SOUTH);
        table.addMouseListener(this);      //给表格对象注册鼠标监视器
        this.setBounds(0,0,512,384);
        this.setVisible(true);
```

```
        }
        public void mouseClicked(MouseEvent e) {
            int row=table.getSelectedRow();         //获得鼠标当前点选的表格单元格行号
            int column=table.getSelectedColumn();//获得鼠标当前点选的表格单元格列号
            String value=(String)table.getValueAt(row,column);
//获得鼠标当前点选的表格单元格数据
            state.setText("行号: "+row+" 列号: "+column+" 值: "+value);
//在标签上显示当前选择的表格行、列以及单元格值
        }
        public void mouseEntered(MouseEvent e) {}
        public void mouseExited(MouseEvent e) {}
        public void mousePressed(MouseEvent e) {}
        public void mouseReleased(MouseEvent e) {}
}
public class Ch9_17{
    public static void main(String arg[]) {
        new TableWin("表格组件示例1");
    }
}
```

【例题 9_18】 改写例题 9_17，初始时表格中只有 10 个数据行，通过单击"增加行"按钮增加新行，单击"删除行"按钮删除当前鼠标点选的行（见图 9.23）。

图 9.23 插入、删除表格行

```
import java.awt.BorderLayout;
import java.awt.GridLayout;
import java.awt.event.ActionEvent;
import java.awt.event.ActionListener;
import javax.swing.*;
import javax.swing.table.DefaultTableModel;
class TableWin1 extends JFrame implements ActionListener{
    String [] title={"班级","学号","姓名","性别","年龄","籍贯"};//定义表格的列标题;
    Object [][]data;                    //定义表格数据单元格
    JTable table;
    DefaultTableModel model;
    JScrollPane scroll;
    JButton insert,delete;
    JPanel panel;
    public TableWin1(String s){
        this.setTitle(s);                //窗体标题
        data=new Object[10][6];          //创建表格数据单元格,初始时只有 10 个数据行。
        for(int i=0;i<10;i++){
            for(int j=0;j<6;j++)
                data[i][j]=null;
        }
        model=new DefaultTableModel(data,title);//创建 DefaultTableModel 对象
        table=new JTable(model);         //用 DefaultTableModel 对象 model 创建 table
        scroll=new JScrollPane(table);
        insert=new JButton("增加行");
        delete=new JButton("删除行");
        panel=new JPanel();
        panel.setLayout(new GridLayout(1,2));
        panel.add(insert);panel.add(delete);
        this.getContentPane().add(scroll,BorderLayout.CENTER);
        this.getContentPane().add(panel,BorderLayout.SOUTH);
```

```java
        insert.addActionListener(this);
        delete.addActionListener(this);              //给按钮对象注册事件监听器
        this.setBounds(0, 0, 512, 384);
        this.setVisible(true);
    }
    //"增加行"、"删除行"按钮事件处理代码
    public void actionPerformed(ActionEvent e) {
        if(e.getSource()==insert) {                  //当事件源为 insert
            Object[] data={null,null,null,null,null};   //定义一个空行
            model.addRow(data);                      //给表格增加新行
        }
        if(e.getSource()==delete)                    //当事件源为 delete
            model.removeRow(table.getSelectedRow());//删除当前选择的行
    }
}
public class Ch9_18{
    public static void main(String arg[]) {
        new TableWin1("表格组件示例2");
    }
}
```

说明：在给表格新增行时，总是在当前表格的末尾进行增加；在删除某个表格行时，要预先点选该行，否则出错。

当程序中的事件源有多个时，可以通过"e.getSource()==事件源" 判断当前触发事件的是哪一个事件源 如 if(e.getSource()==insert) 表示当前触发事件的是 insert 对象，也可以用 if(e.getActionCommand().equals("增加行"))来替代 if(e.getSource()==insert)，表示判断当前触发事件的事件源是否是一个名为"增加行"的按钮对象。这是因为 ActionEvent 类对象 e 调用其 getActionCommand()方法可以用于获取当前事件源的命令字符串（如事件源名称）。同理，也可以用 if(e.getActionCommand().equals("删除行"))来替换 if(e.getSource()==delete)。

9.7.13 JTabbedPane

页标签面板 JTabbedPane 可以用来存放许多标签页，而每一张标签页又可以存放不同的容器或组件，用户只要单击每一张标签页上的标签，便可切换至不同的标签页。

JTabbedPane 的常用构造方法有：

- public JTabbedPane()创建一个具有默认的 JTabbedPane.TOP 选项卡布局的 TabbedPane 对象。
- public JTabbedPane(int tabPlacement)创建一个空的 TabbedPane 对象，使其具有以下指定选项卡布局中的一种：JTabbedPane.TOP、JTabbedPane.BOTTOM、JTabbedPane.LEFT 或 JTabbedPane.RIGHT。

JTabbedPane 的常用方法有：

- public void add(Component component,Object constraints)：将一个 component 添加到页标签面板中。如果 constraints 为 String 或 Icon，则它将作为该标签页标题。
- public void addChangeListener(ChangeListener l)：给 JTabbedPane 对象注册事件监听器。
- public int getSelectedIndex()：返回当前选择的标签页的索引。
- public void setToolTipTextAt(int index,String toolTipText)：将 index 位置的标签页提示文本设置为 toolTipText。

与页签面板 JTabbedPane 关联的事件一般是 ChangeEvent 类型事件，该事件所对应的接口是 ChangeListener，该接口提供了一个方法 stateChanged(ChangeEvent e)，当选择某个标签页时将调用该方法。

【例题 9_19】 通过标签页选择 3 个不同班级的学生信息表格，当选择不同标签页时，在窗体底部的一个标签对象显示该标签页的索引值。同时要求为每个标签页添加提示文本（见图 9.24）。

图 9.24 页标签面板显示学生信息

```java
import java.awt.BorderLayout;
import javax.swing.*;
import javax.swing.event.ChangeEvent;
import javax.swing.event.ChangeListener;
class TabledPaneWin extends JFrame implements ChangeListener{
    String [] title={"班级","学号","姓名","性别","年龄","籍贯"};//定义表格的列标题;
    Object [][]data;                    //定义表格数据单元格
    JTable table1,table2,table3;
    JScrollPane scroll1,scroll2,scroll3;
    JTabbedPane tabbedPane;
    JLabel state;
    public TabledPaneWin(String s){
      this.setTitle(s);              //窗体标题
      data=new Object[50][6];        //创建表格数据单元格
      for(int i=0;i<50;i++){
        for(int j=0;j<6;j++)
          data[i][j]=null;
      }
      /*创建3个表格对象*/
      table1=new JTable(data,title);
      table2=new JTable(data,title);
      table3=new JTable(data,title);
      /*将3个表格对象放在不同的滚动面板中*/
      scroll1=new JScrollPane(table1);
      scroll2=new JScrollPane(table2);
      scroll3=new JScrollPane(table3);
      tabbedPane=new JTabbedPane();   //创建页签面板对象
      /*将3个滚动面板对象放置到页签面板的不同页中*/
      tabbedPane.add(scroll1, "一班学生信息");
      tabbedPane.setToolTipTextAt(0,"单击这里显示一班学生表格");
      tabbedPane.add(scroll2, "二班学生信息");
      tabbedPane.setToolTipTextAt(1,"单击这里显示二班学生表格");
      tabbedPane.add(scroll3, "三班学生信息");
      tabbedPane.setToolTipTextAt(2,"单击这里显示三班学生表格");
      state=new JLabel();
      this.getContentPane().add(tabbedPane,BorderLayout.CENTER);
      this.getContentPane().add(state,BorderLayout.SOUTH);
      tabbedPane.addChangeListener(this);      //给页签面板添加事件监听器
      this.setBounds(0,0,512,384);
      this.setVisible(true);
    }
    public void stateChanged(ChangeEvent arg0) {
```

```
        /*当点选不同页签时,状态栏state中显示对应页签的索引值,
         * 索引值从0开始*/
        int index=tabbedPane.getSelectedIndex();
        state.setText(""+index);}
    }
    public class Ch9_19{
      public static void main(String arg[]) {
        new TabledPaneWin("表格组件示例1");
      }
    }
```

9.7.14 JTree

类似于 Window 资源管理器的目录管理方式,JTree 用于构造树形的分层视图。JTree 由一系列位于不同层次的树结点组成。每个 DefaultMutableTreeNode 类(位于 javax.swing.tree 包中)对象可以充当树的一个结点。创建一个树状视图前首先需要创建出多个具有层次关系的 DefaultMutableTreeNode 类对象(结点),然后以根结点为参数,创建 JTree 类对象。

DefaultMutableTreeNode 的常用构造方法:

- public DefaultMutableTreeNode(Object userObject):创建一个允许有子结点的树结点对象,并使用 userObject 对其进行初始化,如:DefaultMutableTreeNode node1=new DefaulTMutableTreeNode("第1学期");将创建一个名为"第1学期"的树结点。
- public DefaultMutableTreeNode(Object userObject,boolean allowsChildren):创建树结点对象,并使用 userObject 对其进行初始化,用 allowsChildren 指定是否允许该结点有子结点。

DefaultMutableTreeNode 的常用方法:

- public TreeNode getRoot():返回包含此结点的树的根对象。
- public Object getUserObject():返回此结点的用户对象。此处的用户对象是指 DefaultMutableTreeNode 类构造方法参数 Object userObject。
- public boolean isLeaf():判断某结点是否"叶子"结点。

JTree 的常用构造方法为 public JTree(TreeNode root):创建一个以 root 为根结点的树状视图。运用 JTree 类的 public Object getLastSelectedPathComponent()方法可以返回当前选择结点对象。

由于树状视图 JTree 本身没有滚动条,所以也需要将 JTree 加到 JScrollPane 中。

树组件上可以触发 TreeSelectionEvent 类型事件,对应的事件监听器接口为 TreeSelectionListener,JTree 类对象使用 addTreeSelectionListener( ) 方法实现事件监听器的注册。

【例题 9_20】 创建一个树状视图,其根结点名为"学期",包含两个二级结点"第1学期""第2学期",每个二级结点又各自包含"期中前"及"期中后"两个子结点。当点选树状视图中的某个叶结点时,在文本区中显示该叶结点的名称以及其父结点的名称,如图 9.25 所示。

图 9.25 树状视图及其事件处理示例

```java
import java.awt.GridLayout;
import javax.swing.*;
import javax.swing.JTree;
import javax.swing.event.TreeSelectionEvent;
import javax.swing.event.TreeSelectionListener;
import javax.swing.tree.DefaultMutableTreeNode;
public class Ch9_20 {
  public static void main(String args[]){
    JTreeWin treeWin=new JTreeWin();
    treeWin.setBounds(0,0,630,180);
    treeWin.setVisible(true);
  }
}
class JTreeWin extends JFrame implements TreeSelectionListener{
  JTree tree;
  JTextArea area;
  JScrollPane scroll1,scroll2;
  DefaultMutableTreeNode root,node1,node2,node11,node12,node21,node22;
  public JTreeWin(){
      this.setLayout(new GridLayout(1,2));         //将窗体布局为GridLayout
      /**创建树状视图中所需的结点对象*/
      root=new DefaultMutableTreeNode("学期");     //根结点
      node1=new DefaultMutableTreeNode("第1学期");
      node11=new DefaultMutableTreeNode("期中前");
      node12=new DefaultMutableTreeNode("期中后");
      node2=new DefaultMutableTreeNode("第2学期");
      node21=new DefaultMutableTreeNode("期中前");
      node22=new DefaultMutableTreeNode("期中后");
      /**形成结点间的"父子"关系并形成"树"*/
      node1.add(node11);node1.add(node12);
      node2.add(node21);node2.add(node22);
      root.add(node1);root.add(node2);
      tree=new JTree(root);
      tree.addTreeSelectionListener(this);
      //给树对象注册事件监听器
      area=new JTextArea();
      scroll1=new JScrollPane(tree);
      scroll2=new JScrollPane(area);
      //将树对象及文本区对象添加到滚动面板中
      this.add(scroll1);                           //将"树"添加到窗体
      this.add(scroll2);
  }
  //事件处理代码
  public void valueChanged(TreeSelectionEvent e) {
      DefaultMutableTreeNode node= (DefaultMutableTreeNode)tree.getLastSelectedPathComponent();
      //得到当前点选的结点对象
      if(node.isLeaf()){                           //若是叶子结点
          String name=(String)node.getUserObject();
          //得到当前点选结点对象的用户对象
          String parentname=node.getParent().toString();
          //得到当前点选的结点对象的父结点
          area.append("当前点选的树结点名称为: "+name+", 其父结点为: "+parentname+"\n");
      }
```

    }
}

9.7.15 JDialog、JOptionPane 与 JFileChooser

对话框是提供用户与系统进行交互的 GUI 组件，在 Java Swing 中，JDialog、JOptionPane 及 JFileChooser 类都可以创建对话框。

1. JDialog

创建 JDialog 类或其子类对象可以产生一个用于与用户交互的对话框。

JDialog 的常用构造方法 public JDialog(Dialog owner,String title,boolean modal)：创建一个具有指定标题 title、模式 modal 和指定所有者 owner 的对话框，其中，owner 是指该对话框弹出时所依赖的窗体对象，owner 可以为 null；模式 modal 为 true 时表示有模式对话框（当该对话框被显示时，它所依赖的窗体不可以进行任何操作，只有先操作了该对话框，才能操作它所依赖的窗体），默认情况下，JDialog 都是有模式的。模式 modal 为 false 时表示无模式对话框（当该对话框被显示时，还可以对它所依赖的窗体进行操作）。JDialog 的默认布局管理器是 BorderLayout。

【例题 9_21】编程实现：当按下窗体中的"单击我弹出对话框"按钮时，弹出自定义对话框。当单击自定义对话框的"取消"按钮时，窗体中的标签显示"对话框被关闭"字样，如图 9.26 所示。

```java
import java.awt.FlowLayout;
import java.awt.event.ActionEvent;
import java.awt.event.ActionListener;
import javax.swing.*;
public class Ch9_21{
    public static void main(String arg[]) {
      new DialogWin();                   //创建窗体对象
    }
}
//窗体类
class DialogWin extends JFrame implements ActionListener{
    MyDialog mydialog;                   //声明对话框
    JButton button;
    JLabel state;
    public DialogWin(){                  //窗体类构造方法
        this.setTitle("对话框测试");      //窗体标题栏标题
        this.setLayout(new FlowLayout());
        button=new JButton("单击我弹出对话框") ;
        state=new JLabel();
        button.addActionListener(this);  //button 注册事件监听器
        this.getContentPane().add(button);  //将 button 添加到窗体
        this.getContentPane().add(state);   //将 state 添加到窗体
        this.setBounds(200, 200, 300, 200); //给窗体定义位置和大小
        this.setVisible(true);              //设置窗体可见
    }
    //窗体按钮事件处理
    public void actionPerformed(ActionEvent e) {
        mydialog=new MyDialog(this,"自定义对话框",true);
        //创建对话框,它依赖于当前窗体对象弹出,并是有模式对话框
        state.setText(mydialog.str);        //标签上显示对话框被操作的状态
    }
}
```

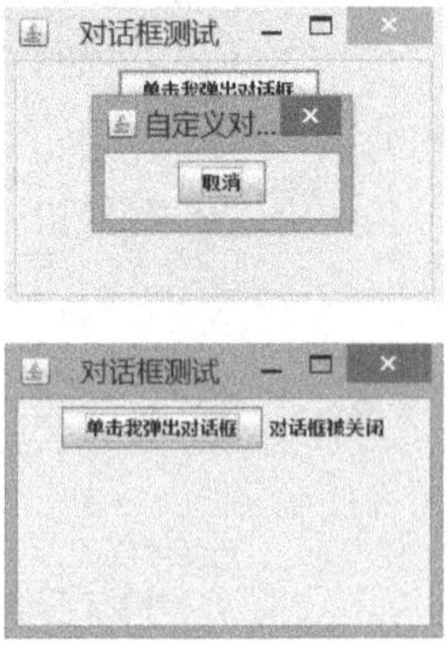

图 9.26 自定义对话框

```java
//对话框类
class MyDialog extends JDialog implements ActionListener{
    JButton button;
    String str;
    public  MyDialog(JFrame f,String t,boolean b){        //对话框含参构造方法
        super(f,t,b);                        //调用父类JDialog的构造方法
        this.setLayout(new FlowLayout());    //将对话框设置为FlowLayout布局
        button=new JButton("取消");          //给对话框创建按钮对象
        button.addActionListener(this);
        //给对话框中的按钮对象注册事件监听器
        this.getContentPane().add(button);   //将按钮对象添加到对话框
        this.setBounds(260, 260, 160, 90);   //给对话框定义位置和大小
        this.setVisible(true);               //设置对话框可见
    }
     //对话框按钮事件处理代码
    public void actionPerformed(ActionEvent e) {
            str="对话框被关闭";
            this.dispose();                   //关闭对话框
    }
}
```

2. JOptionPane

运用 javax.swing.JOptionPane 类提供的以下静态方法可以创建出常用的各种对话框如确认对话框、输入对话框以及消息对话框等，每个静态方法又有多个重载的方法，此处只列出其中较典型、常用的方法。

① public static int showConfirmDialog(Component parentComponent,Object message, String title, int optionType,int messageType)throws HeadlessException 生成一个确认对话框，其各个参数含义如下：

参数 parentComponent 指定该消息对话框所依赖的窗体对象；message 指定对话框上显示的消息文本；title 指定对话框的标题栏标题；optionType 可以取以下 JOptionPane 的常量值：YES_NO_OPTION、YES_NO_CANCEL_OPTION 或 OK_CANCEL_OPTION，用于确定对话框上显示的按钮；messageType 可以取 ERROR_MESSAGE、INFORMATION_MESSAGE、WARNING_MESSAGE、QUESTION_MESSAGE 或 PLAIN_MESSAGE 中的一个，用于确定对话框上显示的图标。

② public static String showInputDialog(Component parentComponent,Object message,String title,int messageType)throws HeadlessException 生成一个可供用户输入内容的对话框，其各个参数含义如下：

参数 parentComponent 指定该消息对话框所依赖的窗体对象；message 指定对话框上显示的消息文本；title 指定对话框的标题栏标题；messageType 可以取值为 ERROR_MESSAGE、INFORMATION_MESSAGE、WARNING_MESSAGE、QUESTION_MESSAGE 或 PLAIN_MESSAGE 之一。

③ public static void showMessageDialog(Component parentComponent,Object message,String title,int messageType)throws HeadlessException 生成一个消息对话框，其各个参数含义如下：

参数 parentComponent 指定该消息对话框所依赖的窗体对象；message 指定对话框上显示的消息文本；title 指定对话框的标题栏标题；messageType 可以其值为 ERROR_MESSAGE、INFORMATION_MESSAGE、WARNING_MESSAGE、QUESTION_MESSAGE 或 PLAIN_MESSAGE 之一。

3. JColorChooser

javax.swing.JColorChooser 类的静态方法 public static Color showDialog(Component component,

String title,Color initialColor)throws HeadlessException 可以创建一个颜色对话框。

其中，参数 component 指定对话框所依赖的窗体对象；title 指定颜色对话框的标题栏标题；initialColor 设置颜色对话框返回的初始颜色。

【例题 9_22】 JOptionPane、JColorChooser 测试。运行结果如图 9.27 所示。

```java
import java.awt.Color;
import java.awt.FlowLayout;
import java.awt.event.ActionEvent;
import java.awt.event.ActionListener;
import javax.swing.*;
public class Ch9_22 {
    public static void main(String args[]){
        new DialogTest();
    }
}
class DialogTest extends JFrame implements ActionListener{
    JButton bt1,bt2,bt3,bt4;
    JLabel label;
    JPanel panel;
    int n;
    String s=null;
    public DialogTest(){
        bt1=new JButton("消息对话框");
        bt2=new JButton("确认对话框");
        bt3=new JButton("输入对话框");
        bt4=new JButton("颜色对话框");
        label=new JLabel();
        panel=new JPanel();
        panel.add(bt1);panel.add(bt2);
        panel.add(bt3);panel.add(bt4);
        panel.add(label);
        this.add(panel);
        bt1.addActionListener(this);
        bt2.addActionListener(this);
        bt3.addActionListener(this);
        bt4.addActionListener(this);
        this.setTitle("对话框测试");
        this.setSize(500, 100);
        this.setVisible(true);
    }
    public void actionPerformed(ActionEvent e) {
     if(e.getSource()==bt1){
        JOptionPane.showMessageDialog(this, "一个消息对话框", "消息", JOptionPane.QUESTION_MESSAGE);
      }
      else if(e.getSource()==bt2){
        n=JOptionPane.showConfirmDialog(this, "一个确认对话框","", JOptionPane.
          YES_NO_CANCEL_OPTION, JOptionPane.INFORMATION_MESSAGE);
        if(n==JOptionPane.YES_OPTION)
           label.setText("您选择了'是(Y)'");
        else if(n==JOptionPane.NO_OPTION)
           label.setText("您选择了'否(N)'");
        else if(n==JOptionPane.CANCEL_OPTION)
           label.setText("您选择了'取消'");
```

```
            }
            else if(e.getSource()==bt3){
                s=JOptionPane.showInputDialog(this,"请输入电子邮件信息: ","Bob_123@126.
                    com");
                label.setText("您在输入对话框中的输入为: "+s);
            }
            else if(e.getSource()==bt4){
                Color newColor=JColorChooser.showDialog(this, "颜色对话框", Color.
                    WHITE);
                //得到颜色对话框中选择的新颜色
                bt4.setBackground(newColor);
                //按钮 bt4 背景色变为颜色对话框中选择的新颜色
            }
        }
    }
```

图 9.27　各类对话框

说明：要重点掌握 JOptionPane 的几个静态方法参数的灵活运用，试改变例题中 messageType 的取值，观察所弹出对话框的变化。注意当操作了各个对话框后事件处理代码的写法。

还有一种重要的对话框是文件打开、保存对话框，它们的使用需要文件输入/输出流操作，在后面章节中介绍。

9.7.16　JSlider

JSlider 组件用于创建滑动条，其滑动杆可以显示主刻度标记以及主刻度之间的次刻度标记，通过拖动其滑块，来改变滑动杆上的取值。

JSlider 的常见构造方法有：

- public JSlider()：创建一个范围在 0 和 100 之间并且初始值为 50 的水平滑动条对象。
- public JSlider(int orientation,int min,int max,int value)：用参数指定方向 orientation、指定的最小值 min、最大值 max 以及初始值 value 创建一个滑动条对象。orientation 可以是 SwingConstants.VERTICAL 或 SwingConstants.HORIZONTAL。

JSlider 的常用方法有：

- public void addChangeListener(ChangeListener l)：给 JSlider 对象注册事件监听器。
- public void setPaintTicks(boolean b)：确定是否在滑动条上显示刻度标记。默认情况下，此属性为 false。
- public void setMajorTickSpacing(int n)：用于设置主刻度标记的间隔。要显示主刻度，setPaintTicks 必须设置为 true。

- public void setMinorTickSpacing(int n)：用于设置次刻度标记的间隔。要显示主刻度，setPaintTicks 必须设置为 true。
- public void setPaintLabels(boolean b)：确定是否在滑动条上显示数字标记。
- public void setMaximum(int maximum)：将滑动条的最大值设置为 maximum。
- public void setMinimum(int minimum)：将滑动条的最小值设置为 minimum。
- public int getValue()：返回滑块的当前值。

【例题 9_23】 编程实现：通过拖动滑动条上的滑块来改变文字颜色（见图 9.28）。

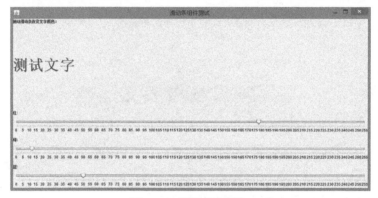

图 9.28 设置文字颜色

```
import javax.swing.*;
import java.awt.*;
import javax.swing.event.*;
class ChangeTextWin extends JFrame implements ChangeListener{
    JSlider slider[]=new JSlider[3];                        //JSlider对象数组
    JLabel label,content;
    JPanel panel1,panel2;
    public ChangeTextWin(){
        this.setTitle("滑动条组件测试");
        label= new JLabel("拖动滑动条改变文字颜色: ");
        content=new JLabel("测试文字");
        content.setFont(new Font("宋体",Font.BOLD,50));   //设置字体
        panel1 = new JPanel();
        panel1=new JPanel();
        panel1.setLayout(new GridLayout(6,2));
        for(int i=0;i<3;i++){
            slider[i]=new JSlider();          //创建JSlider对象并给slider数组赋值
            slider[i].setPaintTicks(true); //显示刻度
            slider[i].setMaximum(255);     //设置滑动条最大刻度值
            slider[i].setMajorTickSpacing(5); //设置主刻度间隔
            slider[i].setMinorTickSpacing(1); //设置次刻度间隔
            slider[i].setPaintLabels(true);   //显示刻度的数字标记
            slider[i].addChangeListener(this); //添加事件监听器
        }
        panel1.add(new JLabel("红:"));panel1.add(slider[0]);
        panel1.add(new JLabel("绿:"));panel1.add(slider[1]);
        panel1.add(new JLabel("蓝:"));panel1.add(slider[2]);
        panel2=new JPanel();
        panel2.setLayout(new GridLayout(2,1));
        panel2.add(content);
```

```
        panel2.add(panel1);
        this.getContentPane().add(label,BorderLayout.NORTH);
        this.getContentPane().add(panel2,BorderLayout.CENTER);
        this.setBounds(0, 0,1200,600);
        this.setVisible(true);
    }
    //事件处理代码
    public void stateChanged(ChangeEvent e) {
        content.setForeground(new
Color(slider[0].getValue(),slider[1].getValue(),slider[2].getValue()));
        //按红、绿、蓝滑动条上各自的取值给标签content 的设置前景色
    }
}
public class Ch9_23{
    public static void main(String[] args){
        new ChangeTextWin();
    }
}
```

9.7.17 JprogressBar

JprogressBar 是用于创建进度条的组件，它以图形化的方式来描述某个任务的进度，在任务完成过程中，进度条显示该任务完成的百分比。

其常用构造方法有：

- public JProgressBar()：创建一个显示边框但不带进度字符串的水平进度条。初始值和最小值都为 0，最大值为 100。
- public JProgressBar(int orient,int min,int max)：创建指定方向 orient、最小值 min 和最大值 max 的进度条。

其常用方法有：

- public void addChangeListener(ChangeListener l)：给 JProgressBar 对象注册事件监听器。
- public void setOrientation(int newOrientation)：将进度条的方向设置为 newOrientation，取值为 SwingConstants.VERTICAL 或 SwingConstants.HORIZONTAL）。默认方向为 SwingConstants.HORIZONTAL。
- public void setPreferredSize(Dimension preferredSize)：设置此组件的合适大小，此方法是 JprogressBar 继承自 javax.swing.JComponent 类的方法。
- public void setBorderPainted(boolean b)：设置 borderPainted 属性，如果进度条应该绘制其边框，则此属性为 true。
- public void setStringPainted(boolean b)：设置 stringPainted 属性的值，该属性确定进度条是否应该呈现进度字符串。
- public void setString(String s)：设置进度字符串的值。
- public void setMaximum(int n)：设置进度条的最大值为 n。
- public void setMinimum(int n)：设置进度条的最小值为 n。

9.7.18 Timer

Timer 是一个定时器类，它可以在指定时间间隔触发一个或多个 ActionEvent 类型事件。

其构造方法为 public Timer(int delay,ActionListener listener)：创建一个 Timer 对象，将触发事件的时间间隔设为 delay 毫秒，同时，给定时器注册事件监听器 listener。

其常用方法有：
- public void addActionListener(ActionListener listener)：给 Timer 对象注册事件监听器。
- public void start()：启动 Timer 对象，每过一个指定的 delay（单位：毫秒）就触发一次 ActionEvent 事件。
- public void stop()：停止 Timer 对象。
- public void restart()：重新启动 Timer 对象。
- public void setDelay(int delay)：设置 Timer 的两次连续的事件间隔的时间（单位：毫秒）。

【例题 9_24】创建进度条以及定时器对象，让进度条随着定时器指定的事件间隔完成进度显示，如图 9.29 所示。

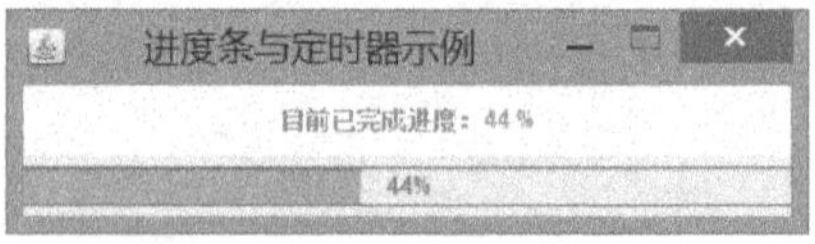

```
import javax.swing.*;
import javax.swing.border.*;
import java.awt.*;
import java.awt.event.*;
import javax.swing.event.*;
class ProgressBarWin extends JFrame implements ActionListener,ChangeListener{
    JProgressBar progressbar;
    JLabel label;
    JPanel panel;
    Timer timer;
    public ProgressBarWin(){
        this.setTitle("进度条与定时器示例");
        this.getContentPane().setBackground(Color.WHITE);     //设置窗体背景
        progressbar = new JProgressBar(SwingConstants.HORIZONTAL,0,100);
        //创建进度条
        progressbar.setStringPainted(true);//设置在进度条上显示其进度百分比字符串
        progressbar.addChangeListener(this);//给进度条注册事件监听器
        progressbar.setPreferredSize(new Dimension(400,20));
        //设置进度条的宽度、高度
        progressbar.setForeground(Color.CYAN);    //设置进度条前景色
        label=new JLabel("",JLabel.CENTER);       //内容在标签上居中显示
        label.setForeground(Color.GREEN);         //设置标签文字颜色
        panel=new JPanel();
        panel.add(progressbar);
        panel.add(label);
        this.add(panel, BorderLayout.SOUTH);
        timer=new Timer(100,this);
        //创建定时器对象，每100 ms触发一次ActionEvent事件，监听器为当前窗体对象
        this.getContentPane().add(label,BorderLayout.CENTER);
        this.setBounds(100,100,400, 100);
        this.setResizable(false);//不可更改窗体大小
        this.setVisible(true);
    }
    //定时器的事件处理代码
    public void actionPerformed(ActionEvent e) {
        int value = progressbar.getValue();      //获取0~100之间的进度值
        if(value<100){
```

图 9.29 带进度条的窗体

```
                value++;
                progressbar.setValue(value);      //设置进度条的值
            }
            else{
                timer.stop();                     //停止timer
                progressbar.setValue(0);
            }
        }
    //进度条事件处理代码
        public void stateChanged(ChangeEvent e) {
            int value = progressbar.getValue();
            label.setText("目前已完成进度: "+Integer.toString(value)+" %");
            //标签上显示对应进度百分比
        }
}
public class Ch9_24{
    public static void main(String[] args){
        ProgressBarWin pbw=new ProgressBarWin();
        pbw.timer.start();                        //启动timer
    }
}
```

9.7.19 键盘事件示例

前面的例题中涉及了最常见的几种事件类型，如 ActionEvent、ItemEvent、MouseEvent 等，实际应用中，键盘事件也经常被使用。键盘事件类是 KeyEvent，与之相关的事件监听器接口为 KeyListener，事件适配器类为 KeyAdapter。触发 KeyEvent 类型事件的事件源通过调用 addKeyListener 方法注册事件监视器。

键盘事件类 KeyEvent 的常用方法有：

- public char getKeyChar()：判断键盘的哪个键被按下、敲击或释放，返回键上的字符。
- public int getKeyCode()：返回与此事件中的键关联的整数 keyCode（键码值），这些键码值取值为 KeyEvent 的常量，如 VK_LEFT 是表示非数字键盘左方向键的键码值，VK_F1 是 F1 键的键码值，等等。
- public static String getKeyText(int keyCode)：返回描述 keyCode 的 String，如"HOME"、"F1" 或 "A"。

以下通过实例说明键盘事件处理方法。

【例题 9_25】通过按键盘的上、下、左、右键来移动按钮，当单击按钮时，按钮变为红色（见图 9.30）。

图 9.30 带进度条的窗体

```
import java.awt.*;
import java.awt.event.*;
import javax.swing.*;
class MoveButtonFrame extends JFrame implements KeyListener,ActionListener{
    JButton bt;
    JPanel panel;
    JComponent c=null;
    public MoveButtonFrame(){
        this.setTitle("移动按钮");
```

```java
        bt=new JButton("可移动的按钮");
        panel=new JPanel();
        panel.add(bt);                                  //将按钮添加到面板
        this.add(panel,BorderLayout.CENTER);  //将panel及添加到窗体
        bt.addKeyListener(this);                        //注册KeyEvent类型事件监听器
        bt.addActionListener(this);                     //注册ActionEvent类型事件监听器
        this.setSize(500, 200);
        this.setVisible(true);
    }
    public void move(JComponent c,String direction){//按钮移动方法
        bt=(JButton)c;
        int x=bt.getBounds().x;
        int y=bt.getBounds().y;
        //得到按钮当前坐标
        if(direction.equals("up"))
           y=y-20;
        else if (direction.equals("down"))
           y=y+20;
        else if (direction.equals("left"))
           x=x-20;
        else if (direction.equals("right"))
           x=x+20;
        bt.setLocation(x,y);
        //将按钮置到新坐标位置
    }
    public void keyTyped(KeyEvent e) {}
    public void keyPressed(KeyEvent e) {
        if(e.getKeyCode()==KeyEvent.VK_UP)              //若按上方向键
           this.move(bt,"up");
        else if(e.getKeyCode()==KeyEvent.VK_DOWN)       //若按下方向键
           this.move(bt,"down");
        else if(e.getKeyCode()==KeyEvent.VK_LEFT)       //若按左方向键
           this.move(bt,"left");
        else if(e.getKeyCode()==KeyEvent.VK_RIGHT)      //若按右方向键
           this.move(bt,"right");
    }
    public void keyReleased(KeyEvent e) {}
    public void actionPerformed(ActionEvent e) {
        bt.setBackground(Color.RED);                    //按钮背景色变为红色
    }   //处理按钮单击事件
}
public class Ch9_25 {
    public static void main(String[] args) {
        new MoveButtonFrame();
    }
}
```

说明:一个事件源可以同时触发多个类型的事件,本例中的按钮对象 bt 就同时实现了两个事件监听器 KeyListener 和 ActionListener。

9.8 其他几个应用

9.8.1 更换窗体标题栏图标

通常，Java GUI 窗体标题栏默认图标为"咖啡杯"图标，若要更换该图标，则需要以下步骤：

（1）需要先得到一个 Toolkit 类对象 tk（java.awt.Toolkit 类是所有 Abstract Window Toolkit 实际实现的抽象父类）。

（2）再用 tk 调用 getImage()方法生成一个 Image 对象 img。

（3）用当前窗体对象调用 setIconImage(img)方法即可。

【例题 9_26】给窗体更换标题栏图标，如图 9.31 所示。

```java
import java.awt.Image;
import java.awt.Toolkit;
import javax.swing.JFrame;
import javax.swing.ImageIcon;
public class Ch9_26{
    public static void main(String args[]){
        new FrameIcon("更换窗体标题栏图标");}
}
class FrameIcon extends JFrame{
    public FrameIcon(String s){
        super(s);
        this.setVisible(true);
        Toolkit tk=this.getToolkit();            //得到一个 Toolkit 对象
        Image img=tk.getImage(getClass().getResource("a.jpg"));
        //从当前目录下加载图片并创建 Image 对象
        this.setIconImage(img);                  //设置窗体标题栏图标
        this.setBounds(200,10,300,200);
    }
}
```

图 9.31 更换窗体标题栏图标

9.8.2 让窗体在屏幕上居中显示

java.awt 包中有两个与显示设备有关的类 GraphicsEnvironment 以及 GraphicsDevice。要想让自己创建的窗体在屏幕上居中显示，则需要通过以下程序语句完成：

```java
GraphicsEnvironment ge = GraphicsEnvironment.getLocalGraphicsEnvironment();
GraphicsDevice gd = ge.getDefaultScreenDevice();
Rectangle rec = gd.getDefaultConfiguration().getBounds();
…
```

【例题 9_27】创建窗体，使之在屏幕中央显示

```java
import java.awt.GraphicsDevice;
import java.awt.GraphicsEnvironment;
import java.awt.Rectangle;
import javax.swing.JFrame;
public class Ch9_27 {
    public static void main(String[] args) {
        JFrame f=new JFrame("在屏幕上居中显示窗体");
        GraphicsEnvironment ge = GraphicsEnvironment.getLocalGraphicsEnvironment();
//得到 GraphicsEnvironment 对象
        GraphicsDevice gd = ge.getDefaultScreenDevice();            //得到屏幕设备
        Rectangle rec = gd.getDefaultConfiguration().getBounds();
```

```
            System.out.println("屏幕宽度: " + rec.getWidth());
            System.out.println("屏幕高度: " + rec.getHeight());
            f.setSize(300, 200);
            f.setLocation((int)(rec.getWidth()-300)/2, (int)(rec.getHeight()-200)/2);
            f.setVisible(true);
        }
    }
```

9.8.3 将窗体显示为任务栏图标

【例题 9_28】编程实现：当窗体最小化时，在任务栏中显示窗体图标；当窗体被关闭时，任务栏图标消失。

```
import java.awt.AWTException;
import java.awt.Image;
import java.awt.SystemTray;
import java.awt.Toolkit;
import java.awt.TrayIcon;
import java.awt.event.WindowAdapter;
import java.awt.event.WindowEvent;
import javax.swing.JFrame;
class SystemTrayWin extends JFrame{
    Image image;
    TrayIcon trayicon;
    public SystemTrayWin(){
        this.setTitle("将窗体显示为任务栏图标");
        this.setBounds(0 , 0, 200,100);
        this.setVisible(true);
        this.addWindowListener(new WindowAdapter(){
        public void windowIconified(WindowEvent e) {//窗体最小化时在任务栏添加图标
            image = Toolkit.getDefaultToolkit().createImage(getClass().getResource
                ("a.jpg"));
            //以当前目录下的a.jpg创建Image对象
            trayicon = new TrayIcon(image);
            //创建任务栏图标
            try {
                SystemTray.getSystemTray().add(trayicon);      //在任务栏添加图标
            }
            catch (AWTException e1) {
                e1.printStackTrace();
            }
        }
        public void windowClosing(WindowEvent e) {            //窗体关闭时将任务栏图标移除
            SystemTray.getSystemTray().remove(trayicon); //从任务栏移除图标
        }
        });
    }
}
public class Ch9_28{
    public static void main(String args[]){
    new SystemTrayWin();
    }
}
```

9.9 字体与颜色

9.9.1 Font 类

在 GUI 编程中，常常要给各个组件对象设置字体。java.awt 包中的 Font 类用于创建字体类对象。其常用构造方法为：

public Font(String name,int style,int size)：根据指定名称 name、样式 style 和磅值 size 创建一个 Font 对象。

例如：Font font= new Font("宋体", Font.BOLD+Font.ITALIC, 24);表示创建字体为"宋体"，风格为倾斜并加粗，字号为 24 磅的字体对象，BOLD、ITALIC 都是 Font 类的常量。

那么，怎样才能知道本机到底支持哪些字体类型呢？下面举例说明：

【例题 9_29】 获取并输出本机支持的字体列表。

```
import java.awt.GraphicsEnvironment;
public class Ch9_29{
    public static void main(String[] args){
        GraphicsEnvironment ge = GraphicsEnvironment.getLocalGraphicsEnvironment();
        String[] fontlist = ge.getAvailableFontFamilyNames();
        System.out.println("本机支持的字体列表: ");
        for(String s:fontlist){
           System.out.println(s);
        }
    }
}
```

9.9.2 Color 类

在 GUI 编程中，也常常要给各个组件对象设置及文本设置颜色（前景色及背景色）。java.awt 包中的 Color 类可以创建颜色类的对象。其常用构造方法有：

- public Color(int r,int g,int b)：创建具有指定红、绿、蓝色值的 Color 对象，其取值范围为 0~255。
- public Color(int r,int g,int b,int a) 指定红、绿、蓝色值 Color 对象，a 表示透明度，取值范围是 0~255。

对应于常见颜色，Color 类中定义了一些颜色常量，如 Color.RED 等。

【例题 9_30】 以不同字体显示文本（见图 9.32）。

```
import java.awt.Font;
import java.awt.GridLayout;
import javax.swing.*;
public class Ch9_30 {
    public static void main(String args[]) {
        new FontTest();
    }
}
class FontTest extends JFrame {
    JPanel panel;
    public FontTest() {
        this.setTitle("显示多种字体");
        panel=new JPanel();
        panel.setLayout(new GridLayout(5,1));          //将面板设为网格布局方式
```

图 9.32　字体设置

```
        /*定义字体数组并初始化*/
        Font[] fonts ={new Font("Times New Roman",Font.ITALIC, 20),
        new Font("宋体",Font.PLAIN, 24), new Font("黑体", Font.BOLD, 24),
        new Font("Calibri", Font.BOLD+Font.ITALIC, 18),new Font("楷体", Font.PLAIN,
           20) };
        /*定义要显示的文本数组*/
        String[] text= {"Times New Roman:Welcome", "宋体:您好",
        "黑体:您好", "Calibri:Welcome","楷体:您好"};
        for (int i=0;i<5;i++){
          JLabel label=new JLabel();       //创建标签对象
          label.setFont(fonts[i]);          //给每个标签设置文本字体
          label.setText(text[i]);           //在每个标签上显示文本
          panel.add(label);                //将各个标签添加到面板中
        }
        this.add(panel);                   //将面板添加到窗体
        this.setSize(300,200);
        this.setVisible(true);
        this.setDefaultCloseOperation(JFrame.EXIT_ON_CLOSE);
        }
}
```

【例题 9_31】编程实现如图 9.33 所示的界面，可以实现对文本字体、字号、风格和颜色的设定和改变。

图 9.33　字体设置

```
import java.awt.*;
import java.awt.event.*;
import javax.swing.*;
public class Ch9_31 {
    public static void main(String[] args) {
    new TextSetDemo();
    }
}
class TextSetDemo extends JFrame implements ActionListener,ItemListener{
    JLabel l1,l2,l3,demo;
    JComboBox<String> combo;
    JRadioButton radio1,radio2,radio3;
    ButtonGroup group;
    JButton bt;
    JTextField jtf;
    JPanel panel;
    /*demo当前的字体、字号、风格、颜色*/
    String fontname=null;         //字体名
    int size;                     //字号
    int style;                    //风格
    Color color=null;             //颜色
    public TextSetDemo(){
    this.setTitle("设置文字示例");
    this.getContentPane().setBackground(Color.WHITE);//窗口背景色为白色
```

```java
        l1=new JLabel("选择字体: ");
        l2=new JLabel("输入字号: ");
        l3=new JLabel("选择风格: ");
        demo=new JLabel("测试文字: 欢迎您,Welcome");
        combo=new JComboBox<String>();
        GraphicsEnvironment ge = GraphicsEnvironment.getLocalGraphicsEnvironment();
        String[] fontlist = ge.getAvailableFontFamilyNames();//获取本机支持的字体列表
          for(int i=0;i<fontlist.length;i++){
                combo.addItem(fontlist[i]);
          }//添加本机支持的所有字体作为下拉列表待选项
        radio1=new JRadioButton("粗体");
        radio2=new JRadioButton("斜体");
        radio3=new JRadioButton("普通");
        group=new ButtonGroup();
        group.add(radio1);group.add(radio2);group.add(radio3);//组合单选按钮
        jtf=new JTextField(10);
        bt=new JButton("文字颜色");
        panel=new JPanel();
        panel.add(l1);panel.add(combo);
        panel.add(l2);panel.add(jtf);
        panel.add(l3);panel.add(radio1);
        panel.add(radio2);panel.add(radio3);
        panel.add(bt);
        /*添加事件监听器*/
        combo.addItemListener(this);
        jtf.addActionListener(this);
        radio1.addItemListener(this);
        radio2.addItemListener(this);
        radio3.addItemListener(this);
        bt.addActionListener(this);
        /*得到demo当前的字体、字号、风格、颜色*/
        fontname=demo.getFont().getFontName();
        size=demo.getFont().getSize();
        style=demo.getFont().getStyle();
        color=demo.getForeground();
        this.getContentPane().add(panel,BorderLayout.NORTH);
        this.getContentPane().add(demo,BorderLayout.CENTER);
        //将panel及demo添加到窗体
        this.setSize(960, 200);
        this.setVisible(true);
        this.setDefaultCloseOperation(JFrame.EXIT_ON_CLOSE);
    }
    /*文本框jtf以及按钮bt的事件处理代码*/
    public void actionPerformed(ActionEvent e) {
        if(e.getSource()==jtf){
        size= Integer.parseInt(jtf.getText());
        }
        else if(e.getSource()==bt){
        color=JColorChooser.showDialog(this, "颜色对话框", this.getContentPane().getBackground());
        //获取颜色对话框中选择的颜色
        }
        demo.setFont(new Font(fontname,style,size));
        //以当前fontname,style,size为参数标签文字的字体
        demo.setForeground(color);         //以当前颜色对话框中的颜色为标签文字的颜色
```

```
        }
    /*下拉列表combo以及单选按钮radio1、radio2、radio3的事件处理代码*/
    public void itemStateChanged(ItemEvent e) {
    if(e.getSource()==combo){
    fontname=(String)combo.getSelectedItem();       //得到下拉列表的当前选项
    }
    else if(e.getSource()==radio1){
        style=Font.BOLD;
    }
    else if(e.getSource()==radio2){
        style=Font.ITALIC;
    }
    else if(e.getSource()==radio3){
        style=Font.PLAIN;
    }
    demo.setFont(new Font(fontname,style,size));    //给demo设置字体
    demo.setForeground(color);                      //给demo设置文本颜色
    }
}
```

注意：本例中，对于"输入字号"文本框，需要在其中输入字号值回车后才可以生效。

9.10 GUI图形绘制

9.10.1 Graphics类

在GUI程序设计中，有时需要进行绘图处理。java.awt包中的Graphics类是所有图形上下文的抽象父类，允许应用程序在组件上进行绘图。

Graphics提供了一系列的方法来实现基本图形的绘制，用于绘制线段、矩形、椭圆、圆弧、多边形等：

- drawLine(int x1,int y1,int x2,int y2)：画一条以(x1,y1)为起点坐标，以(x2,y2)为终点坐标的线段。
- drawRect(int x,int y,int width,int height)：绘制一个以(x,y)为其左上角顶点坐标，以width和height为宽和高的空心矩形。
- fillRect(int x,int y,int width,int height)：绘制一个以(x,y)为其左上角顶点坐标，以width和height为宽和高的实心矩形（得到一个着色的矩形块）。例如：g.setColor(Color.yellow);g.fillRect(50,80,20,30)。
- drawOval(int x,int y,int width,int height)：绘制空心椭圆。其中参数x和参数y指定该椭圆形外接矩形左上角的坐标，width和height是该椭圆横轴和纵轴长度，当width和height相等时，绘制出来的是一个正圆。
- fillOval(int x,int y,int width,int height)：绘制用预定的颜色填充的实心椭圆形。
- drawArc(int x,int y,int width,int height,int startAngle, int arcAngle)：绘制圆弧线，该圆弧是椭圆的一部分。(x,y)是该椭圆的外接矩形左上角的坐标，width和heigh是该椭圆外接矩形的宽和高，startAngle的单位是"度"，起始角度0°是指3点钟方位，参数startAngle和arcAngle表示从startAngle角度开始，逆时针方向画arcAngle度的弧（正值度数是逆时针方向，负值度数是顺时针方向）。

- fillArc(int x,int y,int width, int height, int startAngle, int arcAngle)：用预定的颜色绘制圆弧线。
- drawPolyline(int[] xPoints, int[] yPoints, int nPoints)：绘制一个由 xPoints 和 yPoints 坐标数组定义的闭合多边形，(x, y)坐标定义一个端点坐标。此方法绘制由 nPoints 个线段定义的多边形，该方法并不自动闭合多边形，要画一个闭合的多边形，给出的坐标点的最后一点必须与第一点相同。
- fillPolygon(int xPoints[],int yPoints[],int nPoints)：用方法 setColor()设定的颜色着色多边形。
- drawString(String str,int x,int y)：在 x,y 坐标处开始绘制 string 文本。
- drawImage(Image image,int left,int top,ImageObserver observer)：绘制一个图片。
- setColor(Color color)：设置画笔颜色。
- setFont(Font font)：设置画笔字体。

9.10.2 Canvas 类

java.awt 包中的 Canvas 是一个画布类，它是处于屏幕上的一个空白矩形区域，在该区域可以用 Graphics 类的各个绘图方法进行图形绘制，也可以处理事件源事件。具体的应用中，一般通过创建 Canvas 类的子类来创建画布，然后，将该画布添加到窗体中，从而实现在窗体中绘制图形。Canvas 的常用方法有：

- public void paint(Graphics g)： Canvas 的子类的通过重写此方法实现图形绘制。
- public void repaint()：完成图形的重新绘制，该方法是 Canvas 继承自 java.awt.Component 的方法。
- public void update(Graphics g)：此方法用于响应对 repaint 的调用。首先通过使用背景色填充 canvas 来清理原有图形，然后通过调用此 canvas 的 paint 方法重绘它。

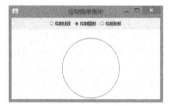

图 9.34 图形绘制

【例题 9_32】编写程序绘制不同的图形（见图 9.34）。

```
import java.awt.*;
import java.awt.event.*;
import javax.swing.*;
//画布类
class DrawCanvas extends Canvas implements MouseListener,MouseMotionListener{
   JRadioButton radio1,radio2,radio3;
   int x1,x2,y1,y2;
   public DrawCanvas(){
      this.addMouseListener(this);
      this.addMouseMotionListener(this);
   }
   public void setRadio1(JRadioButton radio1) {
   this.radio1 = radio1;
   }
   public void setRadio2(JRadioButton radio2) {
   this.radio2 = radio2;
   }
   public void setRadio3(JRadioButton radio3) {
   this.radio3 = radio3;
   }
   public void paint(Graphics g) {       //重写paint方法
   g.setColor(Color.RED);                //设置画笔颜色
```

```java
            if(radio1.isSelected()==true)
            g.drawLine(x1,y1,x2,y2);          //绘制线段
            if(radio2.isSelected()==true){
                int r=(int)Math.sqrt((x2-x1)*(x2-x1)+(y2-y1)*(y2-y1));
                g.drawOval(x1,y1,r,r);        //绘制圆
            }
            if(radio3.isSelected()==true)
                g.drawRect(x1,y1,x2,y2);      //绘制矩形
        }
        //实现MouseListener方法
        public void mousePressed(MouseEvent e){
        x1=e.getX();
        y1=e.getY();
        //获取鼠标按下时的起点坐标
        }
        public void mouseClicked(MouseEvent e){}
        public void mouseEntered(MouseEvent e){}
        public void mouseExited(MouseEvent e){}
        public void mouseReleased(MouseEvent e){}
        //实现MouseMotionListener方法
        public void mouseDragged(MouseEvent e){
        x2=e.getX();
        y2=e.getY();
        //获取鼠标拖动时的终点坐标
        repaint();
        }
        public void mouseMoved(MouseEvent e){}
}
//窗体类
class DrawShape extends JFrame {
    int x1,y1,x2,y2;
    JRadioButton radio1,radio2,radio3;
    ButtonGroup group;
    JPanel panel;
    DrawCanvas canvas=null;
    public DrawShape(){
    this.setTitle("绘制简单图形");
    radio1=new JRadioButton("绘制线段");
    radio2=new JRadioButton("绘制圆形");
    radio3=new JRadioButton("绘制矩形");
        group=new ButtonGroup();
        group.add(radio1);group.add(radio2);group.add(radio3);
        panel=new JPanel();
        panel.add(radio1);
        panel.add(radio2);
        panel.add(radio3);
        this.getContentPane().add(panel,BorderLayout.NORTH);
         //将panel添加到窗体
        canvas=new DrawCanvas();           //创建画布类对象
        canvas.setRadio1(radio1);canvas.setRadio2(radio2);canvas.setRadio3(radio3);
//调用setRadio1、setRadio2和setRadio3方法给DrawCanvas中的按钮对象赋值
        this.getContentPane().add(canvas,BorderLayout.CENTER);
         //将画布添加到窗体
        this.getContentPane().setBackground(Color.WHITE);
```

```
        //将窗体背景色设为白色
        this.setSize(500,300);
        this.setVisible(true);
    }
}
//主类
public class Ch9_32 {
    public static void main(String[] args){
       new DrawShape();
    }
}
```

分析：本例中的画布类中涉及两个事件监听器接口 MouseListener 和 MouseMotionListener，当鼠标在窗体的画布上被按下时，得到起点坐标(x1,y1)，当拖动鼠标并释放鼠标时，得到终点坐标(x2,y2)，而 x1、y1、x2、y2 是绘制各种简单图形所需要的。

在实际的 GUI 绘图编程中，也可以通过给 JPanel 类派生子类，并重写 paint(Graphics g)方法来完成图形绘制，再将该 JPanel 子类添加到窗体，实现在土窗体中显示所绘制的图形。

【例题 9_33】生成验证码并将其显示在窗体中，要求验证码是由 0～9 等数字以及大写的英文字母构成，验证码图片上有干扰点（见图 9.35）。

图 9.35 绘制验证码

```
import java.awt.Color;
import java.awt.Font;
import java.awt.Graphics;
import javax.swing.JFrame;
import javax.swing.JPanel;
//用于在其上绘制图形的 JPanel 子类
class CodePanel extends JPanel {
    public void paint(Graphics g) {         //重写paint方法
      int width=300;
      int height=100;
      //设定验证码显示的宽度、高度
      String chars="0123456789ABCDEFGHIJKLMNOPQRSTUVWXYZ";
       // 定义组成随机验证码的字符表，由 36 个字符组成
      char[] rands=new char[4];
      for(int i=0;i<4;i++) {
         int rand=(int)(Math.random()*36);
         rands[i]=chars.charAt(rand);
      }
      //从构成验证码的 36 个字符中随机取出 4 个字符并存储在字符数组 rands 中
      g.setColor(Color.WHITE);              //绘制用于显示验证码的矩形的填充颜色
      g.fillRect(0,0,width,height);          //绘制用于显示验证码的矩形
      for (int i=0;i<100;i++) {
         int x=(int)(Math.random()*width);
         int y=(int)(Math.random()*height);
         int red=(int)(Math.random()*255);
         int green=(int)(Math.random()*255);
         int blue=(int)(Math.random()*255);
         g.setColor(new Color(red, green, blue));
         //随机生成干扰点的颜色
         g.drawOval(x, y, 1, 0);             //绘制干扰点
      }// 随机产生 100 个干扰点
      g.setColor(Color.BLUE);               //验证码字符的颜色
      g.setFont(new Font(null, Font.ITALIC | Font.BOLD, 100));//验证码字符的字体
      g.drawString("" + rands[0], 0, 80);
```

```java
        g.drawString("" + rands[1], 60, 80);
        g.drawString("" + rands[2], 120, 80);
        g.drawString("" + rands[3], 180, 80);
        // 在指定位置上输出验证码的各个字符
        g.dispose();
        }
}
class CheckCodeWin extends JFrame{
    CodePanel codePane=new CodePanel();
    public CheckCodeWin() {
    this.setTitle("验证码窗体");
    this.add(codePane);                    //将面板对象添加到窗体
    this.setSize(300,150);
    this.setVisible(true);
    }
}
public class Ch9_33 {
    public static void main(String[] args) {
    new CheckCodeWin();
    }
}
```

说明：目前验证码主要出现在 Web 程序中，本例主要目的是示例验证码的生成方法和绘制验证码的基本方法。

小 结

本章介绍了用于 Java GUI 编程中组件与容器的概念和关系；布局管理器类的典型应用；事件编程的几种不同实现方式；重点介绍了常用的 Swing 组件及其事件编程方法；Java GUI 编程中用到的几个工具类的应用；GUI 绘图等内容。由于 Java GUI 的系统类库较庞大、内容丰富，读者还需要在应用实践中不断查阅 JDK 开发文档、积累编程经验，逐步深入了解 Java GUI 编程中涉及的其他系统类、接口的用法。

习 题 9

上机实践题

1. 编写一个用于给学生信息数据库表中录入新记录的窗体程序，效果如图 9.36 所示。要求如下：

（1）在一个 JPanel 对象上 panel1 添加标签对象，标签上显示"请在以下文本框中输入信息并单击'录入'按钮录入新记录"，将标签文字设置为"宋体"、加粗、20 号。

（2）将另一个 JPanel 对象 panel2 一盒式布局方式进行布局，在其中添加"学号""姓名""性别""年龄""籍贯"及各自对应的文本框。

（3）在一个 JPanel 对象上 panel3 添加"录入"按钮对象。

（4）将 panel1、panel2、panel3 分别添加到窗口的 NORTH、Center 及 SOUTH 区域。

（5）程序运行时，窗口在屏幕的中央显示。

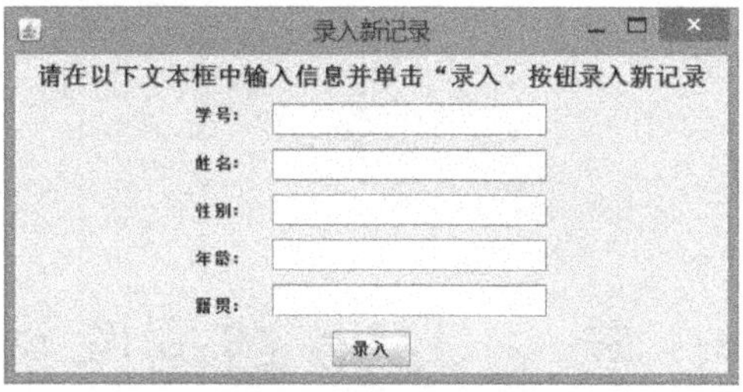

图 9.36　录入学生信息

2. 在窗口中添加一个文本框，一个按钮，一个文本区。要求实现当在文本框中输入一个句子并回车，或单击按钮时，在文本区中显示将该句子中包含的单词及其个数（用不同的事件监听器实现方式实现该题）。

3. 测试下面程序的输出：
```
import javax.swing.*;
public class WindowStyle {
    public static void main(String[] args) {
    try{
        UIManager.LookAndFeelInfo[]
        infos = UIManager.getInstalledLookAndFeels();
        for(UIManager.LookAndFeelInfo info:infos){
            UIManager.setLookAndFeel(info.getClassName());
            System.out.println(info.getClassName());
            JOptionPane.showInputDialog(info.getName()+"风格");
        }
        }catch(Exception ex){}
    }
}
```

4. 编写程序实现一个用于查询数据库表信息并将查询结果显示在表格中的窗口（如图 9.37 所示），不含事件处理代码。

图 9.37　查询学生信息窗口

5. 编写程序，实现一个简单文本编辑器，要求如下：

（1）包含两个菜单"文件"与"编辑""文件"菜单中包括"新建""打开""保存""退出"等菜单项；"编辑"菜单中包括"撤销""剪切""复制""粘贴""全选"等菜单项。

（2）右击文本编辑器的任何位置，都会弹出快捷菜单，快捷菜单包括"剪切""复制""粘贴""全选"等菜单项。

（3）通过以上各种菜单命令，可以实现文本编辑的各种操作。

第 10 章　Java 数据库编程

【本章内容提要】
- JDBC 简介；
- JDBC API；
- MySQL 简介；
- 数据库基本操作的 SQL 语法；
- MySQL 的使用；
- JDBC 数据库基本操作；
- 运用 JavaBean 进行数据库操作；
- 数据库的批处理与事务操作；
- JDBC 操作 Access 数据库。

10.1　JDBC 简介

JDBC（Java DataBase Conectivity）提供了 Java 应用程序访问、操作各种数据库管理系统的 API，它由一组类和接口组成。在用 JDBC API 操作这些数据库管理系统时，还需要配置驱动程序，在实际应用中，JDBC 驱动程序的种类有以下两种：

① JDBC-ODBC 驱动程序：该驱动程序首先将对 JDBC 的调用转化为 ODBC（Open DataBase Conectivity，是微软公司提供的数据库操作编程接口）的调用，然后再利用 ODBC 与数据库进行连接。例如：Java 应用程序在访问、操作 Access 数据库时，就需要通过配置 JDBC-ODBC 驱动程序。

② 专用驱动程序：是由各个数据库厂商提供的，专门用于 JDBC 连接特定数据库管理系统的专用驱动程序。只需要下载这些驱动程序的压缩文件，将其解压并将其中的相关 .jar 文件添加到 Java 项目中，就可以使用 JDBC API 中的类和接口编写用于程序并通过驱动程序连接、操作该数据库。例如：Java 应用程序在访问、操作 MySQL、Oracle、高版本的 SQL Server 数据库时，就需要使用专用驱动程序。

10.2　JDBC API

JDBC API 中的接口和类都包含在 java.sql 包中，利用这些接口和类可以使应用程序方便地连接和操作数据库并返回操作结果。下面对这些接口提供的方法进行详细介绍。

1．DriverManager 类

java.sql.DriverManager 用来连接数据源。DriverManager 类通过调用其静态方法 public static Connection getConnection(String url,String user,String password) throws SQLException 来连接数据源。其中，参数 url 表示数据源的字符串；user 与 password 分别代表数据源的用户名及密码。

2．Connection 接口

java.sql.Connection 接口完成对某一指定数据库的连接。其常用方法有：

- Statement createStatement()：throws SQLException：创建一个 Statement 对象来将 SQL 语句发送到数据库。
- Statement createStatement(int resultSetType,int resultSetConcurrency)throws SQLExceptio n：创建一个 Statement 对象，该对象将生成具有给定类型和并发性的 ResultSet 对象。resultSetType 取值为 ResultSet.TYPE_FORWARD_ONLY（结果集游标只能向下滚动）、ResultSet.TYPE_SCROLL_INSENSITIVE（结果集的游标可以上下滚动，当数据库变化时，当前结果集不变）或 ResultSet.TYPE_SCROLL_SENSITIVE 之一（结果集的游标可以上下滚动，当数据库变化时，当前结果集同步改变）。resultSetConcurrency 取值为 ResultSet.CONCUR_READ_ONLY(不能用结果集更新数据库中的表)或 ResultSet.CONCUR_UPDATABLE（能用结果集更新数据库中的表）之一。
- PreparedStatement prepareStatement(String sql)throws SQLException：创建一个 Prepared Statement 对象来将参数化的 SQL 语句发送到数据库。不带参数的 SQL 语句通常使用 Statement 对象执行。如果多次执行相同的 SQL 语句，使用 PreparedStatement 对象可能更有效。
- PreparedStatementprepareStatement(String sql,int resultSetType,int resultSetConcurrency)throws SQL Exception：创建一个 PreparedStatement 对象，其中，参数 sql 是指 SQL 语句，resultSetType 与 resultSetConcurrency 的含义与 createStatement(int resultSetTyp e,int resultSet Concurrency)方法中这两个参数含义相同。
- void commit()throws SQLException：用于数据库事务操作，使所有上一次提交/回滚后进行的更改成为持久更改，并释放此 Connection 对象当前持有的所有数据库锁。
- void setAutoCommit(boolean autoCommit)throws SQLException：将此连接的自动提交模式设置为 true 或 false。如果连接处于自动提交模式下，则它的所有 SQL 语句将被执行并作为单个事务提交。否则，它的 SQL 语句将聚集到事务中，直到调用 commit 方法或 rollback 方法为止。默认情况下，新连接处于自动提交模式。
- void rollback()throws SQLException：取消在当前事务中进行的所有更改，并释放此 Connection 对象当前持有的所有数据库锁。
- void close()throws SQLException：立即释放此 Connection 对象的数据库和 JDBC 资源。

3．Statement 接口

一个 Statement 对象用于执行 SQL 语句并获得语句执行后产生的结果。其常用方法有：

- ResultSet executeQuery(String sql)throws SQLException：用于数据库查询（Select 语句）操作，执行参数 String sql 给定的 SQL 语句，返回 ResultSet 对象。

- int executeUpdate(String sql)throws SQLException：主要用于数据库增、删、改（Insert、Update 或 Delete 语句）操作，执行参数 String sql 给定 SQL 语句。
- void addBatch(String sql)throws SQLException：将给定的 SQL 命令添加到此 Statement 对象的当前命令列表中。通过调用方法 executeBatch 可以批量执行此列表中的命令。
- int[] executeBatch()throws SQLException：将一批命令提交给数据库来执行，如果全部命令执行成功，则返回更新计数组成的数组。
- void close()throws SQLException：立即释放此 Statement 对象的数据库和 JDBC 资源。

4．PreparedStatement 接口

该接口是 Statement 接口的子接口，与 Statement 接口功能类似，效率较高。同时，支持给 SQL 语句传递参数，PreparedStatement 对象通过调用 setXXX()方法来设置这些参数。

PreparedStatement 也提供了 executeQuery()、executeUpdate()、addBatch(String sql)等方法，其用法与 Statement 接口中这些方法的用法相同。

针对数据库表中各个字段的不同数据类型，PreparedStatement 提供了若干 setXXX()方法，用于给 SQL 语句传递参数：如 setBoolean(int parameterIndex,boolean x)、setInt(int parameterIndex,int x)、setByte(int parameterIndex,byte x)、setFloat(int parameterIndex,float x)、setDouble(int parameterIndex,double x)、setString(int parameterIndex,String x)、setDate(int parameterIndex,Date x)等方法。其中，parameterIndex 取值为 1，2，…。

5．ResultSet 接口

其对象表示 SQL 语句检索结果的结果集。该接口提供了一系列用于获取当前结果集中某条数据库记录每个字段值的方法，如可用以下 getXXX()方法获取 ResultSet 对象（结果集）中当前行数值类型的字段值。

- int getInt(int columnIndex)throws SQLException：以 int 的形式获取此 ResultSet 对象的当前行中由 columnIndex 指定列的值，columnIndex 是指字段所在列编号如 1，2，…。
- int getInt(String columnLabel)throws SQLException：以 int 的形式获取此 ResultSet 对象的当前行中由 columnLabel 指定列的值，columnLabel 是指该字段的名称。

类似的方法还有：

- double getDouble(int columnIndex)throws SQLException 或 double getDouble(String columnLabel)throws SQLException，float getFloat(int columnIndex)throws SQLException 或 float getFloat(String columnLabel)throws SQLException 等，其参数含义与 getInt()方法相同。
- 对于文本类型的字段值，可用 String getString(int columnIndex)throws SQLException 或 String getString(String columnLabel)throws SQLException 获取，其参数含义与 getInt()方法相同。

其他更多的 getXXX()方法，请读者查阅 JDK 开发文档。

ResultSet 接口也提供了多个控制"游标"的方法，通过调用这些方法，使得"游标"移动到当前数据结果集指定的行。

若需要在结果集中前后移动、显示结果集指定的一条记录或随机显示若干条记录等。这时，需要返回一个可滚动的结果集如在调用 createStatement(int resultSetType,in t resultSetConcurrency)方法时，首先将 resultSetType 设为 ResultSet.TYPE_SCROLL _SENSITIVE，其次，通过语句 ResultSet rs=stmt.executeQuery(SQL 语句)产生可滚动结果集，继而用 rs 调用以下各个"游标"来控制游标的移动：

- boolean absolute(int row)throws SQLException：将游标移动到此 ResultSet 对象的 row 给定的行。
- void afterLast()throws SQLException：将游标移动到此 ResultSet 对象的末尾即最后一行之后。
- boolean first()throws SQLException：将游标移动到此 ResultSet 对象的第一行。
- boolean last()throws SQLException：将游标移动到此 ResultSet 对象的最后一行。
- int getRow()throws SQLException：获取当前行编号。
- boolean previous()throws SQLException：将游标移动到此 ResultSet 对象的上一行。
- boolean next()throws SQLException：将游标从当前位置向前移一行。
- public boolean isAfterLast()：判断游标是否在最后一行之后。
- public boolean isBeforeFirst()：判断游标是否在第一行之前
- public boolean ifFirst()：判断游标是否指向结果集的第一行。
- public boolean isLast()：判断游标是否指向结果集的最后一行。

10.3 MySQL 简介

MySQL 是一个小巧、灵活、功能齐全、多用户、多线程关系型数据库管理系统。开发者为瑞典 MySQL AB 公司。由于其体积小、速度快、开放源码等特点，MySQL 被广泛地应用 Java 管理信息系统和中小型 Web 应用中。

1．MySQL 常用数据类型

- CHAR (M)：CHAR 数据类型用于表示固定长度的字符串，可以包含最多达 255 个字符。其中 M 代表字符串的长度。
- VARCHAR (M)：VARCHAR 是一种比 CHAR 更加灵活的数据类型，同样用于表示字符数据，但是 VARCHAR 可以保存可变长度的字符串，其中 M 代表该数据类型所允许保存的字符串的最大长度。
- INT (M) [Unsigned] ：INT 数据类型用于保存− 2 147 483 647～2 147 483 648 范围之内的任意整数数据。如果用户使用 Unsigned 选项，则有效数据范围调整为 0～4 294 967 295。
- FLOAT [(M,D)]：FLOAT 数据类型用于表示数值较小的浮点数据，可以提供更加准确的数据精度。其中，M 代表浮点数据的长度（即小数点左右数据长度的总和），D 表示浮点数据位于小数点右边的数值位数。
- DATE ：DATE 数据类型用于保存日期数据，默认格式为 YYYY-MM-DD。
- TEXT / BLOB ：TEXT 和 BLOB 数据类型可以用来保存 255～65 535 个字符，如果用户需要把大段文本保存到数据库内的话，可以选用 TEXT 或 BLOB 数据类型。TEXT 和 BLOB 这两种数据类型基本相同，唯一的区别在于 TEXT 不区分大小写，而 BLOB 对字符的大小写敏感。

2．MySQL 中创建表的主要参数

- Primary Key：具有 Primary Key 限制条件的字段用于区分同一个数据表中的不同记录。
- NOT NULL：NOT NULL 限制条件规定用户不得在该字段中插入空值。

- Auto_Increment：具有 Auto_Increment 限制条件的字段值从 1 开始，每增加一条新记录，值就会相应地增加 1。

3．MySQL 显示数据表的命令

- mysql > show tables;

该命令将会列出当前数据库下的所有数据表。

- mysql > show columns from tablename;

该命令用于显示数据表的字段。

- mysql > desc tablename;

该命令将会返回指定数据表的所有字段和字段相关信息。

10.4　数据库基本操作的 SQL 语法

在关系型数据库中，数据表中的一行称为一个记录，一列称为一个字段，每列的标题称为字段名。

1．创建数据库

在 SQL 语言中，创建一个新数据库基本语法格式如下：
`CREATE DATABASE 数据库名称;`

2．创建表

在 SQL 语言中，创建一个新表基本语法格式如下：
`CREATE TABLE 表名称 (列名 数据类型 约束, …) ;`

3．增加数据记录

在数据库表中增加数据记录语法形式：
`INSERT INTO 表名 [(字段名表)] VALUES (值表);`

4．查询数据记录

语法形式：
`SELECT [DISTINCT] [别名.]字段名或表达式 [AS 列标题] FROM 表或视图 别名 [ WHERE 条件][ GROUP BY 分组表达式][ ORDER BY 排序表达式[ ASC | DESC ]] ;`

5．删除数据库记录

语法形式：
`delete from 表名 where 条件;`

6．删除数据库

语法形式：
`drop database 数据库名;`

10.5　MySQL 的使用

1．安装与配置

以 MySQL 5.1.40 为例，下载 MySQL 安装包并双击，开始 MySQL 的安装。安装过程（见图 10.1）只需按照安装向导的默认选项一步步完成即可。安装完成后进入 MySQL 的配置阶段（见图 10.2），

在配置阶段需要做一定的修改,需要将字符集设置为手动方式,并选择 GB 2312 字符编码(见图 10.3),以便支持中文字符。为 MySQL 设置密码(见图 10.4),单击 Next 按钮,在弹出的对话框中单击 Execute 按钮,完成配置过程并启动 MySQL 服务(见图 10.5)。

图 10.1　MySQL 安装向导

图 10.2　MySQL 配置向导

图 10.3　设置字符集编码

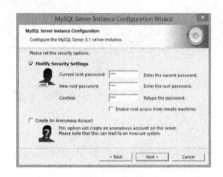
图 10.4　设置密码

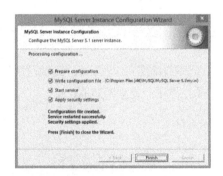
图 10.5　完成配置并启动 MySQL 服务

在完成了 MySQL 的安装、配置并启动 MySQL 服务后,就可以在 MySQL 提供的命令行环境下进行数据库的各种操作。

注意:从开始菜单启动 MySQL 命令行,输入配置过的密码,继而在 mysql> 提示符后键入符合 MySQL 语法规定的相关 SQL 命令即可。

(1)创建数据库 Student
```
mysql>create database Student;
```

（2）打开数据库 Student

mysql＞use Student;

（3）在 Student 库中创建表 studentInfo

mysql＞create table studentInfo(stuno varchar(20) primary key not null,stuname varchar(40),stusex char(4),stuage int,stuaddress varchar(50));

（4）显示 studentInfo 表结构（见图 10.6）

mysql＞desc studentInfo;

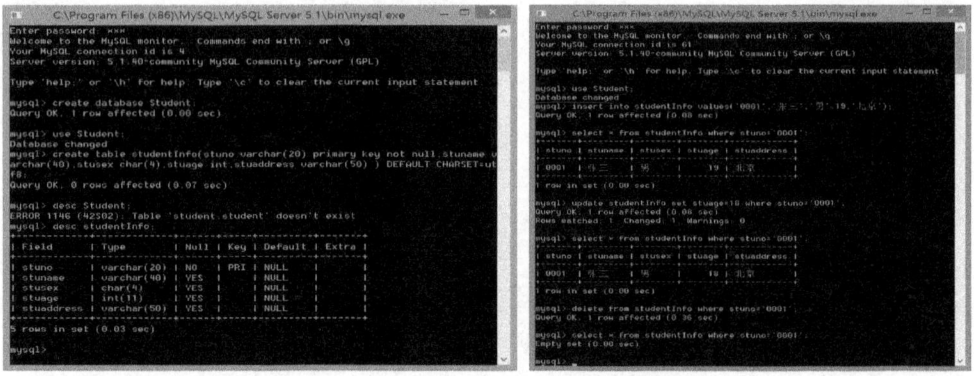

图 10.6　在 MySQL 命令行中进行数据库操作

提示：MySQL 语法要求较严格，在其中使用 SQL 语句时要细心。

2. MySQL 驱动程序

下载与 MySQL 版本吻合的驱动程序，本章以 mysql-connector-java-5.1.8 为例，首先下载该驱动程序，将其解压，在解压后的文件夹中包含 mysql-connector-java-5.1.8-bin.jar，就是我们需要的驱动程序，该驱动程序在 Java 项目开发中需要被导入项目中。

10.6　JDBC 数据库基本操作

做好了以上新建数据库、数据库表的准备工作，就可以编写程序进行数据库操作了。

利用 JDBC API 编写数据库操作程序一般步骤如下：

Step 1：加载驱动程序。

```
try {
Class.forName(driverName);
//不同 DBMS(Database Manager System)所需的驱动程序串 driverName 有不同的写法
    }
catch (ClassNotFoundException e) {…}
```

Step 2：连接数据库（或数据源）。

```
try{
    Connection conn= DriverManager.getConnection(url, user, password);
      /* 不同 DBMS 的 url 有不同的写法，user 和 password 分别表示数据库用户名及密码。对于 MySQL，用户名默认为 root*/
}
catch(SQLException e){…}
```

Step 3：准备向数据库发送 SQL 语句。

```
try{
   Statement stmt=conn.createStatement();
```

}
catch (SQLException e) {…}

Step 4：执行 SQL 语句。
```
try{
    stmt. executeQuery(sql 语句);       // 执行查询操作
```
或
```
stmt. executeUpdate(sql 语句);        //执行更改、删除数据库记录操作
}
catch (SQLException e) {…}
```

Step 5：以查询操作为例，若 SQL 语句执行成功，则会返回查询结果集 rs。例如：
```
try{
ResultSet rs=sql.executeQuery("SELECT * FROM employee");
}
catch (SQLException e) {…}
```

Step 6：关闭数据库连接
```
try {
rs.close();
stmt.close();
conn.close();
}
catch (SQLException e) {…}
```

提示：以上操作步骤中 JDBC API 的 Connection、Statement、ResultSet 的运用是"环环相扣"的，在关闭数据库连接释放资源时，其关闭次序与创建次序反序。

【**例题 10_1**】 编写程序，在 studentInfo 表中中新增一条数据记录。

```
import java.sql.Connection;
import java.sql.DriverManager;
import java.sql.SQLException;
import java.sql.Statement;
public class Ch10_1 {
    public static void main(String[] args) {
    Connection conn=null;
    Statement stmt=null;
    String driverName = "com.mysql.jdbc.Driver";      //MySQL 的 JDBC 驱动程序
    String url="jdbc:mysql://localhost:3306/Student";  //数据库
    String user="root";                                //MySQL 默认的用户名
    String password="123";                             //密码
    try {
        Class.forName(driverName);                     //加载驱动程序
      }
    catch (ClassNotFoundException e) {
        e.printStackTrace();}
    try {
        conn=DriverManager.getConnection(url, user, password);   //连接数据库
        stmt=conn.createStatement();
        String sql="insert into studentInfo
(stuno,stuname,stusex,stuage,stuaddress) values('0001','张三','男',19,'北京')";
        stmt.executeUpdate(sql);
      }
    catch (SQLException e) {
        e.printStackTrace();}
```

```
        finally{
            try {
                if(stmt!=null)
                    stmt.close();
                if(conn!=null)
                    conn.close();
             }
            catch (SQLException e) {
                e.printStackTrace();}
            }
            System.out.println("新记录增加成功! ");
        }
    }
```

注意：本例需要在项目中导入 MySQL JDBC 驱动程序，在 Eclipse 环境下新建项目 Ch10 以及类 Ch10_1.java，在项目名 Ch10 上右击，选择 "Build Path" → "Add External Archives"，在弹出的对话框中选择保存好的 mysql-connector-java-5.1.8-bin.jar 文件并打开，就可以将该驱动程序被包含在项目中了（见图 10.7）。

图 10.7 在项目中导入 MySQL 驱动程序

说明：常见数据库的 JDBC 驱动程序名称和 url 如下。

- **SQLServer**：其 JDBC 驱动程序为：com.microsoft.jdbc.sqlserver.SQLServerDriver，URL 为 jdbc:microsoft:sqlserver://[IP]:1433;DatabaseName=[DBName]。
- **MySQL**：其 JDBC 驱动程序为："com.mysql.jdbc.Driver"，URL 为 jdbc:mysql://localhos t: 3306/[D BName]。
- **Oracle**：其 JDBC 驱动程序驱动程序为 oracle.jdbc.driver.OracleDriver，URL 为 jdbc:oracle: thi n:@[ip]:1521:[sid]。
- **Access**：需要配置 JDBC-ODBC 驱动，其用法在本章稍后的内容中进行介绍。

按照例题 10_1 的方式，当向数据库发送了 SQL 语句，数据库中的 SQL 解释器首先要将该 SQL 语句进行预处理并生成底层的命令，再执行。可以想到，当程序向数据库发送大量 SQL 语句时，无疑会增加数据库的负担，降低执行效率。

为此，Java 提供了 PreparedStatement 接口，可以对要发送给数据库的 SQL 语句进行预处理并生成该数据库底层的内部命令，然后直接让数据库执行该命令即可，从而减轻了数据库的负担，提高了数据库访问速度。例题 10_2 就是运用 PreparedStatement 进行数据库操作的。

【例题 10_2】 用 PreparedStatement 完成例题 10_1 的功能，代码如下：
```
import java.sql.Connection;
import java.sql.DriverManager;
import java.sql.PreparedStatement;
import java.sql.SQLException;
public class Ch10_2 {
    public static void main(String[] args) {
    Connection conn=null;
    PreparedStatement pstmt=null;
    String driverName = "com.mysql.jdbc.Driver";
    String  url="jdbc:mysql://localhost:3306/Student";
```

```java
        String user="root";
        String password ="123";
        String sql=null;
        try {
            Class.forName(driverName);
        } catch (ClassNotFoundException e) {
            e.printStackTrace();
        }
        try {
            conn=DriverManager.getConnection(url, user, password);
            sql="insert into studentInfo(stuno,stuname,stusex,stuage,stuaddress) values(?,?,?,?,?)";
            pstmt=conn.prepareStatement(sql);
            /*给SQL语句中的通配符?赋上具体的值**/
            pstmt.setString(1,"0002");
            pstmt.setString(2,"李四");
            pstmt.setString(3,"男");
            pstmt.setInt(4,20);
            pstmt.setString(5,"山西");
            pstmt.executeUpdate();
            System.out.println("新记录增加成功！");
         }
         catch (SQLException e) {
            System.out.println("新记录增加失败！");
            e.printStackTrace();
        }
        finally{
            try {
                if(pstmt!=null)
                   pstmt.close();
                if(conn!=null)
                   conn.close();
                }
            catch (SQLException e) {
               e.printStackTrace();
            }
        }
    }
}
```

【例题 10_3】 查询数据库表。

```java
import java.sql.Connection;
import java.sql.Statement;
import java.sql.DriverManager;
import java.sql.ResultSet;
import java.sql.SQLException;
public class Ch10_3 {
  public static void main(String[] args) {
     Connection conn=null;
     Statement stmt=null;
     ResultSet rs=null;
     String driverName = "com.mysql.jdbc.Driver";
     String  url="jdbc:mysql://localhost:3306/Student";
     String user = "root";
     String password = "123";
```

```java
    String sql=null;
    try {
        Class.forName(driverName);
    }
    catch (ClassNotFoundException e) {
        e.printStackTrace();
    }
    try {
        conn=DriverManager.getConnection(url, user, password);
        sql="select * from studentInfo";
        stmt=conn.createStatement();
        rs=stmt.executeQuery(sql);           //得到查询结果集
        while(rs.next()){                    //顺序遍历结果集
            System.out.printf("%-8s%-8s%-8s%-4d%-8s",rs.getString(1),rs.getString(2),
                  rs.getString(3),rs.getInt(4),rs.getString(5));
            System.out.println();
        }
    } catch (SQLException e) {
        e.printStackTrace();
    }
    finally{
        try {
            if(rs!=null)rs.close();
            if(stmt!=null)stmt.close();
            if(conn!=null)conn.close();
        } catch (SQLException e) {
            e.printStackTrace();
        }
    }
  }
}
```

执行结果如图 10.8 所示：

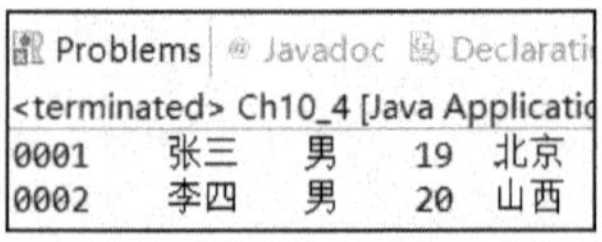

图 10.8　查询结果

```
<terminated> Ch10_4 [Java Application]
0001    张三    男    19    北京
0002    李四    男    20    山西
```

说明：next()是 ResultSet 中的一个游标方法，该游标最初位于结果集 rs 的第一条记录之前，每调用一次 next()，游标下移一行，当移到结果集最后一行之后时，返回 false。

【例题 10_4】 将编号为 0002 的学生年龄修改为 18 并显示修改结果。

```java
import java.sql.Connection;
import java.sql.DriverManager;
import java.sql.PreparedStatement;
import java.sql.ResultSet;
import java.sql.SQLException;
public class Ch10_4 {
    public static void main(String[] args) {
    Connection conn=null;
    PreparedStatement pstmt=null;
    ResultSet rs=null;
    String driverName = "com.mysql.jdbc.Driver";
    String sql=null;
    try {
        Class.forName(driverName);
    } catch (ClassNotFoundException e) {
        e.printStackTrace();
```

```java
        }
        try {
            conn=DriverManager.getConnection("jdbc:mysql://localhost:3306/Student",
"root", "123");
            sql="update studentInfo set stuage=18 where stuno='0002'";
            pstmt=conn.prepareStatement(sql);
            pstmt.executeUpdate();
            System.out.println("记录修改成功！");
            rs=pstmt.executeQuery("select * from studentInfo");
            while(rs.next()){//输出修改后所有记录
                System.out.printf("%-8s%-8s%-8s%-4d%-8s",rs.getString(1),rs.getString(2),
                    rs.getString(3),rs.getInt(4),rs.getString(5));
                System.out.println();
            }
        } catch (SQLException e) {
            System.out.println("记录修改失败！");
            e.printStackTrace();
        }
        finally{
            try {
                if(rs!=null)rs.close();
                if(pstmt!=null)pstmt.close();
                if(conn!=null)conn.close();
            } catch (SQLException e) {
                e.printStackTrace();
            }
        }
    }
}
```

执行结果如图 10.9 所示。

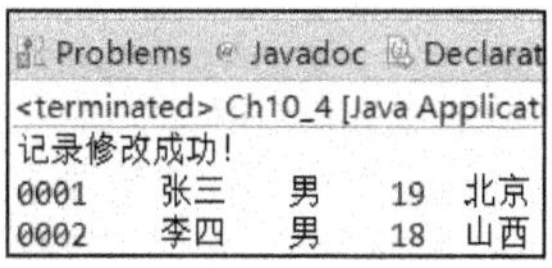

图 10.9 记录修改后结果

【例题 10_5】删除编号为 0002 的学生记录。

与例题 10_4 的编写方法一样，只需将修改例题 6_4 中的 SQL 语句为 sql="delete from studentInfo where stuno='0002' "，输出提示语句相应修改为"记录删除成功"或"记录删除失败"即可。

在以上数据库数据库的增查改删操作中，只有查询操作调用 executeQuery()方法，而增、改、删操作都调用 executeUpdate()方法。

有时，我们需要在一个可滚动的结果集中随机显示若干条记录。可以通过以下语句得到一个可滚动的结果集。

  `Statement stmt=con.createStatement(int resultSetType,int resultSetConcurr ency);`

其中，resultSetType 的取值决定滚动方式，取值可以是 ResultSet.TYPE_FORWORD_ ONLY（表示结果集的游标只能向下滚动）或 ResultSet.TYPE_SCROLL_INSENSITIVE（结果集的游标可以上下移动，当数据库变化时，当前结果集不变）或 ResultSet.TYPE_SCROLL_SENSI TIVE（返回可滚动的结果集，当数据库变化时，当前结果集同步改变）。

resultSetConcurrency 的取值决定是否可以用结果集更新数据库，Concurrency 取值可以是 ResultSet.CONCUR_READ_ONLY(不能用结果集更新数据库中的表)或 ResultSet.CON CUR_UPDATETABLE（能用结果集更新数据库中的表）。

然后通过下面语句得到结果集：

ResultSet rs=stmt.executeQuery(SQL 语句);继而用 rs 调用其各个"游标"控制方法来完成对结果集中数据记录的的随机访问。

假定现在 studentInfo 表的数据记录情况如图 10.10 所示，编写程序，完成对各记录的随机访问。

0001	张三	男	19	北京
0002	李四	男	18	山西
0003	王丽	女	19	湖南
0004	章华	女	18	河南
0005	马文	男	18	安徽

图 10.10 studentInfo 表记录

【例题 10_6】 运用 ResultSet 接口的游标方法随机查询 studentInfo 表记录。

```java
import java.sql.Connection;
import java.sql.DriverManager;
import java.sql.Statement;
import java.sql.ResultSet;
import java.sql.SQLException;
public class Ch10_6 {
  public static void main(String[] args) {
    Connection conn=null;
    Statement stmt=null;
    ResultSet rs=null;
    String driverName = "com.mysql.jdbc.Driver";
    String sql=null;
    try {
        Class.forName(driverName);
    } catch (ClassNotFoundException e) {
        e.printStackTrace();}
    try {
        conn=DriverManager.getConnection("jdbc:mysql://localhost:3306/Student","root", "123");
          stmt=conn.createStatement(ResultSet.TYPE_SCROLL_INSENSITIVE,ResultSet.CONCUR_READ_ONLY);
        //确定可滚动结果集的滚动方式和不能用结果集更新数据库中的表
        sql="select * from studentInfo";
        rs=stmt.executeQuery(sql);
        rs.last();                    //将游标置到结果集最后一行
        int n=rs.getRow();            //获取结果集记录数
        System.out.println("最后一行记录为: ");
        System.out.printf("%-8s%-8s%-8s%-4d%-8s\n",rs.getString(1),rs.getString(2),
              rs.getString(3),rs.getInt(4),rs.getString(5));
        //输出结果集中最后一行记录
        System.out.println("当前记录对应的行号为: "+n);
        //输出当前游标对应的行号
        rs.absolute(2);              //将游标置到第 2 行
        System.out.println("第 2 行记录为: ");
        System.out.printf("%-8s%-8s%-8s%-4d%-8s\n",rs.getString(1),rs.getString(2),
              rs.getString(3),rs.getInt(4),rs.getString(5));
        //输出第 2 行记录
        rs.previous();               //将游标上移一行
        System.out.println("第 1 行记录为: ");
        System.out.printf("%-8s%-8s%-8s%-4d%-8s\n",rs.getString(1),rs.getString(2),
              rs.getString(3),rs.getInt(4),rs.getString(5));
        rs.previous();
        System.out.println("游标是否移到了第 1 行记录之前: "+rs.isBeforeFirst());
    } catch (SQLException e) {
        e.printStackTrace();
```

```
        }
        finally{
            try {
                if(rs!=null)rs.close();
                if(stmt!=null)stmt.close();
                if(conn!=null)conn.close();
            } catch (SQLException e) {
                e.printStackTrace();
            }
        }
    }
}
```
执行结果如图 10.11 所示。

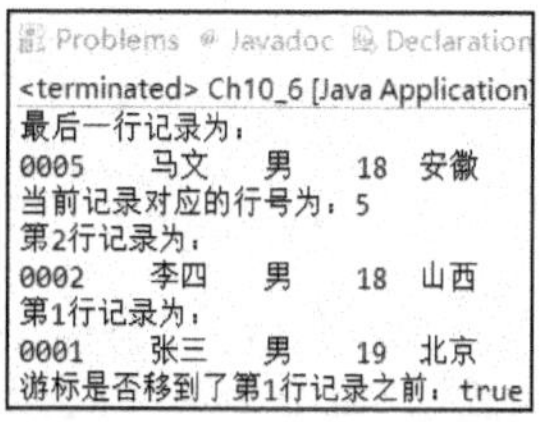

图 10.11　随机访问查询结果集

10.7　运用 JavaBean 进行数据库操作

例题 10_1 到例题 10_6 中虽然完成了数据库的增、查、改、删等基本操作，但其代码中加载驱动程序，连接数据源等代码都是一样的，能不能让这些重复性的代码只编写一次，在需要的地方进行调用呢？

第 3 章中介绍过 JavaBean，它是一种组件技术，可以被共享和重用。在数据库操作中可以运用 JavaBean 达到降低代码重复，将对数据库表的直接操作转化为对类对象的操作（这是对象-关系映射 ORMapping 编程思想的体现）。

JavaBean 有广义和狭义之分，一个操作数据库的 JavaBean 属于广义的 JavaBean，即一个普通的 Java 类，其定义没有严格的规范，这样的 JavaBean 在数据库应用中被称为 DAO（Data Access Object）。与数据库应用相关的另外一种 JavaBean 被称为 VO（Value Object），它是狭义 JavaBean，这就意味着它的构造要满足 JavaBean 的规范。VO 主要用来封装相关数据库表记录：VO 的每个成员变量对应着数据库表的每个字段,将来,对于数据库表字段值得存取就可以转化为用 VO 的 setter 和 getter 方法为 VO 的成员变量赋值或取值操作。

【例题 10_7】运用 DAO+VO+Java 主类的方式操作数据库。

首先，针对 Student 数据库中 studentInfo 表的各个字段，定义 VO 类完成对 studentInfo 表记录的封装，代码如下：

```
package vo;
public class Student {
private String stuno;
private String stuname;
private String stusex;
private int stuage;
private String stuaddress;
public Student(){}//无参构造方法
public String getStuno() {
    return stuno;
}
public void setStuno(String stuno) {
    this.stuno = stuno;
}
public String getStuname() {
    return stuname;
```

```java
    }
    public void setStuname(String stuname) {
        this.stuname = stuname;
    }
    public String getStusex() {
        return stusex;
    }
    public void setStusex(String stusex) {
        this.stusex = stusex;
    }
    public int getStuage() {
        return stuage;
    }
    public void setStuage(int stuage) {
        this.stuage = stuage;
    }
    public String getStuaddress() {
        return stuaddress;
    }
    public void setStuaddress(String stuaddress) {
        this.stuaddress = stuaddress;
    }
}
```

说明：Student 类是一个狭义 JavaBean，在其中定义了与数据库表 studentInfo 中的各个字段对应的成员变量，并给每个成员变量定义了 setter()和 getter()方法，用于操作各个成员变量。通过操作 Student 类的各个成员变量可以达到间接操作 studentInfo 表的目的。

其次，定义 DAO 类，代码如下：

```java
package dao;
import java.sql.Connection;
import java.sql.DriverManager;
import java.sql.PreparedStatement;
import java.sql.ResultSet;
import java.sql.SQLException;
import java.util.ArrayList;
import vo.Student;
public class DatabaseConBean {
    Connection conn=null;
    PreparedStatement pstmt=null;
    ResultSet rs=null;
    String driverName="com.mysql.jdbc.Driver";
    String user="root";
    String password="123";
    String  url="jdbc:mysql://localhost:3306/Student";
    String sql=null;
    ArrayList<Object> list=null;
    Student stu=null;
    public void conn(){                      //用于加载驱动程序，连接数据库的方法
        try {
            Class.forName(driverName);
            conn=DriverManager.getConnection(url, user, password);
        }
        catch (ClassNotFoundException e) {
```

```java
            e.printStackTrace();}
        catch (SQLException e) {
            e.printStackTrace();}
    }
    /*数据库查询方法**/
    public ArrayList<Object> excuteQuery(String sql)//自编的executeQuery()方法
    {
        list=new ArrayList<Object>();        //创建ArrayList对象
        try {
            pstmt=conn.prepareStatement(sql);
            rs=pstmt.executeQuery();         //调用系统的executeQuery()方法
            while(rs.next()){
                stu=new Student(); //创建用于封装studentInfo表记录的Student类对象stu
                stu.setStuno(rs.getString(1));
                stu.setStuname(rs.getString(2));
                stu.setStusex(rs.getString(3));
                stu.setStuage(rs.getInt(4));
                stu.setStuaddress(rs.getString(5));
                //将查询结果集对象rs的各个字段值赋值给stu的各个成员变量
                list.add(stu);        //将stu对象作为一个数据结点加入到list中
            }
        }
        catch (SQLException e) {
            e.printStackTrace();}
        return list;            //返回list
    }
    /*数据库增、删、改方法**/
    public void excuteUpdate(String sql){//自编的excuteUpdate()方法
        try {
            pstmt=conn.prepareStatement(sql);
            int n=pstmt.executeUpdate();//调用系统的executeUpdate()方法
            if(n==1)
                System.out.println("操作成功!");
        }
        catch (SQLException e) {
            e.printStackTrace();}
    }
}
```

编写主类，实现对studentInfo表查、增、改、删操作，代码分别如下：

```java
//查询数据库的类Query
import java.util.ArrayList;
import dao.DatabaseConBean;
import vo.Student;
public class Query {
    public static void main(String[] args) {
        DatabaseConBean bean=new DatabaseConBean();    //创建DAO类对象
        bean.conn();//调用加载驱动程序，连接数据库的conn()方法
        String sql="select * from studentInfo";
        ArrayList<Object> list=bean.excuteQuery(sql);  //调用查询方法得到一个
                                                       ArrayList对象
        for(int i=0;i<list.size();i++){//遍历list并将其每个结点转换为Student对象，
            Student stu=(Student)list.get(i);
            //取得一个list结点(VO类对象)并下转型为Student对象
```

```java
            System.out.printf("%-8s%-8s%-8s%-4d%-8s",stu.getStuno(),stu.getStuname(),
                    stu.getStusex(),stu.getStuage(),stu.getStuaddress());
                //运用Student的getter方法获得数据库表中的每个字段值
            System.out.println();
        }
    }
}
//新增记录的类Insert
import java.util.ArrayList;
import dao.DatabaseConBean;
import vo.Student;
public class Query {
    public static void main(String[] args) {
        DatabaseConBean bean=new DatabaseConBean();  //创建DAO类对象
        bean.conn();//调用加载驱动程序,连接数据库的conn()方法
        String sql="select * from studentInfo";
        ArrayList<Object> list=bean.excuteQuery(sql);     //调用查询方法得到一个
                                                          ArrayList对象
        for(int i=0;i<list.size();i++){//遍历list并将其每个结点转换为Student对象,
            Student stu=(Student)list.get(i);//取得一个list结点并下转型为Student对象

            System.out.printf("%-8s%-8s%-8s%-4d%-8s",stu.getStuno(),stu.getStuname(),
                    stu.getStusex(),stu.getStuage(),stu.getStuaddress());
                //运用Student的getter方法获得数据库表中的每个字段值
            System.out.println();
        }
    }
}
```

提示：在执行完新增记录的程序后，可再次执行 Query，观察查询结果是否已将新记录成功添加到 studentInfo 表中。

```java
//修改数据记录的类Update
import dao.DatabaseConBean;
public class Update {
    public static void main(String[] args) {
        DatabaseConBean bean=new DatabaseConBean();//创建DAO类对象
        bean.conn();                //调用加载驱动程序,连接数据库的conn()方法
        String sql="update studentInfo set stuname='李四' where stuno='0008'";
        bean.excuteUpdate(sql);   //调用DAO类的excuteUpdate方法完成记录修改
    }
}
//删除数据记录的类Delete
import dao.DatabaseConBean;
public class Delete {
    public static void main(String[] args) {
        DatabaseConBean bean=new DatabaseConBean();//创建DAO类对象
        bean.conn();                //调用加载驱动程序,连接数据库的conn()方法
        String sql="delete from studentInfo where stuno='0008' ";
        bean.excuteUpdate(sql);   //调用DAO类的excuteUpdate方法完成记录删除
    }
}
```

提示：在执行完新增记录、修改记录以及删除记录的程序后，可再次执行 Query，观察查询结果是否操作成功。

思考：Query 类中既用到了 DAO（DatabaseConBean 类），也用到了 VO（Student 类），但 Insert、Update 以及 Delete 类中只用到了 DAO（connBean 类），没有用 VO（Student 类），如何改编程序，使得 Insert、Update 及 Delete 类中通过使用 Student 类达到操作数据库表的目的？

10.8 数据库的批处理与事务操作

1. 相关方法

JDBC 允许执行批量操作，批量操作的作用是向数据库同时提交一系列的 SQL 语句。在 Statement 接口中提供了相应批量处理的方法：

- void addBatch(String sql) throws SQLException：将给定的 SQL 命令添加到此 Statement 对象的批处理命令列表中。
- int[] executeBatch()throws SQLException：将一批命令提交给数据库来执行，如果全部命令执行成功，则返回更新计数组成的数组。

PreparedStatement 是 Statement 的子接口，也可以用以上两个方法进行批处理操作。

2. JDBC 的事务操作

事务处理是保证数据库中数据完整性和一致性的重要机制。在 JDBC 的数据库操作中，一个事务指由一个或多个 SQL 语句组成的一个不可分割的工作单元，具有原子性的特点即当组成事务的每一个操作要么全做，要么一个都不做。

当一个连接对象被创建时，默认情况下事务被设置为自动提交状态。即每执行一条 SQL 语句就会自动对该语句调用 commit()方法进行提交，为了将多条 SQL 语句作为一个事务执行，Connection 对象可以通过调用 setAutoCommit(false)方法禁止自动提交，在整个事务操作完成后，再调用 commit()方法整体提交。若事务中一个操作失败，可在异常捕获语句中调用 rollback()方法对事务进行回滚操作，即撤销本次事务的所有操作，回到什么都没做的状态，从而保证数据的一致性。

【例题 10_8】 运用批处理及事务处理在数据库表 studentInfo 中新增多条数据记录。

```java
import java.sql.Connection;
import java.sql.DriverManager;
import java.sql.PreparedStatement;
import java.sql.ResultSet;
import java.sql.SQLException;
import java.sql.Statement;
public class Ch10_8 {
 public static void main(String[] args) {
    Connection conn=null;
    Statement stmt=null;
    PreparedStatement pstmt=null;
    ResultSet rs=null;
    String driverName = "com.mysql.jdbc.Driver";
    try {
       Class.forName(driverName);
    } catch (ClassNotFoundException e) {
       e.printStackTrace();
```

```java
        }
        try {
            conn=DriverManager.getConnection("jdbc:mysql://localhost:3306/Student",
            "root","123");
            stmt=conn.createStatement();
            /*查询显示事务操作前的数据记录*/
            rs=stmt.executeQuery("select * from studentInfo");
            System.out.println("事务操作前 studentInfo 表中的记录为: ");
            while(rs.next()){
                System.out.printf("%-8s%-8s%-8s%-4d%-8s",rs.getString(1),rs.getString(2),
                  rs.getString(3),rs.getInt(4),rs.getString(5));
                System.out.println();
            }
            /*批处理及事务操作*/
            conn.setAutoCommit(false);       //关闭事务自动提交
            pstmt=conn.prepareStatement("insert into studentInfo(stuno,stuname,stusex,stuage,stuaddress) values(?,?,?,?,?)");
            pstmt.setString(1,"0006");
            pstmt.setString(2,"李莉");
            pstmt.setString(3,"女");
            pstmt.setInt(4,20);
            pstmt.setString(5,"上海");
            pstmt.addBatch();                //添加批处理
            pstmt.setString(1,"0007");
            pstmt.setString(2,"赵扬");
            pstmt.setString(3,"男");
            pstmt.setInt(4,18);
            pstmt.setString(5,"甘肃");
            pstmt.addBatch();
            pstmt.setString(1,"0008");
            pstmt.setString(2,"孙兰兰");
            pstmt.setString(3,"女");
            pstmt.setInt(4,19);
            pstmt.setString(5,"陕西");
            pstmt.addBatch();
            pstmt.executeBatch();            //执行批处理
            conn.commit();                   //提交事务
            conn.setAutoCommit(true);        //开启事务自动提交
            /*查询显示事务操作后的数据记录*/
            rs=stmt.executeQuery("select * from studentInfo");
            System.out.println("事务操作后 studentInfo 表中的记录为: ");
            while(rs.next()){
                System.out.printf("%-8s%-8s%-8s%-4d%-8s",rs.getString(1),rs.getString(2),
                  rs.getString(3),rs.getInt(4),rs.getString(5));
                System.out.println();
            }
        }
        catch (SQLException e) {
            try {
              conn.rollback();               //事务回滚操作
              }
            catch (SQLException e1) {
              e1.printStackTrace();}
            e.printStackTrace();
```

```
            }
        finally{
            try {
                if(rs!=null)stmt.close();
                if(stmt!=null)stmt.close();
                if(pstmt!=null)pstmt.close();
                if(conn!=null)conn.close();
                }
            catch (SQLException e) {
                e.printStackTrace();
                }
            }
        }
```

运行程序，执行结果如图10.12所示。

【例题10_9】综合示例—简单的学生信息管理：要求编写GUI程序实现以下功能对Student库中studentInfo表的增、查、改、删。

编程思路：

- 分别编写用于查询数据库表的窗体类QueryWin，增加记录的类InsertWin，修改记录的类UpdateWin以及删除记录的类DeleteWin并分别实现各自的事件处理。

图10.12 事务处理前后的数据库记录变化

- 编写主窗体类DatabaseMainFrame，在其中创建QueryWin、InsertWin、UpdateWin及DeleteWin类的对象并进行相应的事件处理；
- 编写主类DatabaseMainClass，在其中创建DatabaseMainFrame的对象。

说明：本例中用到了例题10_7中定义的VO类Student以及DAO类DatabaseConBean，只是要将DatabaseConBean类的excuteUpdate方法修改为如下形式：

```
public int excuteUpdate(String sql){     //自编的excuteUpdate()方法
    try {
        pstmt=conn.prepareStatement(sql);
        n=pstmt.executeUpdate();          //调用系统的executeUpdate()方法
    }
    catch (SQLException e) {
        e.printStackTrace();}
    return n;
}
```

（1）DatabaseMainClass类代码如下：

```
package myproject.stumanager;
public class DatabaseMainClass {
    public static void main(String[] args) {
        new DatabaseMainFrame();
    }
}
```

（2）DatabaseMainFrame类代码如下：

```
package myproject.stumanager;
import java.awt.*;
```

```java
import java.awt.event.ActionEvent;
import java.awt.event.ActionListener;
import javax.swing.*;
public class DatabaseMainFrame extends JFrame implements ActionListener{
    JLabel label1,label2;
    JButton query,insert,update,delete;
    JPanel panel1,panel2,panel3;
    public DatabaseMainFrame(){
        this.setTitle("学生信息管理系统");
        label1=new JLabel("***欢迎使用学生信息管理系统***");
        label2=new JLabel("@版权属于×××团队");
        label1.setFont(new Font("宋体",Font.BOLD,30));
        label1.setForeground(Color.BLUE);
        label2.setFont(new Font("宋体",Font.BOLD,20));
        query=new JButton("查询");
        query.setSize(40,15 );
        insert=new JButton("录入");
        update=new JButton("修改");
        delete=new JButton("删除");
        panel1=new JPanel();
        panel1.add(label1);
        panel2=new JPanel();
        panel2.setLayout(new GridLayout(1,4,20,20));
        panel2.add(query);panel2.add(insert);
        panel2.add(update);panel2.add(delete);
        panel3=new JPanel();
        panel3.add(label2);
        this.add(panel1,BorderLayout.NORTH);
        this.add(panel2,BorderLayout.CENTER);
        this.add(panel3,BorderLayout.SOUTH);
        GraphicsEnvironment ge = GraphicsEnvironment.getLocalGraphicsEnvironment();
        //得到 GraphicsEnvironment 对象
        GraphicsDevice gd = ge.getDefaultScreenDevice();//得到屏幕设备
        Rectangle rec = gd.getDefaultConfiguration().getBounds();
        this.setSize(700, 350);
        this.setLocation((int)(rec.getWidth()-700)/2, (int)(rec.getHeight()-
           350)/2);
        //窗口在屏幕上居中显示
        query.addActionListener(this);
        insert.addActionListener(this);
        update.addActionListener(this);
        delete.addActionListener(this);           //给各个按钮对象添加事件监听器
        this.setResizable(false);                 //窗体不可调整大小
        this.setVisible(true);
    }
    //事件处理代码
    public void actionPerformed(ActionEvent e) {
        if(e.getSource()==query)
            new QueryWin();                       //创建查询窗体类 QueryWin 对象
        else if(e.getSource()==insert)
            new InsertWin();                      //创建新增记录窗体类 InsertWin 对象
        else if(e.getSource()==update)
            new UpdateWin();                      //创建修改记录窗体类 UpdateWin 对象
        else if(e.getSource()==delete)
            new DeleteWin();                      //创建删除记录窗体类 DeleteWin 对象
```

}
}
（3）查询窗体类 QueryWin 类代码如下：
```java
package myproject.stumanager;
import java.awt.*;
import java.awt.event.ActionEvent;
import java.awt.event.ActionListener;
import java.util.ArrayList;
import javax.swing.*;
import dao.DatabaseConBean;
import vo.Student;
public class QueryWin extends JFrame implements ActionListener {
    JLabel label;
    JTextField text;
    JButton bt1,bt2;
    JPanel panel1,panel2;
    JTable table;
    JScrollPane jsppanel;
    String field[]={"学号","姓名","性别","年龄","籍贯"};       //表格组件的列标题
    Object record[][]=new Object[10000][5];                    //表格组件的数据区域
    public QueryWin(){
        this.setTitle("查询窗口");
        this.setLayout(new GridLayout(3,1));
        label=new JLabel("请输入学号");
        text=new JTextField(20);
        bt1=new JButton("显示所有学生信息");
        bt2=new JButton("查询");
        bt1.addActionListener(this);
        bt2.addActionListener(this);
        panel1=new JPanel();
        panel1.add(bt1);
        panel2=new JPanel();
        panel2.add(label);panel2.add(text);panel2.add(bt2);
        table=new JTable(record,field);
        jsppanel=new JScrollPane(table);
        this.add(panel1);this.add(panel2);this.add(jsppanel);
        GraphicsEnvironment ge = GraphicsEnvironment.getLocalGraphicsEnvironment();
        GraphicsDevice gd = ge.getDefaultScreenDevice();
        Rectangle rec = gd.getDefaultConfiguration().getBounds();
        this.setSize(500, 300);
        this.setLocation((int)(rec.getWidth()-500)/2, (int)(rec.getHeight()-300)/2);
        this.setVisible(true);
    }
    //事件代码
    public void actionPerformed(ActionEvent e) {
        if(e.getSource()==bt1){
            DatabaseConBean bean=new DatabaseConBean();    //创建 DAO 类对象
            bean.conn();                    //调用加载驱动程序，连接数据库的 conn()方法
            String sql="select * from studentInfo";
            ArrayList<Object> list=bean.excuteQuery(sql);
            /*遍历 list 并将其每个结点转换为 Student 对象，
             * 运用 Student 的 getter 方法获得数据库表中的每个字段值**/
            if(list.size()!=0){        //记录不为空
              for(int i=0;i<list.size();i++){
```

```
                Student stu=(Student)list.get(i);
                //取得一个list结点并下转型为Student对象
                record[i][0]=stu.getStuno();
                record[i][1]=stu.getStuname();
                record[i][2]=stu.getStusex();
                record[i][3]=stu.getStuage();
                record[i][4]=stu.getStuaddress();
                //将数据库查询结果赋值给表格组件的对应单元格
            }
            table.repaint();              //刷新表格组件对象
        }
        else{
            JOptionPane.showMessageDialog(null,"数据库记录为空!");
        }
    }
    else if(e.getSource()==bt2){
        String no=text.getText().trim();
        DatabaseConBean bean=new DatabaseConBean();//创建DAO类对象
        bean.conn();                 //调用加载驱动程序,连接数据库的conn()方法
        String sql="select * from studentInfo where stuno="+no;
        ArrayList<Object> list=bean.excuteQuery(sql);
        if(list.size()!=0){          //记录不为空
        Student stu=(Student)list.get(0);
          record[0][0]=stu.getStuno();
          record[0][1]=stu.getStuname();
          record[0][2]=stu.getStusex();
          record[0][3]=stu.getStuage();
          record[0][4]=stu.getStuaddress();
        //将数据库查询结果赋值给表格组件的对应单元格
          table.repaint();
        }
        else{
            JOptionPane.showMessageDialog(null,"查询的数据库记录不存在!");
        }
    }
  }
}
```

分析：该类用到JTable对象来显示数据库查询结果,要注意如何将数据库查询结果置于表格组件对象的各个单元格中;同时,该类中同时使用了DAO以及VO类。

(4) 新增记录窗体类InsertWin类代码如下

```
package myproject.stumanager;
import java.awt.*;
import java.awt.event.ActionEvent;
import java.awt.event.ActionListener;
import javax.swing.*;
import dao.DatabaseConBean;
public class InsertWin extends JFrame implements ActionListener {
    JLabel label,l1,l2,l3,l4,l5;
    JTextField t1,t2,t3,t4,t5;
    JButton bt;
    JPanel panel1,panel2,panel3;
    Box Hbox,Vbox1,Vbox2;
    public InsertWin(){
        this.setTitle("录入新记录");
```

```java
            label=new JLabel("请在以下文本框中输入信息并单击"录入"按钮录入新记录");
            label.setFont(new Font("宋体",Font.BOLD,20));
            label.setForeground(Color.BLUE);
            bt=new JButton("录 入");
            bt.addActionListener(this);
            panel1=new JPanel();
            panel1.add(label);
            t1=new JTextField(20);t2=new JTextField(20);
            t3=new JTextField(20);t4=new JTextField(20);
            t5=new JTextField(20);
            Vbox1=Box.createVerticalBox();
            Vbox1.add(new JLabel("学 号: "));Vbox1.add(Box.createVerticalStrut(20));
            Vbox1.add(new JLabel("姓 名: "));Vbox1.add(Box.createVerticalStrut(20));
            Vbox1.add(new JLabel("性 别: "));Vbox1.add(Box.createVerticalStrut(20));
            Vbox1.add(new JLabel("年 龄: "));Vbox1.add(Box.createVerticalStrut(20));
            Vbox1.add(new JLabel("籍 贯: "));
            Vbox2=Box.createVerticalBox();
            Vbox2.add(t1);Vbox2.add(Box.createVerticalStrut(10));
            Vbox2.add(t2);Vbox2.add(Box.createVerticalStrut(10));
            Vbox2.add(t3);Vbox2.add(Box.createVerticalStrut(10));
            Vbox2.add(t4);Vbox2.add(Box.createVerticalStrut(10));
            Vbox2.add(t5);
            Hbox=Box.createHorizontalBox();
            Hbox.add(Vbox1);
            Hbox.add(Box.createHorizontalStrut(20));
            Hbox.add(Vbox2);
            panel2=new JPanel();
            panel2.add(Hbox);
            panel3=new JPanel();
            panel3.add(bt);
            this.add(panel1, BorderLayout.NORTH);
            this.add(panel2, BorderLayout.CENTER);
            this.add(panel3, BorderLayout.SOUTH);
            GraphicsEnvironment ge = GraphicsEnvironment.getLocalGraphicsEnvironment();
            GraphicsDevice gd = ge.getDefaultScreenDevice();
            Rectangle rec = gd.getDefaultConfiguration().getBounds();
            this.setSize(600, 300);
            this.setLocation((int)(rec.getWidth()-600)/2,
(int)(rec.getHeight()-300)/2);
            this.setVisible(true);
    }
    public void actionPerformed(ActionEvent e) {
        String no=t1.getText().trim();
        String name=t2.getText().trim();
        String sex=t3.getText().trim();
        int age=Integer.parseInt(t4.getText().trim());
        String address=t5.getText().trim();
        DatabaseConBean bean=new DatabaseConBean();        //创建DAO类对象
        bean.conn();                    //调用加载驱动程序,连接数据库的conn()方法
        String sql="insert into studentInfo(stuno,stuname,
stusex,stuage,stuaddress) values('"+no+"','"+name+"','"+sex+"',"+age+",'
"+address+"')";
        int n=bean.excuteUpdate(sql);      //调用excuteUpdate方法完成新增记录
        if(n>=1){
            JOptionPane.showMessageDialog(null, "新记录录入成功! ");
```

```
            t1.setText(null);t2.setText(null);t3.setText(null);
            t4.setText(null);t5.setText(null);
        }
        else{
            JOptionPane.showMessageDialog(null, "新记录录入失败，该学号已经存在！");
        }

    }
}
```

（5）修改记录窗体类 UpdateWin 类代码如下：

```
package myproject.stumanager;
import java.awt.*;
import java.awt.event.ActionEvent;
import java.awt.event.ActionListener;
import java.util.ArrayList;
import javax.swing.*;
import dao.DatabaseConBean;
import vo.Student;
public class UpdateWin extends JFrame implements ActionListener {
    JLabel label,l1,l2,l3,l4,l5;
    JTextField t,t1,t2,t3,t4,t5;
    JButton bt1,bt2;
    JPanel panel1,panel2,panel3;
    Box Hbox,Vbox1,Vbox2;
    public UpdateWin(){
        this.setTitle("修改记录");
        label=new JLabel("请输入要修改记录的学号: ");
        t=new JTextField(20);
        bt1=new JButton("查询");
        bt2=new JButton("修改");
        bt1.addActionListener(this);
        bt2.addActionListener(this);
        panel1=new JPanel();
        panel1.add(label);
        panel1.add(t);
        panel1.add(bt1);
        t1=new JTextField(20);t2=new JTextField(20);
        t3=new JTextField(20);t4=new JTextField(20);
        t5=new JTextField(20);
        Vbox1=Box.createVerticalBox();
        Vbox1.add(new JLabel("学 号: "));Vbox1.add(Box.createVerticalStrut(20));
        Vbox1.add(new JLabel("姓 名: "));Vbox1.add(Box.createVerticalStrut(20));
        Vbox1.add(new JLabel("性 别: "));Vbox1.add(Box.createVerticalStrut(20));
        Vbox1.add(new JLabel("年 龄: "));Vbox1.add(Box.createVerticalStrut(20));
        Vbox1.add(new JLabel("籍 贯: "));
        Vbox2=Box.createVerticalBox();
        Vbox2.add(t1);Vbox2.add(Box.createVerticalStrut(10));
        Vbox2.add(t2);Vbox2.add(Box.createVerticalStrut(10));
        Vbox2.add(t3);Vbox2.add(Box.createVerticalStrut(10));
        Vbox2.add(t4);Vbox2.add(Box.createVerticalStrut(10));
        Vbox2.add(t5);
        Hbox=Box.createHorizontalBox();
        Hbox.add(Vbox1);
        Hbox.add(Box.createHorizontalStrut(20));
        Hbox.add(Vbox2);
```

```java
            panel2=new JPanel();
            panel2.add(Hbox);
            panel3=new JPanel();
            panel3.add(bt2);
            this.add(panel1, BorderLayout.NORTH);
            this.add(panel2, BorderLayout.CENTER);
            this.add(panel3, BorderLayout.SOUTH);
            GraphicsEnvironment ge = GraphicsEnvironment.getLocalGraphicsEnvironment();
            GraphicsDevice gd = ge.getDefaultScreenDevice();
            Rectangle rec = gd.getDefaultConfiguration().getBounds();
            this.setSize(600, 300);
            this.setLocation((int)(rec.getWidth()-600)/2, (int)(rec.getHeight()-300)/2);
           this.setVisible(true);
        }
        public void actionPerformed(ActionEvent e) {
            if(e.getSource()==bt1){
                String no=t.getText();
                DatabaseConBean bean=new DatabaseConBean();        //创建DAO类对象
                bean.conn();                 //调用加载驱动程序,连接数据库的conn()方法
                String sql="select * from studentInfo where stuno="+no;
                ArrayList<Object> list=bean.excuteQuery(sql);//查询要修改的记录
                if(list.size()!=0){           //记录不为空
                    Student stu=(Student)list.get(0);            //得到VO对象
                    t1.setText(stu.getStuno());
                    t2.setText(stu.getStuname());
                    t3.setText(stu.getStusex());
                    t4.setText(String.valueOf(stu.getStuage()));
                    t5.setText(stu.getStuaddress());//在各个文本框中显示本记录的各个字段值
                }
            }
            else if(e.getSource()==bt2){
                String no=t1.getText().trim();
                String name=t2.getText().trim();
                String sex=t3.getText().trim();
                int age=Integer.parseInt(t4.getText().trim());
                String address=t5.getText().trim();
                //获取各个文本框值
                DatabaseConBean bean=new DatabaseConBean();//创建DAO类对象
                bean.conn();//调用加载驱动程序,连接数据库的conn()方法
                String sql="update studentInfo set stuno='"+no+"',
                 stuname='"+name+"',stusex="+ "'"+sex+"',stuage="+age+",
                 stuaddress='"+address+"'"+" where stuno='"+no+"'";
                int n=bean.excuteUpdate(sql);
                if(n>=1)
                    JOptionPane.showMessageDialog(null, "记录修改成功!");
                else
                    JOptionPane.showMessageDialog(null, "记录修改失败!");
            }
        }
    }
}
```

(6)删除记录窗体类DeleteWin类代码如下:

```java
package myproject.stumanager;
import java.awt.*;
import java.awt.event.ActionEvent;
```

```java
import java.awt.event.ActionListener;
import javax.swing.*;
import dao.DatabaseConBean;
public class DeleteWin extends JFrame implements ActionListener {
    JLabel label;
    JTextField t;
    JButton bt;
    JPanel panel;
    public DeleteWin(){
        this.setTitle("删除记录");
        label=new JLabel("请输入要删除记录的学号: ");
        t=new JTextField(20);
        bt=new JButton("删除");
        bt.addActionListener(this);
        panel=new JPanel();
        panel.add(label);
        panel.add(t);
        panel.add(bt);
        this.add(panel, BorderLayout.CENTER);
        GraphicsEnvironment ge = GraphicsEnvironment.getLocalGraphicsEnvironment();
        GraphicsDevice gd = ge.getDefaultScreenDevice();
        Rectangle rec = gd.getDefaultConfiguration().getBounds();
        this.setSize(600, 300);
        this.setLocation((int)(rec.getWidth()-600)/2,
(int)(rec.getHeight()-300)/2);
        this.setVisible(true);
    }
    public void actionPerformed(ActionEvent e) {
        String no=t.getText().trim();
        DatabaseConBean bean=new DatabaseConBean();   //创建DAO类对象
        bean.conn();                  //调用加载驱动程序，连接数据库的conn()方法
        String sql="delete from studentInfo where stuno='"+no+"'";
        int n=bean.excuteUpdate(sql);
        if(n>=1)
            JOptionPane.showMessageDialog(null, "记录删除成功！");
        else
            JOptionPane.showMessageDialog(null, "记录删除失败！");
    }
}
```

主窗体运行结果如图 10.13 所示。

单击主窗体中的"查询"按钮，进入学生信息查询窗体，效果如图 10.14 和图 10.15 所示。其中，当在"请输入学号"文本框中输入要查询学生的学号，并单击"查询"按钮时，表格中显示该条学生记录；当单击"显示所有学生信息"按钮时，表格中显示所有学生信息。

图 10.13　主窗体　　　　图 10.14　查询某条记录　　　　图 10.15　显示所有学生记录

单击主窗体中的"录入"按钮，进入学生信息录入窗体，运行效果如图 10.16 所示。当录入

成功时，对话框提示录入成功，否则，提示录入失败。

图 10.16 录入新的学生记录

单击主窗体中的"修改"按钮，进入学生信息修改窗体，运行效果如图 10.17 所示。首先在"请输入要修改记录的学号"文本框中输入学生学号，单击"查询"按钮，则在其下方的各个文本框中显示该条记录各个字段值，直接在这些文本框中修改字段值如将 0009 号学生的籍贯由原来"陕西"改为"山东"，单击"修改"按钮，若修改成功，弹出修改成功对话框，否则，弹出修改失败对话框。

单击主窗体中的"删除"按钮，进入学生信息删除窗体，运行效果如图 10.18 所示。

图 10.17 修改学生记录　　　　　　　　图 10.18 删除学生记录

提示：在完成记录的添加、修改、删除操作后，可以再次执行查询，以验证这些操作是否被正确执行。

10.9 JDBC 操作 Access 数据库

在 Java 应用开发中，使用 Microsoft Access 数据库可以满足小型应用项目的需要。本节介绍 Java 操作 Access 的目的有两个：其一是 Java 连接、操作 Access 是通过 JDBC-ODBC 桥接的方式，不需要专用的驱动程序，这种方式具有一定的代表性；其二是相对而言，Access 具有简单、易用的特点。对于刚接触 JDBC 数据库编程的读者，可以将注意力集中在 JDBC 数据库程序设计本身，而不是 DBMS，Access 就成了一个不错的选择。

Java 操作 Access 数据库时，需要通过配置 JDBC-ODBC 数据源的方式进行，同时，需要给数据源添加 Windows 自带的 ODBC 驱动程序。

首先，在 Access 中新建数据库 Student，在其中新建表 studentInfo(stuno 文本类型 primary key not null,stuname 文本类型,stusex 文本类型,stuage 数字类型,stuaddress 文本类型)，其各个字段类型定义以及初始记录如图 10.19 所示。

图 10.19 studentInfo 表

按照以下步骤配置 JDBC-ODBC 数据源及驱动程序［Windows 7(32 位)操作系统下调试的］：

（1）在控制面板选择"管理工具"→"数据源(ODBC)"并双击鼠标打开，弹出"ODBC 数据源管理器"对话框（见图 10.20）。

（2）单击"添加"按钮，打开"创建新数据源"对话框（见图 10.21），在其中选择 Microsoft Access Driver(*.mdb)。注意：若 Access 版本在 2007 以上，则选择 Microsoft Access Driver(*.mdb,*.accdb)。

（3）单击"完成"按钮，在弹出的"ODBC Microsoft Access 安装"对话框中输入数据源名 STUDENT（见图 10.22），单击"选择"按钮，在弹出的"选择数据库"对话框中选择 Student.mdb，单击"确定"按钮（见图 10.23）。

图 10.20　ODBC 数据源管理器

图 10.21　选择驱动程序

（4）在"ODBC Microsoft Access 安装"对话框中单击"高级"按钮，在弹出的"设置高级选项"对话框中输入登录名称如 admin，密码如 123，单击"确定"按钮（见图 10.24）。返回 ODBC Microsoft Access 安装"对话框并按"确定"按钮，继而返回"ODBC 数据源管理器"对话框，按"确定"按钮完成数据源的配置。

图 10.22　设定数据源

图 10.23　选择要连接的数据库

图 10.24　设定用户名和密码

【例题 10_10】编写程序，在 studentInfo 表中新增记录。

```java
import java.sql.*;
public class Insert {
    public static void main(String[] args) {
        Connection conn=null;
        Statement stmt=null;
    String driverName = "sun.jdbc.odbc.JdbcOdbcDriver";    //驱动程序
        String url="jdbc:odbc:STUDENT";                    //数据源
        String user = "admin";                             //用户名
        String password = "123";                           //密码
        try {
            Class.forName(driverName);                     //加载驱动程序
        } catch (ClassNotFoundException e) {
          e.printStackTrace();}
```

```
        try {
            conn=DriverManager.getConnection(url, user, password);//连接数据库
            stmt=conn.createStatement();
               String sql="insert into studentInfo(stuno,stuname,stusex,
stuage,stuaddress) values('2015009','赵小刚','男',19,'辽宁')";
            stmt.executeUpdate(sql);
         }
         catch (SQLException e) {
            e.printStackTrace();}
         finally{
            try {
               if(stmt!=null)
               stmt.close();
               if(conn!=null)
               conn.close();
            } catch (SQLException e) {
               e.printStackTrace();}
             }
         System.out.println("新记录增加成功！");
      }
}
```

由新增记录的程序可以看出，JDBC 操作 Access 数据库时，首先需要配置 JDBC-ODBC 数据源；其次，要将驱动程序 String driverName 赋值为 "sun.jdbc.odbc.JdbcOdbcDriver"，数据源 String url 赋值为"jdbc:odbc:数据源名"。其他运用 JDBC API 的核心代码与操作其他数据库的一样。

思考并完成编写查询记录、修改记录以及删除记录的程序；用VO+DAO的方式改写上面增、查、改、删程序。

小　　结

本章介绍了 JDBC 的作用；JDBC API 提供的常用接口、类的应用方法；简介了 MySQL 提供的数据类型和常用命令的用法；介绍了 MySQL 的安装、配置步骤；重点以实例说明了运用 JDBC 操作 MySQL 数据库的方法，运用 JavaBean 操作数据库的方法，数据库的批处理以及事务处理方法以及一个较综合的简单学生信息管理系统等。

习　题　10

上机实践题

1. 在 Access 中新建数据库 Student，在其中创建表 studentInfo(stuno 文本类型 primary key not null,stuname 文本类型,stusex 文本类型,stuage 数字类型,stuaddress 文本类型)。

编写程序实现对该数据库表的增、查、改、删操作并输出结果。

2. 试用 VO+DAO 的方式改写上机实践题1。

3. 仿照例题"简单的学生信息管理"，编写完整的程序并上机调试运行。

第 11 章　Java 输入、输出流

【本章内容提要】

- Java 输入、输出流概述；
- File 类的应用；
- 输入、输出流类；
- 文件字节输入、输出流类；
- 文件字符输入、输出流类；
- 字节数组输入、输出流类；
- 过滤流类；
- 随机访问文件；
- Serializable 接口与对象序列化；
- 标准输入、输出流；
- 文件对话框；
- 用 Desktop 类打开文件。

11.1　Java 输入、输出流概述

在 Java 中，输入、输出流可以形象地理解为执行"读"和"写"操作的数据通信管道，如图 11.1 所示。其中，"源"与"目的地"是广义的概念。

在标准输入输出过程中，"源"是指标准输入设备键盘，"目的地"是指标准输出设备显示器；在文件读写操作中，是指从磁盘"源"文件读数据，写到磁盘中的"目的地"文件中；在网络通信程序中，"源"是指网络中数据发送端主机，数据发送端主机通过"写"操作将数据发往目的主机，"目的地"是指网络中数据接收端主机，数据接收端主机通过"读"操作把发送端主机发来的数据进行接收等。

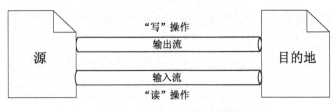

图 11.1　输入输出流示意图

要产生一个"输入流",只需要给 Java 的某个输入流类创建对象即可,该对象就是"输入流"这个数据管道;若要产生一个"输出流",只须创建 Java 的某个输出流类对象即可,该对象就是"输出流"这个数据管道。

11.2 File 类的应用

File 类包含在 java.io 包中,File 是对磁盘文件或目录(文件夹)的抽象,即创建一个 File 类对象,可以代表磁盘中的某文件,也可以代表磁盘中的某个目录(文件夹)。

1. File 类的常用构造方法

- File(String pathname),pathname 表示 File 类所代表的磁盘文件或目录的路径及名称。
- File(String path, String name),path 表示 File 类所代表的磁盘文件或目录的路径,name 表示 File 类所代表的磁盘文件或目录的名称。
- File(File dir, String name),dir 是一个已经存在的表示某磁盘文件或目录路径的 File 类对象,name 表示 File 类所代表的磁盘文件或目录的名称。

2. File 类中获取文件、目录属性的常用方法

- public boolean exists():判断该 File 对象所代表的文件或目录是否存在。
- public boolean isDirectory():判断该 File 对象是否代表一个目录。
- public boolean isFile():判断该 File 对象是否代表一个文件。
- public boolean canRead():判断对 File 对象所代表的文件能否进行读操作。
- public boolean canWrite():判断对 File 对象所代表的文件能否进行写操作。
- public String getName():返回文件或目录名。
- public String getPath():返回文件或目录路径。
- public long length():获取文件的字节数。

3. File 类操作文件、目录的常用方法

- public boolean createNewFile()throws IOException:若该 File 对象代表文件,并且该文件不存在,则创建该文件。
- public boolean mkdir(),在文件系统中创建由该 File 对象表示的目录。
- public boolean delete():删除该 File 对象表示的目录表示的文件或目录。
- public String[] list():将该 File 对象类表示的文件或目录名称以字符串数组形式返回。
- public File[] listFiles():将该 File 对象类表示的文件以 File 对象数组形式返回。

【例题 11_1】测试 File 类的常用方法。

```
import java.io.File;
import java.io.IOException;
public class Ch11_1 {
  public static void main(String args[]) {
    File dir=new File("E:\\Java Pro");
    File file = new File("E:\\Java Pro\\Ch11\\src","Ch11_1.java");
    /*测试File的一些常用 方法**/
    System.out.println("dir是否存在: "+dir.exists());
    System.out.println("dir是否代表目录: "+dir.isDirectory());
```

```
        System.out.println("file是否存在: "+file.exists());
        System.out.println("file是否代表文件: "+file.isFile());
        System.out.println("file的长度为: "+file.length()+"字节");
        File dir1=new File("E:\\aaa");
        if(!dir1.exists())
          { dir1.mkdir();//若aaa目录不存在,则创建它
            System.out.println("aaa创建成功! ");
          }
        File dir2=new File(dir1,"\\bbb");
        if(!dir2.exists())
          { dir2.mkdir();//若bbb文件夹不存在,则创建它,aaa是父目录,bbb是子目录
            System.out.println("bbb创建成功! ");
          }
        File f1=new File(dir1,"new1.txt");
        if(!f1.exists())
        { try {
            f1.createNewFile();//在dir1中创建文件new1.txt
            System.out.println("bbb创建成功! ");
          } catch (IOException e) {}
        }
        File[] files=dir1.listFiles();
        System.out.println("列出dir1下所有文件和目录");
        for(File f:files)
            System.out.println(f);
        //dir2.delete();//删除dir2目录,先将该语句注释起来,
    }
}
```

运行结果如图 11.2 所示。

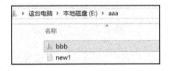

图 11.2　控制台输出及 aaa 文件夹

分析：创建 File file = new File("E:\\Java Pro\\Ch11\\src","Ch11_1.java");就是代表当前源程序 Ch11_1.java 本身。

一个 File 对象既可以抽象代表一个文件如 new1.txt，也可以代表一个目录（文件夹）如 E 盘下的 aaa 文件夹，aaa 文件夹的子文件夹 bbb 等。一个 File 类对象的创建并不意味着它所代表的文件或文件夹就一定存在，可以调用 mkdir()创建出文件夹，调用 createNewFile()方法创建文件等。在程序运行时，现将 dir2.delete();注释起来，等到观察完 E 盘下的 aaa 文件夹、其子文件夹 bbb 以及 new1.txt 是否成功被创建后再执行该语句，再次观察结果。

11.3　输入、输出流类

在 java.io 包中，InputStream 与 OutputStream 分别是表示字节输入流和字节输出流的抽象类，

不能直接实例化。InputStream 与 OutputStream 作为其他字节输入流和字节输出流类的父类。Reader 与 Writer 分别是表示字符输入流和字符输出流的抽象类，不能直接实例化。Reader 与 Writer 作为其他字符输入流、输出流类的父类。

1．InputStream 的常用方法

- public abstract int read() throws IOException：从输入流中读取数据的下一个字节。返回 0～255 范围内的 int 字节值。如果因为已经到达流末尾，则返回值 –1。
- public int read(byte[] b) throws IOException：从输入流中按字节读取数据，并将其存储在字节数组 b 中。以整数形式返回实际读取的字节数。
- public int read(byte[] b,int off,int len) throws IOException：从输入流中按字节读取数据，并将读到的数据存储在字节数组 b 中，以 off 指定的下标开始存储，存储数据长度为 len 个字节。以整数形式返回实际读取的字节数。
- public int available()throws IOException：返回此输入流下一个方法，调用可以不受阻塞地从此输入流读取（或跳过）的估计字节数。
- public void close()throws IOException:关闭此输入流并释放与该流关联的所有系统资源。
- public void reset()throws IOException：将此流重新定位到最后一次对此输入流调用 mark 方法时的位置。
- public long skip(long n)throws IOException：跳过此输入流中数据的 n 个字节。

2．OutputStream 的常用方法

- public abstract void write(int b)throws IOException：将指定的字节写入此输出流。
- public void write(byte[] b)throws IOException：将 b.length 个字节从指定的 byte 数组写入此输出流。
- public void write(byte[] b,int off,int len)throws IOException：将指定 byte 数组中从偏移量 off 开始的 len 个字节写入此输出流。
- public void flush()throws IOException：刷新此输出流并强制写出所有缓冲的输出字节。
- public void flush()throws IOException：刷新此输出流并强制写出所有缓冲的输出字节。

3．Reader 的常用方法

- public int read()throws IOException：读取单个字符。返回值为 0～65 535，如果已到达流的末尾，则返回 –1 。
- public int read(char[] cbuf) throws IOException：读取的字符数，如果已到达流的末尾，则返回 –1 。
- public abstract int read(char[] cbuf,int off,int len)throwsIOException：从输入流中按字符读取数据，并将读到的数据存储在字符数组 b 中，以 off 指定的下标开始存储，存储数据长度为 len 个字节。以整数形式返回实际读取的字符数。
- public abstract void close()throws IOException: 关闭该流并释放与之关联的所有资源。

4．Writer 的常用方法

- public void write(char[] cbuf)throws IOException：写入字符数组。

- public abstract void write(char[] cbuf,int off,int len)throws IOException：将 cbuf 数组中从偏移量 off 开始的 len 个字符写入此输出流。
- public void write(String str,int off,int len)throws IOException：将 cbuf 数组中从偏移量 off 开始的 len 个字符写入此输出流。
- public Writer append(char c)throws IOException：将指定字符添加到此 writer。
- public abstract void flush()throws IOException：刷新此输出流并强制写出所有缓冲的输出字符。
- public abstract void close()throws IOException：关闭该流并释放与之关联的所有资源。

11.4 文件字节输入、输出流类

1. FileInputStream 类

FileInputStream 是文件字节输入流类，是 InputStream 类的子类。其常用构造方法如下：
- FileInputStream(File file) throws FileNotFoundException；
- FileInputStream(String name)throws FileNotFoundException。

分别以 File 类对象 file 或 String 对象 name 所代表的文件创建文件字节输入流对象。当文件不存在时，引发 FileNotFoundException 异常。

2. FileOutputStream 类

FileOutputStream 是文件字节输出流类，是 OutputStream 类的子类。其常用构造方法如下：
- public FileOutputStream(File file) throws FileNotFoundException；
- public FileOutputStream(File file,boolean append)throws FileNotFoundException；
- FileOutputStream(String name) throws FileNotFoundException；
- FileOutputStream(String name, boolean append) throws FileNotFoundException。

分别以 File 类对象 file 或 String 对象 name 所代表的文件创建文件字节输出流对象。若有 boolean append 参数为 true 时，表示在文件原有内容的基础上中追加写，没有 boolean append 参数时，表示擦除文件原有内容并写入写内容。

【例题 11_2】把 Ch11_1.java 文件内容读到控制台输出。

```java
import java.io.File;
import java.io.FileInputStream;
import java.io.FileNotFoundException;
import java.io.IOException;
public class Ch11_2 {
  public static void main(String[] args){
   File source=new File("E:\\Java Pro\\Ch11\\src","Ch11_1.java");
   try {
     FileInputStream fin=new FileInputStream(source);
//以 Ch11_1.java 为读取的源头，创建文件字节输入流类对象
     byte b[]=new byte[(int)source.length()];//数组大小与文件大小相同
     while(fin.read(b)!=-1)//未读完
     {
      String str=new String(b,0,b.length);//将 b 转化为字符串
      System.out.println(str);
     }
```

```
    } catch (FileNotFoundException e) {}
     catch (IOException e){}
  }
}
```

分析：以 E:\\Java Pro\\Ch11\\src 文件夹下的 Ch11_1.java 为数据读取的"源"，用文件字节输入流类对象 fin 调用 read()方法按字节进行读取，并将读到的内容存储在字节数组 b 中，将数组 b 转化为字符串并输出在控制台。

【例题 11_3】将"我在学习文件输入输出流类的用法"写入 E:\\a.txt 文件中。

```java
import java.io.File;
import java.io.FileNotFoundException;
import java.io.FileOutputStream;
import java.io.IOException;
public class Ch11_3 {
  public static void main(String[] args) {
  File destination=new File("E:\\a.txt");
  try {
    FileOutputStream fout = new FileOutputStream(destination);
    //创建文件字节输出流类对象,以 destination 代表的 E:\\a.txt 作为 fout 输出的目的地
    byte b[]="我在学习文件输入输出流类的用法".getBytes();
    //将字符串转化为字节数组
    fout.write(b);//通过输出流管道将 b 的内容写到 E:\\a.txt
    System.out.println("文件写入成功! ");
  } catch (FileNotFoundException e) {}
   catch (IOException e) {}
  }
}
```

【例题 11_4】将 E:\\a.txt 文件内容转存到 E:\\b.txt。

```java
import java.io.File;
import java.io.FileInputStream;
import java.io.FileNotFoundException;
import java.io.FileOutputStream;
import java.io.IOException;
public class Ch11_4 {
  public static void main(String[] args) {
  try {
    File source=new File("E:\\a.txt");           //读取的源头
    File destination=new File("E:\\b.txt");    //写入的目的地
    FileInputStream fin=new FileInputStream(source);//输入流对象
    FileOutputStream fout=new FileOutputStream(destination,true);//输出流对象
    byte b[]=new byte[(int)source.length()];
    while(fin.read(b)!=-1){                     //读文件 a.txt
      fout.write(b);                            //写到 b.txt
    }
    fout.flush();                               //将缓冲区中的内容送到输出流
    fin.close();
    fout.close();                               //关闭输入、输出流对象,释放资源
    System.out.println("文件转存成功! ");
  } catch (FileNotFoundException e) {}
   catch (IOException e) {}
```

        }
    }

11.5 文件字符输入、输出流类

1. FileReader 类

文件字符输入、输出流与文件字节输入输出流的基本编程方法一样，所不同的是文件字符输入输出流读写数据的基本单位为字符，而文件字节输入输出流读写数据的基本单位为字节。

FileReader 是文件字符输入流类。其常用构造方法如下：

- FileReader(String fileName)throws FileNotFoundException：以给定从中读取数据的文件名 fileName 的创建一个新 FileReader 对象。
- FileReader(File file)throws FileNotFoundException：以给定从中读取数据的 File 对象 file 创建一个新 FileReader 对象。

2. FileWriter 类

FileWriter 是文件字符输出流类。其常用构造方法如下：

- FileWriter(File file,boolean append)throws IOException：根据给定的 File 对象 file 创建一个 FileWriter 对象。若参数 append 为 true，则将字符以追加方式写入文件末尾处。若为 false，则擦除方式（原有文件内容被清除）写入新文件。
- FileWriter(String fileName,boolean append)throws IOException：根据给定的文件名 fileName 创建 FileWriter 对象。参数 append 指明追加或擦除写入的方式。

【例题 11_5】生成九九乘法表并将其保存在 E:\\乘法表.txt 中，并将"乘法表.txt"文件内容读取到控制台输出。

```java
import java.io.File;
import java.io.FileNotFoundException;
import java.io.FileReader;
import java.io.FileWriter;
import java.io.IOException;
public class Ch11_5 {
    public static void main(String[] args)  {
        File f = new File("E:\\乘法表.txt");
        FileReader freader;
        FileWriter fwriter;
        try {
            fwriter = new FileWriter(f);//以 f 创建文件字符输出流类对象
            for (int i=1;i<=9;i++) {
              for (int j=1;j<=i;j++){
                fwriter.write(i+"*"+j+"="+(i*j)+"\t");//将乘法表写入E:\\乘法表.txt
              }
              fwriter.write("\r\n");//写入回车换行
            }
            fwriter.flush();
            fwriter.close();
            /*读取乘法表*/
            freader=new FileReader(f);//以 f 创建文件字符输入流类对象
```

```
            char[] c=new char[(int)f.length()];
            while(freader.read(c)!=-1){  //读取乘法表
                String str=new String(c,0,c.length);
                System.out.print(str);//输出到控制台
            }
        } catch (FileNotFoundException e) {}
        catch (IOException e) {}
    }
}
```

程序执行结果如图 11.3 所示。

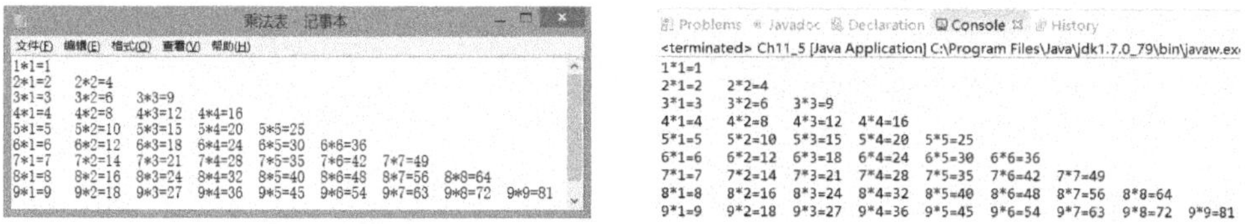

图 11.3 乘法表.txt 及控制台输出

11.6 字节数组输入、输出流类

ByteArrayInputStream 类也是 InputStream 类的子类，用于从内存中的字节数组中读取数据。其常用构造方法：public ByteArrayInputStream(byte[] buf) 及 public ByteArray InputStream(byte[] buf,int offset,int length)：都是以字节数组 buf 为数据读取的"源"。Offset 是指明了在 buf 中读取时的偏移量，length 表示读取的字节个数。

ByteArrayOutputStream 类也是 OutputStream 类的子类，用于给内存中的字节数组写入数据。其常用构造方法：

- public ByteArrayOutputStream()：创建一个新的字节数组输出流。缓冲区的容量最初是 32 字节，如有必要可增加其大小。
- public ByteArrayOutputStream(int size)：创建一个新的字节数组输出流，以 size 指定缓冲区容量（以字节为单位）。

11.7 过 滤 流 类

前面的各种输入输出流类可以视为"初级流"，过滤流可以视为"高级流"，通过过滤流，可以达到对"初级流"的功能增强或提高其输入输出效率的目的。

FilterInputStream 与 FilterOutputStream 分别是 InputStream 与 OutputStream 的子类，FilterInputStream 与 FilterOutputStream 各自又派生出多个子类。其中，BufferedInputStream 与 BufferedOutputStream 就是它们的直接子类。

1. BufferedInputStream 与 BufferedOutputStream 类

为了给初级流 FileInputStream 增加缓冲输入功能，一般格式如下：
BufferedInputStream bufin= new BufferedInputStream(FileInputStream 对象);
或
BufferedInputStream bufin= new BufferedInputStream(FileInputStream 对象,int size);

int size 用于指定缓冲区大小。

为了给初级流 FileOutputStream 增加缓冲输出功能，一般格式如下：

`BufferedOutputStream bufout=new BufferedOutputStream(FileOutputStream 对象);`

或

`BufferedOutput Stream bufout= new BufferedOutputStream(FileOutputStream 对象,int size);`

BufferedReader 与 BufferedWriter 具有与 BufferedInputStream 和 BufferedOutputStream 相似的用法，不再赘述。

【例题 11_6】 将 "E:\\乘法表.txt" 转存为 "D:\\chengfabiao.txt"。

```java
import java.io.BufferedReader;
import java.io.BufferedWriter;
import java.io.FileReader;
import java.io.FileWriter;
import java.io.IOException;
public class Ch11_6 {
    public static void main(String[] args){
        try {
        FileReader fr=new FileReader("E:\\乘法表.txt");
        BufferedReader br=new BufferedReader(fr);
        //用 BufferedReader 过滤初级流 FileReader 对象 fr,
        //以便使用 BufferedReader 的按行读方法
        FileWriter fw=new FileWriter("D:\\chengfabiao.txt");
        BufferedWriter bw=new BufferedWriter(fw);
        //用 BufferedWriter 过滤初级流 FileWriter 对象 fw
        String s=br.readLine();//按行读取文件内容
            while(s!=null){
              bw.write(s);
              bw.newLine();//换行
              s=br.readLine();  //读入下一行
            }
            bw.flush();
            br.close();
            bw.close();
        }
        catch (IOException e)  {}
    }
}
```

说明：由于 BufferedReader 的 rendLine()方法并不读入换行符，所以写入换行时须用 newLine()方法。

2. DataInputStream 与 DataOutputStream

为了增强"初级流"灵活读取各种数据类型数据的能力，可以用 DataInputStream 与 DataOutputStream。

DataInputStream 构造方法 DataInputStream（InputStream in）创建的数据输入流指向一个由参数 in 指定的底层输入流。

DataOutputStream 的构造方法 DataOutputStream（OutnputStream out）创建的数据输出流指向

一个由参数 out 指定的底层输出流。

DataInputStream 类提供了 readBoolean()、readByte()、readChar()、readInt()、readFloat()、read Double() 等常用方法，用于灵活读取各种基本数据类型数据。相应地，DataOutputStream 类提供了 writeBoolean(boolean v)、writeByte(int v)、writeChar(int v)、writeInt(int v)、writeFloat(float v)、writeDouble(double v)等常用方法。

此外，DataInputStream 的 readUTF(DataInput in)方法能从流 in 中读取用 UTF-8 格式编码的 Unicode 字符格式的字符串；DataOutputStream 的 writeUTF(String str)方法能以与机器无关方式使用 UTF-8 编码方式将一个字符串写入基础输出流，常常用于读写中文字符。

【例题 11_7】测试 DataInputStream 与 DataOutputStream 的常用方法。

```java
import java.io.*;
public class Ch11_7 {
  public static void main(String[] args){
  try
    {
     DataOutputStream out=new DataOutputStream(new BufferedOutputStream
            (new FileOutputStream("E:\\data.txt")));
     //建立输出流
    out.writeInt(100);
    out.writeLong(100L);
    out.writeFloat(12.3F);
    out.writeDouble(3.1425926);
    out.writeBoolean(true);
    out.writeUTF("您好");
    out.close();
    DataInputStream in=new DataInputStream(new BufferedInputStream
            (new FileInputStream("E:\\data.txt")));
    //建立输入流
    System.out.println(in.readInt());
    System.out.println(in.readLong());
    System.out.println(in.readFloat());
    System.out.println(in.readDouble());
    System.out.println(in.readBoolean());
    System.out.println(in.readUTF());
    }
    catch(IOException e){
    System.out.println(e.toString());}
    }
}
```

11.8 随机访问文件

前面介绍的各个输入、输出流类对文件的操作是顺序的。要实现随机访问文件，即要能在文件中随意地移动读取位置，需要用到 java.io.RandomAccessFile 类。

RandomAccessFile 类的构造方法：

RandomAccessFile(File file, String mode) 或 RandomAccessFile(String name, String mode)：以

File 对象 file（或文件名 name）指定了要操作的文件，参数 mode 指明了访问模式，其取值有 "r"（只读）、"rw"（读写）等。

要想实现对文件内容进行随机的读、写，需要对文件指针（读写头）进行移动，Java 中提供了如下的方法对文件指针进行操作。

- public long getFilePointer( )：获得当前的文件指针。
- public void seek(long pos)：移动文件指针到 pos 指定的位置。
- public int skipBytes(int n)：使文件指针向前移动指定的 n 个字节。

除了文件指针操作的方法外，RandomAccessFile 类还提供了一系列读写数据的方法，读者可以查阅 JDK 文档。

【例题 11_8】RandomAccessFile 方法测试。

```
import java.io.IOException;
import java.io.FileNotFoundException;
import java.io.RandomAccessFile;
public class Ch11_8 {
    public static void main(String[] args) {
    RandomAccessFile raf;
    try {
        raf = new RandomAccessFile("E:\\random.txt","rw");
        //以可读写方式创建RandomAccessFile对象
        String str ="ABCDEFGHIJKLMNOPQRSTUVWXYZ";
        raf.write(str.getBytes());
        System.out.println(raf.getFilePointer());//输出当前文件指针位置
        raf.seek(10);//移动文件指针到指定的位置
        String str1=raf.readLine();
        System.out.println(str1);
        raf.close();
    }
    catch (FileNotFoundException e) {}
    catch (IOException e) {}
    }
}
```

说明：RandomAccessFile 类的对象既可以充当输入流，也可以充当输出流，这取决于当前 RandomAccessFile 类的对象调用的是 readXXX()方法还是 writeXXX()方法，调用 readXXX()方法时充当输入流，调用 writeXXX()方法时充当输出流。

11.9　Serializable 接口与对象序列化

在一个程序运行的时候，其中的对象的相关信息是被保存在内存中的，程序一旦结束，对象就会被回收，对象的状态信息就会丢失。

但有时候，我们可能需要将对象的状态信息保存下来，然后在需要时再将对象恢复过来。如当两个进程在进行远程通信时，彼此可以发送各种类型的数据包括一个类对象。这时，就需要在发送前先将对象的状态信息转换为字节序列保存下来，以免丢失，该过程被称为对象的序列化过程，反之，接收方则需要把字节序列再恢复为 Java 对象，这一过程被称为对象的反序列化过程。

并非所有的对象都需要或者可以被序列化,但所有需要实现序列化的对象必须首先实现 java.io.Serializable 接口,这个接口中不含有任何的方法声明,是个空接口,其定义如下: public interface Serializable,因此,实现 Serializable 接口时,不需要编写任何的实现代码,只是说明该类可以被序列化。如果一个类可以序列化,它的所有子类都可以被序列化。

对象序列化和反序列化过程需要利用对象输出流类 ObjectOutputStream 和对象输入流类 ObjectInputStream,通过对象输出流将对象状态保存下来(保存到文件或者通过网络传送到其他地方),再通过对象输入流将对象状态恢复。

ObjectOutputStream 类的常用构造方法为 public ObjectOutputStream(OutputStream out) throws IOException:创建将序列化流写入底层流 OutputStream out 的 ObjectOutputStream 对象。当创建了 ObjectOutputStream 对象后,继而用 ObjectOutputStream 对象调用 public final void writeObject(Object obj)throws IOException 方法将指定的对象写入 ObjectOutputStream。

ObjectInputStream 的常用构造方法为 public ObjectInputStream(InputStream in)throws IOException:创建从指定 InputStream in 中读取序列化流的 ObjectInputStream 对象。当创建了 ObjectInputStream 对象后,继而用 ObjectInputStream 对象调用 public final Object readObject()throws IOException,ClassNotFoundException 方法从 ObjectInputStream 读取对象。

【例题 11_9】用对象输入、输出流读写序列化对象。

```java
import java.io.File;
import java.io.FileInputStream;
import java.io.FileNotFoundException;
import java.io.FileOutputStream;
import java.io.IOException;
import java.io.ObjectInputStream;
import java.io.ObjectOutputStream;
import java.io.Serializable;
class Student implements Serializable{//可被序列化的类
  private String stuno;
  private String stuname;
  private int stuage;
  public String getStuno() {
    return stuno;}
  public void setStuno(String stuno) {
    this.stuno = stuno;}
  public String getStuname() {
    return stuname;}
  public void setStuname(String stuname) {
    this.stuname = stuname;}
  public int getStuage() {
    return stuage;}
  public void setStuage(int stuage) {
    this.stuage = stuage;}
}
public class Ch11_9 {
    public static void main(String[] args) {
        ObjectInputStream objin;
```

```
    ObjectOutputStream objout;
    File f=new File("E:\\student.txt");
    try {
      objout=new ObjectOutputStream(new FileOutputStream(f));
      //以 E:\\student.txt 为保存对象序列化信息的对象输出流的目的地
      Student stu=new Student();//创建可被序列化的类对象
      stu.setStuno("0001");stu.setStuname("张三");stu.setStuage(18);
      objout.writeObject(stu);//保存 stu 对象状态
      objin=new ObjectInputStream(new FileInputStream(f));
      //创建对象输入流类对象
      try {
        Student stu1=(Student)objin.readObject();//还原对象状态
        System.out.println("学号: "+stu1.getStuno()+" 姓名: "+stu1.getStuname()+" 性别: "+stu1.getStuage());
        //输出被还原后的对象信息
      } catch (ClassNotFoundException e) {}
    }
    catch (FileNotFoundException e) {}
    catch (IOException e) {}
    }
}
```

程序执行结果如图 11.4 所示。

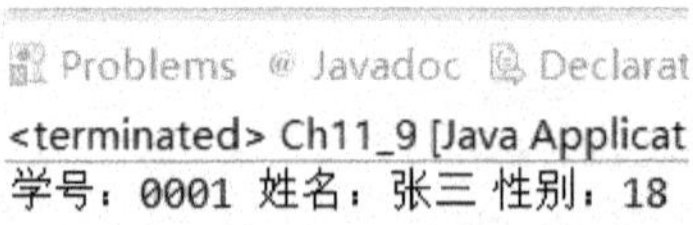

图 11.4 反序列化后对象信息

11.10 标准输入、输出流

Java 程序从键盘读入数据,或向屏幕输出数据是的操作是非常频繁的。为此,java.lang.System 类中定义了 3 个有用的静态类字段:public final static InputStream in、public final static PrintStream out、 public final static PrintStream err。这三个静态字段就是系统的标准流,其中,System.in 表示系统的标准输入流,此流对应于键盘输入或由主机环境或由用户指定的另一个输入源。System.out 表示系统的标准输出流,此流对应于显示器输出或由主机环境或由用户指定的另一个输出目标;System.err 表示系统的标准错误输出流,此流对应于显示器输出或由主机环境或由用户指定的另一个输出目标。

1. 标准输入

Java 的标准输入 System.in 是 InputStream 类的对象,当程序中需要从键盘读入数据的时候,只须调用 System.in 的 read()方法即可。

标准输入的例子请参看【例题 2_2】。

2. 标准输出

Java 的标准输出 System.out 是打印输出流 PrintStream 类的对象。PrintStream 是过滤输出类流 FilterOutputStream 的一个子类,System.out 可以调用 println、print 以及 printf 方法进行输出。

标准输出的例子请参看【例题 2_1】。

3．标准输入、输出重定向

在默认情况下，标准输入流从键盘读取数据，标准输出流和标准错误输出流向控制台输出数据。Java 的 System 类提供了一些以下静态方法完成对标准输入、输出和错误 I/O 进行重定向：

- public static void setIn(InputStream in)：对标准输入流重定向。
- public static void setOut(PrintStream out)：对标准输出流重定向。
- public static void setErr(PrintStream err)：对标准错误输出流重定向。

【例题 11_10】标准输入、输出流重定向测试。

```
import java.io.*;
public class Ch11_10 {
    public static void main(String[] args) throws IOException {
        BufferedInputStream in = new BufferedInputStream(new FileInputStream(
                "E:\\Java Pro\\Ch11\\src\\Ch11_10.java"));
        PrintStream out = new PrintStream(new BufferedOutputStream(new FileOutputStream("E:\\redirect.txt")));
        System.setIn(in);
        //对标准输入流重定向
        System.setOut(out);
        //对标输出流重定向
        BufferedReader br = new BufferedReader(new InputStreamReader(System.in));
        //包装标准输入流
        String s;
        while ((s = br.readLine()) != null)   //读取一行数据
            System.out.println(s);
        out.close();
    }
}
```

分析：本例中 System.setIn(in);将标准输入流重定向为从 E:\\Java Pro\\Ch11\\src\\Ch11_10.java 文件中读取数据，此处，E:\\Java Pro\\Ch11\\src\\Ch11_10.java 是指例题 11_10 源文件。System.setOut(out);是将标准输出流重定向为 E:\\redirect.txt 文件即用 System.out.println(s);

输出时，实际上是将读到的数据写入 E:\\redirect.txt 文件。试将程序中 System.setOut(out);一句注释掉，再次运行程序，观察结果的变化。

11.11 文件对话框

文件对话框是 Java 提供的对文件进行打开、保存操作时出现的对话框。javax.swing 包中的 JFileChooser 类可以帮助我们快速创建文件打开与文件保存对话框，并通过文件输入输出流的相关处理方法对文件对话框进行打开与保存操作。

JFileChooser 类的常用构造方法：

- public JFileChooser()：创建一个指向用户默认目录的 JFileChooser 对象。此默认目录取决于操作系统，如 C:\Users\lenovo。
- public JFileChooser(String currentDirectoryPath)：使用 currentDirectoryPath 给定的路径创建 JFileChooser 对象。

- public JFileChooser(File currentDirectory)：使用 File 对象 currentDirectory 给定的路径来创建 JFileChooser 对象。

JFileChooser 类的常用方法：

- public int showOpenDialog(Component parent)throws HeadlessException：显示用于打开文件的对话框。
- public int showSaveDialog(Component parent)throws HeadlessException：显示用于保存文件的的对话框。

当用户操作文件对话框的按钮或关闭对话框时会返回一个整数值,这个整数值的类型有三种，分别是 JFileChooser.CANCEL_OPTION（表示用户按下取消按钮）；JFileChooser.APPROVE_OPTION（表示用户按下确定按钮）；JFileChooser.ERROR_OPEION（表示有错误产生或是对话框不正常关闭）。利用这三个整数值我们就能判断用户到底在对话框中做了什么操作，并加以处理。如当用户选择了文件并按下确定按钮后，就可以利用以下方法获得文件对话框中所做的文件相关属性，进而进行文件输入输出流操作。

- public File getCurrentDirectory()：返回在文件对话框中选择的目录。
- public File getSelectedFile()：返回在文件对话框中选中的文件对象。
- public String getName(File f)：返回文件对象对应的文件名。

【例题 11_11】编写窗体程序，通过打开文件的对话框打开文件并将文件内容显示在文本区中，通过保存文件对话框将文本区中编辑的内容进行保存。

```java
import javax.swing.*;
import java.awt.BorderLayout;
import java.awt.event.ActionEvent;
import java.awt.event.ActionListener;
import java.io.*;
public class Ch11_11 {
    public static void main(String[] args) {
        new FileChooserTest();
    }
}
class FileChooserTest extends JFrame implements ActionListener{
    JButton open,save;
    JTextArea content;
    JScrollPane scroll;
    JLabel state;
    JPanel panel;
    JFileChooser fileChooser;
    FileReader fr=null;
    FileWriter fw=null;
    BufferedReader bin=null;
    BufferedWriter bout=null;
    public FileChooserTest()
    {   this.setTitle("文件对话框测试");
        open=new JButton("打开文件");
        save=new JButton("保存文件");
        panel=new JPanel();
        panel.add(open);panel.add(save);
```

```java
            content=new JTextArea();
            scroll=new JScrollPane(content);
            open.addActionListener(this);
            save.addActionListener(this);
            state=new JLabel("状态栏: ");
            this.add(panel,BorderLayout.NORTH);
            this.add(scroll,BorderLayout.CENTER);
            this.add(state,BorderLayout.SOUTH);
            fileChooser =new JFileChooser("E:\\");
            //创建一个 FileChooser 对象,并指定默认文件对话框路径。
            this.setSize(320,180);
            this.setVisible(true);
            this.setDefaultCloseOperation(JFrame.EXIT_ON_CLOSE);
        }
        /*文件对话框事件处理代码**/
        public void actionPerformed(ActionEvent e){
            if(e.getActionCommand().equals("打开文件")){
            content.setText(null);//清空文本区
            int n=fileChooser.showOpenDialog(this);//依托当前窗体弹出文件打开对话框
            try {
                if(n==JFileChooser.APPROVE_OPTION){
                File path=fileChooser.getCurrentDirectory();//得到用户在对话框中选择的路径
                String fileName=fileChooser.getSelectedFile().getName();
                File f=new File(path,fileName);
                bin=new BufferedReader(new FileReader(f));
                String str=null;
                while((str=bin.readLine())!=null){
                  content.append(str+"\n");
                  state.setText("您选择打开的文件路径为: "+path+"您选择打开的文件名称为: "+fileName);
                }
                }
                else if(n==JFileChooser.CANCEL_OPTION)
                    state.setText("您没有选择任何文件");
                    bin.close();
                }
            catch (FileNotFoundException exp) {}
            catch (IOException exp) {}
        }
            /*保存文件**/
            if(e.getActionCommand().equals("保存文件")){
                int n=fileChooser.showSaveDialog(this);
                try {
                    if(n==JFileChooser.APPROVE_OPTION){
                    File path=fileChooser.getCurrentDirectory();
                    String fileName=fileChooser.getSelectedFile().getName();
                    File f=new File(path,fileName);
                    bout=new BufferedWriter(new FileWriter(f));
                    bout.write(content.getText());
```

```
                state.setText("您选择保存文件的路径为: "+path+"您保存的文件名称为:
"+fileName);
                }
                else if(n==JFileChooser.CANCEL_OPTION)
                    state.setText("您没有保存任何文件");
                bout.close();
            }
            catch (IOException exp) {}
        }
    }
}
```

程序运行结果如图 11.5 所示。

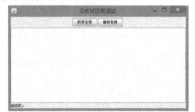 

图 11.5　文件的打开与保存

11.12　用 Desktop 类打开文件

java.awt.Desktop 类是 jdk1.6 后新增的类，Desktop 类允许 Java 应用程序启动已在本机桌面上注册的关联应用程序，完成打开系统默认的浏览器浏览网页，打开系统默认的邮件客户端，打开，编辑，打印本地文件等功能。此处只介绍其打开文件的功能。

可以用 public static boolean isDesktopSupported()方法测试当前平台是否支持此类；可以用 public static Desktop getDesktop()方法返回 Desktop 实例；用 Desktop 类 public void open(File file)throws IOException 方法启动关联应用程序来打开文件。

【例题 11_12】Desktop 类测试。
```
import java.awt.Desktop;
import java.io.File;
public class Ch11_12 {
    public static void main(String[] args) throws Exception{
        Desktop.getDesktop().open(new File("E:\\aaa.pdf"));
    }
}
```

小　　结

本章主要介绍了 Java 输入、输出流的概念；File 类的常用方法的应用；文件字节、字符输入、输出流等初级流类的典型应用方法；过滤流类等高级流类对初级流功能增强的方法；随机访问文件流类的应用；对象序列化的概念及方法；标准输入输出流的应用；文件对话框的实现等。正确理解各种流类的功能并将其分类，掌握其各自的常用方法及典型应用是学习的重点。

习 题 11

上机实践题

1. 编写程序，输出 E 盘根目录下的所有文件和文件夹的名称。

2. 在 E 盘根目录下下有一个文本文件目录.txt，在其中有如下表示章节标题的 4 行文本：

Java 简介与 Java SE 程序开发环境；

Java 语法基础；

Java 面向对象程序设计基础；

Java 继承与多态。

编写程序现将文件中每行文本读出，并给每行文本前加上第 1 章、第 2 章等并将修改后的文本写入目录 1.txt 中。

3. 编程求解正整数 m 到 n 之间的所有素数并将这些素数按照每行 k 个的形式写入文件 E:\\prime.txt 中，要求 m，n，k 通过命令行参数得到。再将 E:\\prime.txt 中的内容读取到控制台输出。

4. 自定义 Customer 类，让该类实现 Serializable 接口，该类中包含成员变量 String name，int age；一个含参构造方法 public Customer（String name, int age）；同时，重写 Object 类的 toString()方法，用于返回 Customer 对象的姓名及年龄。定义主类 XITi11_1_4，在 main()方法中，运用 ObjectOutputStream 将 Customer 对象、一个字符串"欢迎您！"及一个 Date 类对象（已知系统类 String 和 Date 实现过 Serializable 接口）分别写入文件 E:\\objectFile.txt，然后用 ObjectOutputStream 从 E:\\objectFile.txt 中读出这些对象并输出在控制台。

5. 已知 javax.swing 包中的 ProgressMonitorInputStream 类是一个带进度条的文件输入流类。请编程实现读取某文件如第 3 小题中的 E:\\prime.txt，并显示读取该文件的进度条。提示，为了更加清楚地看到读取进度，运用多线程类 Thread 调用其静态方法 sleep（单位：毫秒）来延缓读取过程。

第 12 章　Java 多线程编程

【本章内容提要】

- 程序、进程与线程；
- Java 多线程机制；
- Java 多线程实现的方法；
- 线程的生命周期与状态转换；
- Java 多线程调度机制；
- Thread 类；
- 线程的让步；
- 线程的联合；
- 多线程的互斥与同步；
- 守护线程；
- 线程之间的通信流类。

在前面各个章节中，我们编写的程序都是单线程的，即一个程序只有一条从头至尾的执行线索。然而现实问题中，很多过程都需要多条执行线索同时进行如一个网络服务器可能需要同时处理多个客户端的请求等。Java 的多线程机制支持用户编写多线程程序，允许用户程序中同时存在几条不同的执行线索共同执行。

12.1　程序、进程与线程

程序是一段静态的代码，它是应用软件执行的脚本，程序是一个静态概念。

进程是操作系统调度和执行的基本单位，是程序的一次动态执行过程，对应了从代码加载、执行到执行完毕的一个完整的过程。线程是比进程更小的调度和执行单位，一个进程可以划分为多个线程。

每个进程都有独立的代码和数据空间(进程上下文)，进程间的切换会有较大的开销。线程可以看成时轻量级的进程，同一类线程共享代码和数据空间，每个线程有独立的运行栈和程序计数器（PC），线程切换的开销小。

多线程是这样一种机制，它允许在程序中并发执行多个指令流，每个指令流都称为一个线程，彼此相互独立。

12.2　Java 多线程机制

　　Java 支持用户编写多线程应用程序，即一个程序中安排多条执行线索，每个执行线索对应一个线程，多个线程独立、并发地执行各自的任务。

　　每个 Java 应用程序都有一个默认的主线程。Java 应用程序总是从主类的 main 方法开始执行。当 JVM（Java 虚拟机）加载代码并发现 main 方法之后，就会启动一个线程，这个线程称作"主线程"，该线程负责执行 main 方法。当然，用户可以在程序中创建其他线程。如果 main 方法中没有创建其他的线程，那么当 main 方法执行完最后一个语句，即 main 方法返回时，JVM 就会结束 Java 应用程序。如果 main 方法中又创建了其他线程，那么 JVM 就要在主线程和其他线程之间轮流切换，保证每个线程都有机会使用 CPU 资源，main 方法即使执行完最后的语句（主线程结束），JVM 也不会结束程序，JVM 一直要等到程序中的所有线程都结束之后，才结束 Java 应用程序。

12.3　Java 多线程实现的方法

　　Java 中多线程的实现方法有两种：一是继承 java.lang.Thread 类；二是实现 java.lang.Runnable 接口。

　　通过重写 Thread 类的 run()方法或 Runnable 接口中的 run()方法实现该线程的具体功能。基本的格式如下：

1. 继承 Thread 类

```
class MyThread extends Thread {
  public void run(){//重写 Thread 的 run()方法
    …//线程的具体任务代码
  }
}
public class ThreadTest{
    public static void main(String[] args) {
        MyThread t=new MyThread();//创建线程对象 t
        t.start();//使得线程对象插入就绪队列排队
    }
}
```

2. 实现 Runnable 接口

```
class MyThread implements Runnable{
public void run(){//实现 Runnable 接口的 run()方法
    …//线程的具体任务代码
  }
}
public class ThreadTest
{ public static void main(String[] args) {
      Thread t=new Thread(new MyThread());
      /*以实现过 Runnable 接口的 MyThread 类对象作为
      Thread 构造方法参数，创建线程对象 t*/
      t.start();//使得线程对象插入就绪队列排队
```

        }
    }

注意：以上两种创建 Java 多线程的方法也可以描述为：方法一，给 Thread 类的子类创建对象；方法二，给 Thread 类创建对象，但其构造方法参数必须是一个实现过 Runnable 接口的类对象。

【例题 12_1】 通过给 Thread 类的子类创建对象实现多线程。

```java
class MyThread1 extends Thread {      // Thread类的子类
    public void run() {
        for(int i=1;i<=3;i++)
            System.out.println("MyThread1第"+i+"次执行");
    }
}
class MyThread2 extends Thread {      // Thread类的子类
    public void run() {
        for(int i=1;i<=3;i++)
            System.out.println("MyThread2第"+i+"次执行");
    }
}
public class Ch12_1{
   public static void main(String args[]){
     MyThread1 t1=new MyThread1();
     MyThread2 t2=new MyThread2();
//创建线程对象t1,t2
     t1.start();
     t2.start();
//t1,t2插入就绪队列排队
     for(int i=1;i<=3;i++){              //main线程的任务
         System.out.println("main线程第"+i+"次执行");
     }
   }
}
```

分别执行该程序两次，得到以下结果，如图 12.1 所示。

图 12.1 多线程输出

若多次执行该程序，会发现每次的执行结果不尽相同，输出结果依赖于当前 CPU 资源的使用情况。

【例题 12_2】 改写例题 12_1，通过创建 Thread 类对象实现多线程。

```java
class MyThread3 implements Runnable {
    public void run() {
```

```
      for(int i=1;i<=3;i++)
         System.out.println("MyThread1 第"+i+"次执行");
   }
}
class MyThread4 implements Runnable {
   public void run() {
      for(int i=1;i<=3;i++)
         System.out.println("MyThread2 第"+i+"次执行");
   }
}
public class Ch12_2{
public static void main(String args[]){
   Thread  t1=new Thread(new MyThread3());
   Thread  t2=new Thread(new MyThread4());
   //创建线程对象t1,t2
   t1.start();
   t2.start();
   //t1,t2 插入就绪队列排队
   for(int i=1;i<=3;i++){
      System.out.println("main 线程第"+i+"次执行");
   }
}
}
```

12.4 线程的生命周期与状态转换

一个线程产生后，就处于整个线程生命周期中的某个状态。一个线程对象生命周期中的状态有：

（1）新建状态。当创建一个线程对象后，该线程即处于新建状态，但此时该线程并未开始执行。

（2）就绪状态。当线程对象调用 start()方法后，该线程插入就绪队列排队，等待被调度获得 CPU 资源，称该线程对象对象处于就绪状态。

（3）运行状态。一旦某线程对象获得 CPU 资源，线程就调用自己的 run()方法，进入运行状态。直到完成整个任务进入死亡状态或由于其他原因进入阻塞状态。

（4）阻塞状态。阻塞状态又称为不可运行状态。引起线程进入阻塞状态的原因很多。例如：

- 通过调用 sleep(milliseconnds)使线程进入休眠状态，线程在指定的时间内不会运行。
- 通过调用 wait()方法使线程挂起，直到线程得到了 notify()或 notifyAll()消息，线程才会进入就绪状态。
- 线程在等待某个输入、输出完成。
- 线程试图在某个对象上调用同步控制方法，但是对象锁不可用，因为另一个对象已经获取了这个锁。

（5）死亡状态。死亡状态是当线程完成了所有任务而自然死亡，或被强制性地终止，如调用 Thread 类中的 destroy()或 stop()方法等。

线程生命周期中涉及的各个方法的用法将在后续小节中详细介绍。

12.5　Java 多线程调度机制

Java 中每个线程都有一个优先级，Java 线程的优先级由 Thread 类 setPriority (int newPriority) 方法进行设定，int newPriority 的取值为 1～10，取值越大表明线程的优先级越高。Thread 类的常量 MAX_PRIORITY，MIN_PRIORITY 以及 NORM_PRIORITY 分别对应 10、1 和 5，若不给某个线程具体指定优先级，则该线程默认有限级别为 NORM_PRIORITY。需要注意的是有些操作系统只能识别 3 个级别：1、5、10。

在 Java 中，JVM（Java 虚拟机）的职能之一就是负责管理和调度 Java 多线程，多个线程处在线程池中等待被调度。在采用时间片轮转调度的操作系统中，每个线程都有机会获得 CPU 的使用权，执行线程中的操作。当线程使用 CPU 资源的时间到达后，即使线程没有完成自己的全部操作，JVM 也会中断当前线程的执行，把 CPU 的使用权切换给线程池中下一个排队等待的线程，当前线程将等待下一次 CPU 得到时间片，然后从中断处继续执行，直到线程池中的所有就绪线程执行完毕。

从优先级高低的角度看，JVM 的调度策略似乎是抢占式调度，即先保证高优先级线程执行完毕，再执行低优先级的线程任务。但事实上，只能保证较高优先级的线程得到相对较多的执行机会，低优先级线程较少的执行机会。从时间片轮转调度的角度看，一旦时间片有空闲，则使具有同等优先级的线程以轮流的方式使用时间片。

Java 多线程的调度过程中，不仅受到 JVM 调度策略的影响，也受限于当前操作系统调度策略和 CPU 使用情况。

【例题 12_3】给线程设置优先级。
```
class PriorityThread implements Runnable{
  public void run() {
   for(int i=1;i<=3;i++){
       System.out.println("优先级为"+Thread.currentThread().getPriority()+
"的线程第"+i+"次执行");
    }
   }
}
public class Ch12_3 {
public static void main(String args[]){
   Thread t1=new Thread(new PriorityThread());
   Thread t2=new Thread(new PriorityThread());
   Thread t3=new Thread(new PriorityThread());
   t1.setPriority(Thread.MAX_PRIORITY);
   t2.setPriority(Thread.NORM_PRIORITY);
   t3.setPriority(Thread.MIN_PRIORITY);
   t1.start();
   t2.start();
   t3.start();
  }
}
```
程序的几种可能执行结果如图 12.2（a）、(b)、(c) 所示。

| (a) | (b) | (c) |

图 12.2 不同优先级的多线程执行情况

分析：图 12.2（a）是按照高优先级线程先执行，中等优先级次之，低优先级线程最后执行的理想情况；图 12.2（b）、（c）显示的执行次序则由随机性。因此，仅通过设定线程的优先级无法达到精准控制线程执行顺序的目的。

12.6 Thread 类

Thread 类包含在 java.lang 中，其原型为：public class Thread extends Object implements Runnable。

Thread 类的构造方法：

- public Thread()：创建一个线程对象。
- public Thread(Runnable target)：创建一个线程对象，target 为一个实现过 Runable 接口的类对象。
- public Thread(String name)：创建一个名为 name 的线程对象。
- public Thread(Runnable target,String name)：以一个实现过 Runable 接口的类对象 target 创建一个名为 name 的线程对象。

Thread 类的常用方法：

- public static Thread currentThread()：返回对当前正在执行的线程对象的引用。
- public final String getName()：返回该线程的名称。
- public void interrupt()：intertupt()方法经常用来"吵醒"处于休眠状态的线程。但要注意，该"吵醒"行为是处于休眠状态的线程对象自己调用 interrupt()方法来吵醒自己，重新排队等待 CPU 资源。
- public final boolean isAlive():判断是否处于活动状态。
- public final void join()throws InterruptedException：等待该线程终止。
- public void run()：若以创建 Thread 类的子类对象方式创建线程对象，则该子类应重写 Thread 类的 run()方法；若以创建实现 Runnable 接口的方式创建线程对象，则实现 Runnable 接口的类应实现 Runnable 接口中的 run()方法。run()方法用以定义线程对象被调度之后所执行的操作，run()方法是由系统自动调用的。
- public final void setDaemon(boolean on)：将该线程标记为守护线程或用户线程。
- public final void setName(String name)：给线程设定新名称
- public final int getPriority()：返回线程的优先级。

- public final void setPriority(int newPriority)：给线程设定优先级。
- public void start()：将线程对象插入就绪队列排队。
- public static void sleep(long millis)throws InterruptedException ：让当前线程休眠 millis（ 单位：毫秒 ）。
- public static void yield()：主动让出运行权，让其他线程得以执行。

【例题 12_4】设置线程的名字。

每个线程都具有默认的名字。在不给某个线程对象设置名称时，系统会给用户创建的每个线程对象默认的名称，如第一个线程名为"Thread-0"，第二个线程名为"Thread-1"等。主线程的默认名称为"main"。

```
public class Ch12_4{
    public static void main(String args[]){
        MyThread mt = new MyThread() ;
        /*创建线程对象t1,t2,系统自动给其设置名称**/
        Thread t1=new Thread(mt);
        t1.start() ;
        Thread t2=new Thread(mt);
        t2.start();
        /*创建线程对象t3,t4 并在Thread构造方法中给线程命名**/
        Thread t3=new Thread(mt,"线程A");
        t3.start();
        Thread t4=new Thread(mt,"线程B");
        t4.start();
        /*创建线程对象t5,用 setName()方法给线程命名**/
        Thread t5=new Thread(mt);
        t5.setName("线程C");
        t5.start();
    }
}
class MyThread implements Runnable{
    public void run(){
        for(int i=0;i<2;i++){
            System.out.println(Thread.currentThread().getName()+ "第"+i+"次运行") ;
        }
    }
}
```

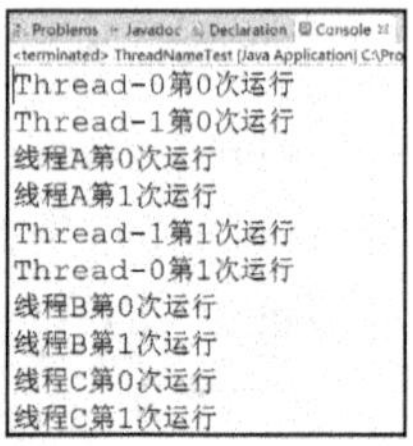

一种可能的执行结果如图 12.3 所示。

图 12.3 设置了名字的线程

说明：在创建一个线程对象的时候，可以在 Thread 构造方法中给线程对象命名如 Thread t3=new Thread(mt,"线程 A"); 也可以用 setName("线程名")方法给一个线程对象命名，如 t5.setName("线程 C"); Thread.currentThread().getName()表示获得当前运行线程的名字。

【例题 12_5】通过 sleep()方法调节线程执行次序。

```
public class Ch12_5 {
    public static void main(String[] args) {
        Thread t1 = new Thread(new SleepThread1());
        Thread t2 = new Thread(new SleepThread2());
        t1.setName("线程1");
```

```
        t2.setName("线程2");
        t1.start();
        t2.start();
    }
}
class SleepThread1 implements Runnable {
    public void run() {
        for (int i=1; i<=5; i++) {
            System.out.println(Thread.currentThread().getName()+"第"+i+"次执行");
            try {
                Thread.sleep(10);//休眠10 ms
            } catch (InterruptedException e) {
                e.printStackTrace();
            }
        }
    }
}
class SleepThread2 implements Runnable {
    public void run() {
        for (int i=1; i<=5;i++) {
            System.out.println(Thread.currentThread().getName()+"第"+i+"次执行");
            try {
                Thread.sleep(10);//休眠10 ms
            } catch (InterruptedException e) {
                e.printStackTrace();
            }
        }
    }
}
```

执行结果如图 12.4 所示。

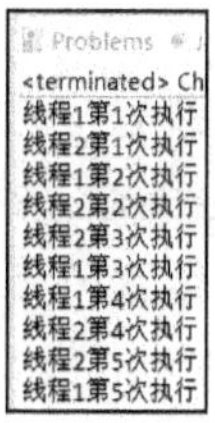

图 12.4 休眠的线程执行

分析：由于线程 1、线程 2 在各自的任务体中都进行了休眠 10 ms 的操作，即每进行一次输出，就休眠，从而让其他线程得以执行，达到调节线程执行次序的目的。

【例题 12_6】 线程的休眠与休眠中断。

```
class SleepThread implements Runnable{
    public void run(){
        System.out.println(Thread.currentThread().getName()+"准备先休眠10秒,再工作。");
        try{
            Thread.sleep(10000);// 线程休眠10 s
            System.out.println("提前结束了休眠,开始工作。") ;
        }
        catch(InterruptedException e){
        System.out.println("休眠被中断...") ;}
        for(int i=0;i<2;i++){
            System.out.println(Thread.currentThread().getName()+"第"+i+"次执行");
        }
    }
}
public class Ch12_6{
```

```
        public static void main(String args[]){
            SleepThread thread = new  SleepThread();
            Thread t=new Thread(thread,"偷懒的线程");
            t.start();
            try {
            Thread.sleep(1000);//主线程休眠1 s
            }
            catch (InterruptedException e) {}
            t.interrupt() ;     //中断线程的休眠
        }
    }
```

执行结果如图 12.5 所示。　　　　　　　　　　　　　　　　图 12.5　线程休眠与中断

说明：运行该程序时发现线程 t 在 1 s 后立刻就执行完毕了，而不是休眠 10 s 后才执行。这是因为当主线程休眠 1 s 后，立即中断了线程 t 的休眠。

sleep 方法的调用可能发生 InterruptedException 异常，顾名思义，即当一个线程在休眠过程中，如被 interrupt()方法"吵醒"，则发生 InterruptedException 异常并中断本次休眠。

要注意的是，休眠中断动作是线程自己"吵醒"了自己如 t.interrupt()。

12.7　线程的让步

当前运行的线程调用 yield()方法，使它暂时放弃 CPU，给其他线程一个执行的机会，线程调用 yield()，并不一定保证它会放弃 CPU 给其他线程。Java 的线程调度器根据调度算法，仍可能使它继续占用 CPU 运行。若该线程的优先级高，即使它调用 yield()，低优先级的线程仍不会占用 CPU 运行。因为 yield()方法只给相同或更高优先级的线程以执行的机会，如果当前系统中没有相同或更高优先级的线程，该方法调用不会产生任何效果，当前线程继续执行。

【例题 12_7】 通过 yield()方法调节线程执行次序。

```
public class Ch12_7 {
    public static void main(String[] args) {
      Thread t1 = new Thread(new YieldThread());
      Thread t2 = new Thread(new YieldThread());
      t1.setName("线程 1");
      t2.setName("线程 2");
      t1.start();
      t2.start();
    }
}
class YieldThread  implements Runnable {
    public void run() {
      for (int i=1; i<=5; i++) {
        System.out.println(Thread.currentThread().getName()+"第"+i+"次执行");
        Thread.yield();//暂停当前正在执行的线程
      }
    }
}
```

两种可能的执行结果如图 12.6 所示。

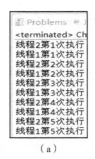
（a）

（b）

图 12.6 让步的线程执行

分析：图 12.6（a）的执行结果显示了使用 yield()方法使得当前线程暂停执行，从而让其他线程有机会得到执行的理想情况。但当一个线程对象用 yield()方法后，仍有可能继续得到执行，如图 12.6（b）所示。

12.8 线程的联合

一个线程 A 在占有 CPU 资源期间，可以让其他线程调用 join()和本线程联合，如 B.join();称 A 在运行期间联合了 B。如果线程 A 在占有 CPU 资源期间一旦联合 B 线程，那么 A 线程将立刻中断执行，一直等到它联合的线程 B 执行完毕，A 线程再重新排队等待 CPU 资源，以便恢复执行。

【例题 12_8】 线程联合测试。

```
public class Ch12_8{
    public static void main(String args[]){
        JoinThread mt=new JoinThread() ;
        Thread t=new Thread(mt);
        t.start();
        for(int i=1;i<=5;i++){
            System.out.println("main 线程运行第"+i+"次输出") ;
            if(i>2){
                try{
                    t.join();//主线程联合了线程 t
                }
                catch(InterruptedException e){}
            }
        }
    }
}
class JoinThread implements Runnable{
    public void run(){
        for(int i=1;i<=5;i++){
            System.out.println(Thread.currentThread().getName()+"第"+i+"次输出") ;
        }
    }
}
```

一种可能的执行结果如图 12.7 所示。

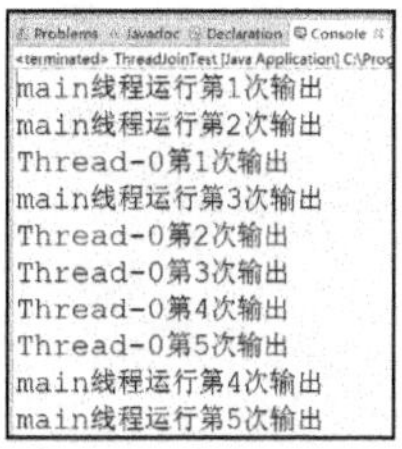

图 12.7 线程联合

说明：由于 main 线程在执行过程中当 i>2 时联合了 t 线程，导致的结果是 main 线程暂停执行，等待 t 线程执行完毕后再继续执行。

12.9 多线程的互斥与同步

12.9.1 线程的互斥

当多个线程竞争使用一个共享资源如内存中的变量时，需要保证在一个时刻只能有一个线程操作该共享变量即互斥地访问和操作该变量。

若所有的线程都是独立地与其他线程没有任何联系地运行，则问题就会简单得多，以至不用加以特别考虑，但实际情况要复杂得多。最常见的典型情况之一是，两个或多个线程共享同一个数据区，有些线程对该数据区写，有些线程对该数据区读，还有些线程对该数据区既读又写，这个共享的数据区就是一个共享变量。

在线程（进程）互斥问题上，涉及一个"临界区"的概念。所谓临界区，就是读、写同一个共享变量的程序段。

Java 提供了 synchronized（同步关键字）关键字实现了线程互斥访问共享资源，确保任何时刻只能有一个线程对共享资源进行访问或操作。

synchronized 关键字可以修饰需要互斥操作的方法，或修饰一段代码形成 synchronized 语句块，相当于给"临界区"加上一把锁，保证在任何时刻只能有一个线程对该方法或代码块进行操作，当任务执行结束时，自动解锁。

synchronized 有两种格式：

格式 1：
```
synchronized(任何对象){
   //访问共享变量的临界区(程序段)，又称同步代码块
}
```
格式 2：同步化方法。在方法的前面加上 synchronized，如：
```
public synchronized void add() {
    //临界区
}
```

【例题 12_9】模拟两个售票窗口同时售票。
```
class SaleTicket implements Runnable{
    private int ticketNum =10;//有10张票
    public void run(){
        while(ticketNum>0){
            if(Thread.currentThread().getName().equals("售票口1"))
                saleTicket();
            else if(Thread.currentThread().getName().equals("售票口2"))
                saleTicket();
        }
    }
    public synchronized void saleTicket(){
      String tName = Thread.currentThread().getName();
      if(ticketNum<=0)
          System.out.println(tName+":票已售完");
```

```java
        else{
            ticketNum-=1;  //票的数量减一
            try {
                Thread.sleep(200);// 当前线程休眠200 ms
                } catch (Exception ex) {}
            System.out.println(tName+"售出一张票,还剩"+ticketNum+"张票");
            }
        }
    }
}
public class Ch12_9 {
    public static void main(String[] args){
        SaleTicket st = new SaleTicket();
        Thread t1 = new Thread(st,"售票口 1");
        Thread t2 = new Thread(st,"售票口 2");
        t1.start();
        t2.start();
    }
}
```

程序执行可能结果如图12.8和图12.9所示。

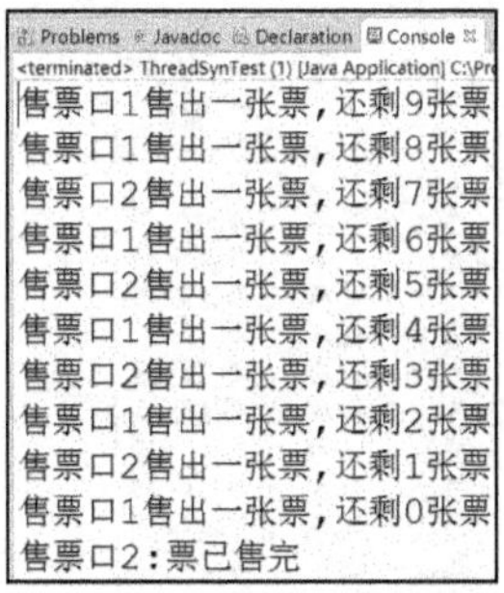

图 12.8　线程的同步 　　　　　　　图 12.9　没有被同步的线程

分析：程序中售票窗口1线程t1和窗口2线程t2共享变量ticketNum，对ticketNum应保证互斥地操作，因此在操作ticketNum变量的方法saleTicket()前加上关键字synchronized进行修饰，正确的结果如图12.8所示。

若去掉saleTicket()方法前的synchronized关键字，则程序的运行结果会发生混乱，如图12.9所示。

将上面程序改写如下并测试：

```java
class SaleTicket implements Runnable {
    private int ticketNum =10;
    public void run() {
        while (true) {
            String tName = Thread.currentThread().getName();
            synchronized(this) {
                if (ticketNum <=0) {
                    System.out.println(tName+":票已售完");
                    break;
                } else {
                    ticketNum-=1;                       //票的数量减1
```

```
                    try {
                        Thread.sleep(200);        // 程序休眠200 ms
                    } catch (Exception ex) {
                    }
                    System.out.println(tName+"售出一张票,还剩"+ticketNum+"张票");
                }
            }
        }
    }
}
public class Ch12_9 {
    public static void main(String[] args) {
        SaleTicket st = new SaleTicket();
        Thread th1 = new Thread(st, "售票口 1");
        Thread th2 = new Thread(st, "售票口 2");
        th1.start();
        th2.start();
    }
}
```

程序执行可能结果如图 12.10 和图 12.11 所示。

图 12.10 线程的同步   图 12.11 没有被同步的线程

12.9.2 互斥线程的协调

线程间常常存在相互协调配合的关系，如当一个线程使用的同步方法中用到某个变量，而此变量又需要其他线程修改后才能符合本线程的需要，那么可以在同步方法中使用 wait()方法。使用 wait()方法可以中断方法的执行，使本线程等待，暂时让出 CPU 的使用权，并允许其他线程使用这个同步方法。其他线程如果在使用这个同步方法时不需要等待，那么它使用完这个同步方法的同时，应当用 notify()或方法 notifyAll()方法通知其他等待线程结束等待。如果使用 notifyAll()方法，就会通知所有的由于使用这个同步方法而处于等待的线程结束等待。曾中断的线程就会从刚才的中断处继续执行这个同步方法，并遵循"先中断先继续"的原则。如果使用 notify()方法，那么只是通知处于等待中的某一个线程结束等待。

wait()、notify()和 notifyAll()都是 Object 类中的 final 方法，被所有的类继承且不允许重写的方法。

其中，wait()方法有两种格式。

格式 1：
public final void wait() throws InterruptedException

只有拥有该对象的"对象锁"的线程才能调用该对象的 wait()方法。该方法的功能是,使调用者(线程)释放该"对象锁",并进入"阻塞"状态,Java 系统将这个调用者(线程)放入该对象的 wait 等待队列中。当另外一个线程调用该对象的 notify()、notifyAll()方法时,唤醒处于这个对象的 wait 等待队列中的线程,进入运行态。线程唤醒后能否沿原来断点处继续执行,取决于该线程能否重新得到该对象的"对象锁"。若得不到对象锁,则根据 synchronized 的获取对象锁的机制,该线程将进入"阻塞"状态,并被放入该对象的对象锁等待队列中。当其他线程归还对象锁时会自动唤醒它。wait()方法使该线程只释放这个对象的对象锁并进入这个对象的 wait 等待队列中,若该线程同时还拥有其他对象的对象锁,这些对象锁不会被释放。

格式 2:
```
public final void wait(long timeout) throws InterruptedException
```
含义以格式一样,只是增加了当指定的时间一到,线程被唤醒,进入运行态。然后线程试图重新获取对象锁。只有获取到对象锁,才能继续原先的断点往下执行。

notify()方法的格式为:public final void notify(),只有拥有该对象的"对象锁"的线程才能调用该对象的 notify()方法方法。该方法的功能是,从该对象的 wait 等待队列中选择一个线程唤醒它,选择的算法由具体实现者决定,可简单认为是从队列中任意选择一个线程。大部分情况下,wait()与 notify()或 notifyAll()是配套成对使用的。若对一个 wait(),程序员忘记用相应的 notify()或 notifyAll()来唤醒,则极大地增加产生死锁的概率。

考虑到尽可能降低死锁产生的潜在可能性,通常建议使用 notifyAll(),其格式为:
```
public final void notifyAll()
```
在使用 wait()、notify()或 notifyAll()时要注意:
- 必须保证,每一个 wait()都有相应的 notify()或 notifyAll()与之配合使用。
- 只有拥有该对象的对象锁的线程才能调用 wait()、notify()或 notifyAll()方法。
- wait()、notify()或 notifyAll()方法必须且只能放在 synchronized 代码块或方法中。

若一个 Java 程序的所有线程都因为申请不到它们所需要的资源而全部进入"阻塞"状态时,该 Java 程序将被挂起,程序再被不能继续执行,这种现象称为死锁。要线程间有协作关系的编程实践中,要避免死锁现象的发生。

【例题 12_10】用多线程模拟"生产者-消费者"问题。

生产者-消费者问题描述的是生产者在生产产品,并将这些产品提供给消费者去消费,在两者之间设置了仓库,生产者将它所生产的产品放入一个仓库中,消费者可从仓库中取走产品去消费,但它们之间必须保持同步。

生产者和消费者之间的关系如下:
- 生产者生产前,若产品没有被消费(仓库满),则生产者等待;生产者生产产品后,通知消费者消费。
- 消费者消费前,若产品已经被消费完(仓库空),则消费者等待;消费者消费后,通知生产者生产。
- 生产者向仓库存放产品与消费者从仓库取走产品的操作不能在同一时刻同时进行。

编程思路:定义生产者、消费者线程分别模拟生产者与消费者,设置一个共享的存储空间模拟仓库;运用等待、通知的线程协调机制,等待使用 wait()方法,通知使用 notifyAll()或者 notify()方法;运用 synchronized 关键字同步生产者、消费者线程即在生产者向仓库存放产品时,消费者不能同时从仓库取走产品，反之亦然。

【例题 12_11】"生产者-消费者"问题模拟：以自定义的"栈"模拟仓库，入栈操作模拟生产者线程将生产的产品放入仓库，出栈操作模拟消费者线程从仓库取走产品。

```java
public class ProducerConsumerTest {
    public static void main(String []args){
        Stack s = new Stack();
        Thread p=new Thread(new Producer(s),"生产者1");
        Thread c=new Thread(new Consumer(s),"消费者1");
        p.start();
        c.start();
    }
}
class Stack{///定义栈类
    private int index=0;//栈顶
    private int s[]=new int[5];//栈空间
    public synchronized void push(int i)//入栈操作
    {
        while(index>4){//栈满
            try {
              System.out.println("栈满"+Thread.currentThread().getName()+"等待……");
                this.wait();//生产者等待
            } catch (InterruptedException e) {
                e.printStackTrace();
            }
        }
        s[index]=i;//将 i 压入栈
        index++;//栈顶增 1
        System.out.println(Thread.currentThread().getName()+"生产了: "+i);
        this.notifyAll();//当生产者在栈中压入数据后通知消费者从栈中取数据
    }
    public synchronized int pop()//出栈操作
    {
        while(index<=0){//栈空
            try {
              System.out.println("栈空"+Thread.currentThread().getName()+"等待……");
                this.wait();//消费者等待
            } catch (InterruptedException e) {
                e.printStackTrace();
            }
        }
        index--;//栈顶减 1
        System.out.println(Thread.currentThread().getName()+"消费了: "+s[index]);
        this.notifyAll();//当消费者取走栈中的数据后通知生产者在栈中压入数据
        return s[index];
    }
}
class Producer implements Runnable{///生产者线程类
    Stack stack;
    public Producer(Stack stack){
```

```
            this.stack=stack;
        }
        public void run(){
            for(int i=0;i<7;i++){  //生产7个产品
               stack.push(i);
               try {
                   Thread.sleep(100);
               } catch (InterruptedException e) {}
            }
        }
    }
    class Consumer implements Runnable{//消费者线程类
        Stack stack;
        public Consumer(Stack stack){
            this.stack=stack;
        }
        public void run(){
            for(int i=0;i<7;i++){//消费7个产品
                stack.pop();
                try {
                    Thread.sleep(100);
                } catch (InterruptedException e) {}
            }
        }
    }
```

一种可能的输出结果如图12.12所示。　　　　　图12.12　生产者-消费者线程输出

说明：本例生产者线程和消费者线程生产、消费的产品是数字，如生产7个产品0、1、2、3、4、5、6，将它们压入栈，消费产品就是从栈中弹出这些数字。注意：并不要求生产者线程将栈压满，只要栈中有产品，就可以通知消费者线程来取走。

本例中生产者和消费者线程共享栈空间和index变量，因此要用synchronized关键字修饰对栈和index操作的方法push()和pop()，以便保证同步。

12.10　守　护　线　程

Java线程可以分为"守护线程"与"非守护线程"两种。默认情况下，线程是非守护线程，我们之前看到的例子都是非守护线程。

一个线程对象可以通过调用setDaemon(boolean on)方法将自己设置成一个守护线程，例如：thread.setDaemon(true);

当程序中的所有非守护线程都已结束运行时，即使守护线程的run()方法中还有需要执行的语句，守护线程也会立刻结束运行。因此，可以用守护线程做一些不很严格的工作，当该守护线程随时结束时也不会产生什么不良的后果。

【例题12_12】 守护线程示例。
```
    class UserThread implements Runnable{///非守护线程类
        public void run() {
            for(int i=1;i<=3;i++){
```

```
            System.out.println(Thread.currentThread().getName()+"第"+i+"次执行");
            try {
              Thread.sleep(100);
            } catch (InterruptedException e) {
              e.printStackTrace();
            }
          }
        }
      }
      class DaemonThread implements Runnable{//守护线程类
          public void run() {
            for(int i=1;i<=10;i++){
              System.out.println(Thread.currentThread().getName()+"第"+i+"次执行");
              try {
                  Thread.sleep(100);
               } catch (InterruptedException e) {
                  e.printStackTrace();
              }
            }
          }
      }
      public class Ch12_12{
          public static void main(String args[]){
              Thread user=new Thread(new UserThread());        //非守护线程对象
              Thread daemon=new Thread(new DaemonThread());    //守护线程对象
              user.setName("用户线程 ");
              daemon.setName("守护线程");
              daemon.setDaemon(true);                           //设置为守护线程
              user.start();
              daemon.start();
          }
      }
```

程序运行结果如图 12.13 所示。

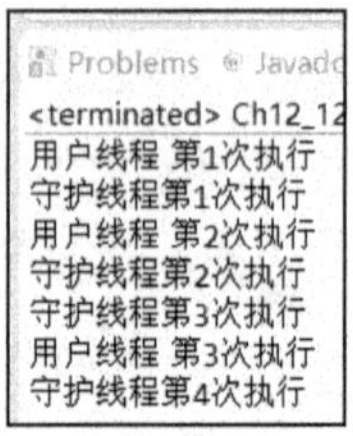

分析：当非守护线程对象 user 执行完 3 次输出任务后，守护线程对象 daemon 虽然还没有执行完 10 次输出任务就提前结束了执行。可将程序中的 daemon.setDaemon(true);一句注释掉，再运行程序，观察结果。

图 12.13 守护线程提前结束执行

12.11 线程之间的通信流类

Java 线程间的通信是通过管道流来实现的。一个线程发送数据到输入管道，另一个线程从输出管道中读数据。管道用来把一个线程或代码块的输出连接到另一个线程或代码块的输入。

Java 提供了 PipedInputStream 类和 PipedOutputStream 这两个专门的类用于充当管道流类，它们位于 java.io 包中。PipedInputStream 类是 InputStream 的子类，表示一个通信管道的接收端，一个线程通过它读取数据；PipedOutputStream 类是 OutputStream 的子类，表示一个通信管道的发送端，一个线程通过它发送数据。

【例题 12_13】线程间的数据通信。
```
import java.io.IOException;
```

```java
import java.io.PipedInputStream;
import java.io.PipedOutputStream;
class Sender implements Runnable {   //负责发送数据的线程
    private PipedOutputStream out = new PipedOutputStream();//声明管道输出流类对象
    public PipedOutputStream getPipedOutputStream() {//获取管道输出流类对象的方法
        return out;
    }
    public void run() {
        String s = "这是通过管道输出流发送过来的数据！";
        try {
            out.write(s.getBytes());//写入数据
            out.close();
        } catch (Exception e) {
            e.printStackTrace();
        }
    }
}
class Receiver implements Runnable {  //负责接收数据的线程
    private PipedInputStream in;//声明管道输入流类对象
    public Receiver(Sender sender) {//构造方法，以 Sender 类对象 sender 为其形参
        try {
            in = new PipedInputStream(sender.getPipedOutputStream());
            //创建管道输入流类对象，实现管道流的连接
        } catch (IOException e) {
            e.printStackTrace();
        }
    }
    public void run() {
        try {
            byte[] b = new byte[1024];
            while ((in.read(b))!= -1){
                System.out.println(new String(b));
            }
            in.close();
        } catch (Exception e) {
            e.printStackTrace();
        }
    }
}
public class Ch12_13{
    public static void main(String args[]) throws Exception {
        Sender sender = new Sender();
        Receiver receiver = new Receiver(sender);
        Thread senderthread=new Thread(sender);
        Thread receiverthread=new Thread(receiver);
        senderthread.start();
        receiverthread.start();
    }
}
```

程序运行结果如图 12.14 所示。

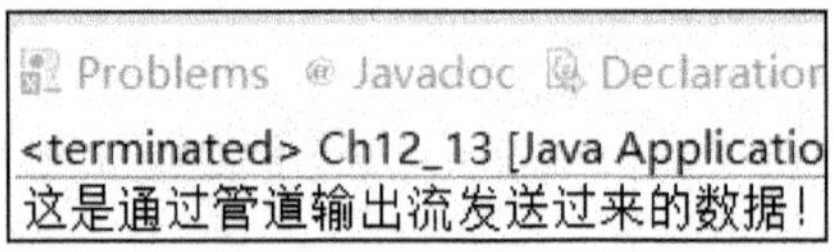

```
                        Problems  @ Javadoc  Declaration
                        <terminated> Ch12_13 [Java Applicatio
                        这是通过管道输出流发送过来的数据!
```
图 12.14 通过管道流读到发送端线程发来的数据

分析：语句 in = new PipedInputStream(sender.getPipedOutputStream());是在创建管道输入流类对象 in 时，指明它要读取数据的源头是管道输出流类对象 out，从而完成管道输入流与输出流的连接。

小　　结

本章主要介绍了线程与多线程的概念；Java 对多线程编程的支持机制；Java 中实现多线程的不同方法；Java 多线程的生命周期及其状态转换；Java 多线程的调度机制；Thread 类常用方法的典型应用；Java 线程的让步与联合的应用；Java 多线程的互斥与同步机制、方法与典型应用；守护线程的特点及应用；运用管道流进行线程间通信等。

习　题　12

上机实践题

1. 下面程序模拟三个售票窗口同时售票，运行该程序，指出其中的错误并改正。

```java
class TicketRunnable implements Runnable {
    private int ticketNum =100;
    public void run() {
        while (true) {
            String tName = Thread.currentThread().getName();
            if (ticketNum <= 0) {
                System.out.println(tName + "无票");
                break;
            } else {
                try {
                    Thread.sleep(100);
                } catch (Exception ex) {}
                ticketNum--;
                System.out.println(tName + "卖出一张票,还剩" + ticketNum + "张票");
            }
        }
    }
}
public class XiTi12_1_1 {
    public static void main(String[] args) {
        TicketRunnable tr = new TicketRunnable();
        Thread t1 = new Thread(tr, "售票窗口 1");
        Thread t2 = new Thread(tr, "售票窗口 2");
        Thread t3 = new Thread(tr, "售票窗口 3");
        t1.start();
```

```
            t2.start();
            t3.start();
        }
    }
```

2. 编写程序模拟分时系统进程（线程）调度执行。要求：

（1）用三个 JButton 对象作为三个进程（线程）实体。

（2）代表每个进程（线程）的按钮对象每被调度一次移动的距离为 100。

（3）假设该分时调度系统的时间片为 100 ms，进程（线程）0，进程（线程）1，进程（线程）2 各自最大用时分别为 900 ms、500 ms、1 200 ms。

3. 运用 join()方法，编写程序实现：主线程（main 方法对应的线程）等待另外一个线程 t 的执行结果（1 到 100 的和），当 t 执行完毕后，将执行结果返回给主线程，主线程负责输出该结果。

4. 编写程序解决"生产者-消费者"问题。

第 13 章　Java 网络编程

【本章内容提要】
- Java 网络编程概述；
- InetAddress 类的应用；
- URL 类的应用；
- URLConnection 类的应用；
- TCP、UDP、端口与套接字；
- 基于 TCP 的 Socket 网络编程；
- 基于 UDP 的网络编程；
- 基于组播的网络编程。

13.1　Java 网络编程概述

由于 Java 具有跨平台的特性，从而使得 Java 成为最流行的网络编程工具之一。Java 网络编程的内涵很广，如包括直接面向网络通信协议的网络编程、面向 Java EE 开发的网络编程、分布式网络编程、Applet 等。

在直接面向网络通信协议的网络编程方面主要有：基于应用层 HTTP（Hypertext Transfer Protocol）协议的 URL 通信程序，它使用 java.net.URL 类来获取 Web 资源；基于传输层 TCP 协议的 Socket 通信程序，它使用 java.net.Socket 类和 java.net.ServerSocket 类，实现基于 TCP 套接字的面向连接的可靠的网络数据报通信编程；基于传输层 UDP 协议的 Datagram 通信程序，它使用 java.net.DatagramPacket 类和 java.net.DatagramSocket 类，提供基于 UDP 的面向无连接的不可靠的网络数据报通信编程等。

面向 Java EE 开发的网络编程：JSP（Java Server Pages），JSP 页面由 HTML 代码和嵌入其中的 Java 代码所组成。服务器在页面被客户端所请求以后，对这些 Java 代码进行处理，然后将生成的 HTML 页面返回给客户端的浏览器；Java Servlet，是一种运行在 Web 服务器端的 Java 程序，是比 JSP 更加底层的组件，完成的功能和 JSP 类似；EJB（Enterprise JavaBean），提供了一个框架来开发和实施分布式商务逻辑，显著地简化了具有可伸缩性和高度复杂的企业级应用的开发等。

分布式网络编程技术主要有 Java RMI（Remote Method Invocation，远程方法调用）的分布式应用程序，它使用 java.rmi.*包中的类实现各种分布式计算等。

Applet 程序：又称为 Java 小程序，它嵌套在超文本标记语言 HTML 文件中，通过网络下载其代码到本地浏览器的 JVM 中执行。

本章重点介绍包含在 java.net 包中的 Java 类提供的网络编程功能。

13.2 InetAddress 类的应用

IP 地址是网络通信中的重要概念，IPv4 版本的 IP 地址地址由 32 个比特来表示，IPv6 版本的 IP 地址地址由 128 bit 来表示。

一个 IP 地址可代表 Internet 上某台计算机，根据 IP 地址就可以与其对应的计算机进行通信。IPv4 地址采用点分十进制的形式表示，即由 4 个 0~255 的数字组成如 192.168.1.1 等。IP 地址由专门的机构负责其定义和分配使用，目前 IP 地址分为 A、B、C、D、E 五类。

由于数字形式的 IP 地址难记难用，因此在 IP 地址的基础上又发展出一种符号化的地址方案，来代替数字型的 IP 地址。每一个符号化的地址都与特定的 IP 地址对应，这样网络上的资源访问起来就容易得多了。这个与网络上的数字型 IP 地址相对应的字符型地址，就被称为域名。域名服务器提供域名与 IP 地址之间相互转换服务。

java.net 包中的 InetAddress 类是对网络中计算机的{IP 地址，域名}的抽象。InetAddress 类通过调用其以下静态方法获得 InetAddress 类对象：

- public static InetAddress getByName(String host)throws UnknownHostException：以给定的主机名得到 InetAddress 对象。
- public static InetAddress[] getAllByName(String host)throws UnknownHostException：以给定的主机名返回 InetAddress 对象数组。
- public static InetAddress getByAddress(byte[] addr)throws UnknownHostException：以给定原始 IP 地址返回 InetAddress 对象。
- public static InetAddress getByAddress(String host,byte[] addr)throws UnknownHost Exception：以给定的主机名和 IP 地址得到 InetAddress 对象。
- public static InetAddress getLocalHost()throws UnknownHostException：返回本机的 InetAddress 对象。

InetAddress 类的其他常用方法如下：

- String getHostAddress()：获取 InetAddress 所含的 IP 地址。
- String getHostName()：获取 InetAddress 所含的域名。
- public byte[] getAddress()：返回此 InetAddress 对象的原始 IP 地址。

【例题 13_1】InetAddress 类常用方法。

```
import java.net.InetAddress;
import java.net.UnknownHostException;
public class Ch13_1 {
    public static void main(String[] args) {
        try {
            InetAddress inet1=InetAddress.getByName("www.sina.com.cn");
            String name=inet1.getHostName();
            String address=inet1.getHostAddress();
            System.out.println("inet1 的输出: "+inet1);
```

```
            System.out.println("inet1 包含的主机域名: "+name+"\ninet1 包含的 IP 地址: "+address);
            InetAddress[] inet2=InetAddress.getAllByName("www.sina.com.cn");
            System.out.println("inet2 的输出: ");
            for(InetAddress inet:inet2)
              System.out.println(inet);
            InetAddress inet3=InetAddress.getLocalHost();
            System.out.println("inet3 的输出: "+inet3);
        }
        catch (UnknownHostException e) {e.printStackTrace();}
    }
}
```

程序运行结果如图 13.1 所示：

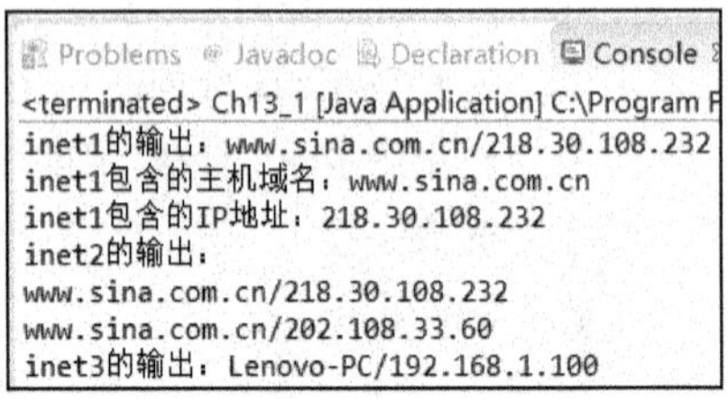

图 13.1　获取 IP 地址和域名

说明：程序执行时，要保证计算机已经正确连接到 Internet。

13.3　URL 类的应用

13.3.1　URL 简介

URL（Uniform Resource Locator，统一资源定位器）的值表示网络上某个资源（如文件等）的地址，实现了对网络资源的定位，其格式如下：

URL 的一般语法格式为：

```
protocol :// hostname[:port] / path / [;parameters][?query]#fragment
```

其中，带方括号的部分为可选项。protocol 指定使用的传输协议，如 http、ftp、file 等；hostname 指定资源所在的计算机，它既可以是 IP 地址，也可以是主机名或域名；port（端口号）为可选，省略时使用方案的默认端口，各种传输协议都有默认的端口号，如 http 的默认端口为 80；path 由零或多个"/"符号隔开的字符串，一般用来表示主机上的一个目录或文件地址；";parameters" 是用于指定特殊参数的可选项；"?query" 是用于给动态网页（如使用 PHP/JSP/ASP/ASP.NET 等技术制作的网页）传递参数的可选项，可有多个参数，参数间用"&"符号隔开，每个参数的名和值用"="连接；fragment 用于指定网络资源中的片断。实际中并非每个 URL 都包含这些元素，对于多数的 URL，协议、主机名和文件名是必需的，而端口号和文件内部的引用等都是可选的。

13.3.2　URL 类的常用方法

URL 类是 java.net 包中的一个重要的类，实现了对 URL 的封装。
URL 的常用构造方法：

- public URL(String spec) throws MalformedURLException：根据字符串 spec 指定的 URL 创建 URL 对象。
- public URL(String protocol,String host,String file)throws MalformedURLException：根据指定的 protocol 名称、host 名称和 file 名称创建 URL 对象。

URL 的常用方法：
- public String getFile()：获取此 URL 对应的网络资源文件名。
- public final InputStream openStream() throws IOException：打开到此 URL 的连接并返回一个 InputStream 对象。
- public URLConnection openConnection() throws IOException：返回一个 URLConnection 对象，它表示到 URL 所引用的远程对象的连接。
- public final Object getContent() throws IOException：获取此 URL 的内容。
- public String getHost()：获取此 URL 的主机名。
- public String getPath()：获取此 URL 的路径部分。
- public int getPort()：获取此 URL 的端口号。
- public String getProtocol()：获取此 URL 的协议名称。

【例题 13_2】读取远程主机的资源。

```
import java.io.*;
import java.net.MalformedURLException;
import java.net.URL;
public class Ch13_2 {
    public static void main(String[] args) {
        URL url;
        InputStream in;
        BufferedReader br;
        try {
            url=new URL("http://www.crphdm.com");//创建URL对象
            in=url.openStream();//打开连接此URL的输入流
            System.out.println("读到的网络资源信息为: ");
            br = new BufferedReader(new InputStreamReader(in, "utf-8"));
            //将字节流转为以utf-8编码的字符流,并将其包装为BufferedReader对象
            String str=null;
            while((str=br.readLine())!=null){//读取一行信息
                System.out.println(str);//输出到控制台
            }
        }
        catch (MalformedURLException e)
        { e.printStackTrace();}
        catch (IOException e)
        { e.printStackTrace();}
    }
}
```

程序运行结果如图 13.2 所示。

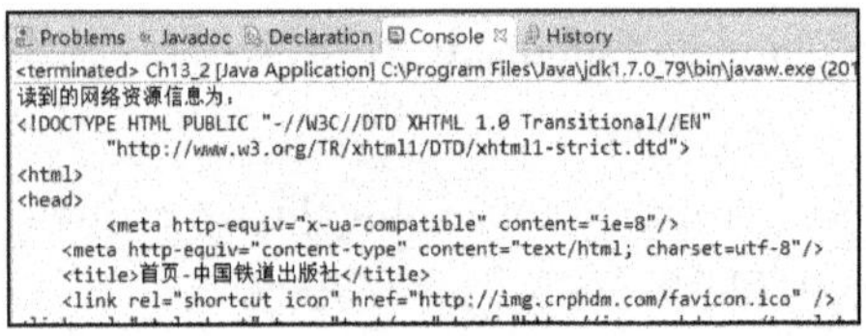

图 13.2　通过 URL 读取远程主机信息（局部）

13.4　URLConnection 类

URLConnection 类也是定义在 java.net 包中，URLConnetion 类的对象可以与指定 URL 建立动态连接。可以向服务器发送请求读取数据，同时也能将数据写回服务器。一般通过 URL 对象调用 openConnection()方法得到一个 URLConnection 类对象。

URLConnection 类的常用方法：
- public abstract void connect()throws IOException：打开到此 URL 引用的资源的通信链接。
- public Object getContent()throws IOException：获取此 URL 连接的内容。
- public InputStream getInputStream()throws IOException：返回从此打开的连接读取的输入流。
- public OutputStream getOutputStream()throws IOException：返回写入到此连接的输出流。

【例题 13_3】编写 GUI 程序，运用 URLConnection 读取远程主机信息并将读到的信息显示在窗体的文本区中。

```java
import java.net.*;
import javax.swing.*;
import java.awt.BorderLayout;
import java.awt.event.*;
import java.io.*;
public class Ch13_3 {
    public static void main(String[] args) {
        new GetRemoteServerInfo();
    }
}
class GetRemoteServerInfo extends JFrame implements ActionListener,Runnable{
    JLabel label;
    JTextField text;
    JButton bt;
    JPanel panel;
    JTextArea area;
    JScrollPane jspanel;
    URL url=null;
    URLConnection urlcon=null;
    String urlstr=null;
    public GetRemoteServerInfo(){
        this.setTitle("获取远程服务器资源");
        label=new JLabel("请输入一个合法的 URL: ");
        text=new JTextField(40);
        bt=new JButton("获取资源");
        bt.addActionListener(this);//注册事件监听器
        panel=new JPanel();
```

```java
        panel.add(label);panel.add(text);panel.add(bt);
        area=new JTextArea(10,40);
        jspanel=new JScrollPane(area);
        this.add(panel, BorderLayout.NORTH);
        this.add(jspanel, BorderLayout.CENTER);
        this.setSize(800, 300);
        this.setVisible(true);
        this.setDefaultCloseOperation(JFrame.EXIT_ON_CLOSE);
    }
    public void run() {
        try {
            url=new URL(urlstr);
            urlcon=url.openConnection();//得到 URLConnection 对象 urlcon
            urlcon.connect();//发起与服务器的连接
            InputStream in=urlcon.getInputStream(); //得到 InputStream 对象 in
            BufferedReader br = new BufferedReader(new InputStreamReader(in,
"utf-8"));
            //包装初级流对象并设定编码方式
            String str=null;
            while((str=br.readLine())!=null){ //读取一行信息
                area.append(str+"\n");           //显示在文本区
                System.out.println(str);
            }
        }
        catch (MalformedURLException e) {
            JOptionPane.showMessageDialog(this, "非法的 URL! "+e.getMessage());
            //提示对话框
        }
        catch (IOException e) {e.printStackTrace();}
    }
    //事件处理代码
    public void actionPerformed(ActionEvent e) {
        area.setText(null);                    //清空文本区
        urlstr=text.getText();                 //获取文本框输入的 URL
        new Thread(this).start();
        //创建读取服务器资源的线程并使之插入就绪队列排队
    }
}
```

程序运行结果如图 13.3 所示。

图 13.3　通过 URLConnection 读取远程主机信息

说明：在读取远程主机信息时，由于网速或其他因素，URL 资源的读取可能会引起阻塞，因此，程序中将读取远程信息的任务放在一个线程中，以免阻塞主线程。

13.5 TCP、UDP、端口与套接字

1．TCP 协议

传输控制协议 TCP（Transmission Control Protocol）是提供面向连接的、可靠的、基于字节流的传输层通信协议。面向连接是指应用进程（或程序）在使用 TCP 协议通信之前，必须先在发送方与接收方之间建立 TCP 连接，在通信完毕后，断开已经建立的连接。保证可靠性上，采用超时重传和捎带确认等机制。

2．UDP 协议

用户数据报协议 UDP（User Datagram Protocol）是面向无连接的、不可靠地的传输层通信协议。在应用进程（或程序）在使用 UDP 协议之前，不必先建立连接，自然也没无须断开连接。这样带来的好处是减少了开销和发送数据的时延，适合于即时通信。

3．端口和 Socket（套接字）

虽然通过 IP 地址或域名可以实现对网络中特定的计算机的寻址，但一般情况下，网络通信双方的主机上都会运行多个进程。那么，到底用哪一个进程来接收和发送数据呢？通常借助于端口号来解决这个问题。

TCP 和 UDP 都采用端口来区分进程。端口是一种抽象的软件结构(包括一些数据结构和 I/O 缓冲区)。端口用一个整数型标识符来表示，即端口号。端口号采用 16 个比特编码，可提供 64×2^{10}（0~65535）个端口号。1024 以下的端口号保留给预定义的服务，用于常用的网络服务和应用如 HTTP 协议使用 80 端口，FTP 协议的默认端口号是 21 等。应用程序与某端口建立连接后，传输层传给该端口的数据都被相应的进程所接收，相应进程发给传输层的数据都通过该端口输出。

区分不同应用程序进程间的网络通信和连接,主要有 3 个参数：通信的目的 IP 地址、使用的传输层协议（TCP 或 UDP）和使用的端口号。通过将这 3 个参数结合起来，与一个套接字 Socket 绑定，应用层就可以和传输层通过套接字接口，区分来自不同应用程序进程或网络连接的通信，实现数据传输的并发服务。

Socket（套接字）是支持 TCP/IP 的网络通信的基本操作单元，可以看做不同主机之间的进程进行双向通信的端点。可将 Socket 表示为：Socket=IP 地址+端口号+TCP/UDP 协议。

13.6 基于 TCP 的 Socket 网络编程

基于 TCP 的 Socket 通信是通过指定 IP 地址以及端口号，采用 C/S（Client/Server）模式建立 TCP 协议下的两个通信进程之间的连接，实现可靠的双向通信，任何一方既可以接受请求，也可以向另一方发送请求。

Java 中提供了用于实现客户端套接字的 Socket 类和用于实现服务器端套接字的 ServerSocket 类，它们封装了网络数据通信的底层细节，可以方便地实现基于 TCP 的 Socket 编程。

13.6.1　ServerSocket 类

java.net.ServerSocket 类用于实现服务器套接字。其常用构造方法为 public ServerSocket(int port)throws IOException：创建绑定到特定端口 port 的服务器套接字对象。

ServerSocket 类的常用方法：
- public Socket accept()throws IOException：等待并侦听客户端的请求，返回与该客户端进行通信用的 Socket 对象。
- public void close()throws IOException：关闭此套接字。
- public InetAddress getInetAddress()：返回此服务器套接字的本地地址。
- public int getLocalPort()：返回此套接字在其上侦听的端口。
- public SocketAddress getLocalSocketAddress()：返回此套接字绑定的端点的地址。

13.6.2　Socket 类

java.net.Socket 类实现客户端套接字。其常用构造方法为：
- public Socket(InetAddress address,int port)throws IOException：创建一个流套接字并将其连接到指定 IP 地址的指定端口号。
- public Socket(String host,int port)throws UnknownHostException,IOException：创建一个流套接字并将其连接到指定主机上的指定端口号。
- public Socket(String host,int port,InetAddress localAddr,int localPort)throws IOException：创建一个套接字并将其连接到指定远程主机上的指定远程端口。Socket 会通过调用 bind()方法来绑定提供的本地地址及端口。host 表示远程主机名，port 表示远程端口号，localAddr 表示要将套接字绑定到的本地地址，localPort 表示要将套接字绑定到的本地端口。
- public Socket(InetAddress address,int port,InetAddress localAddr,int localPort)throws IOException：创建一个套接字并将其连接到指定远程地址上的指定远程端口。Socket 会通过调用 bind()方法来绑定提供的本地地址及端口。

Socket 类的常用方法有：
- public void bind(SocketAddress bindpoint)throws IOException：将套接字绑定到本地地址。
- public void close()throws IOException：关闭此套接字。
- public void connect(SocketAddress endpoint)throws IOException：将此套接字连接到服务器。
- public InetAddress getInetAddress()：返回套接字连接的地址。
- public InetAddress getLocalAddress()：获取套接字绑定的本地地址。
- public InputStream getInputStream()throws IOException：返回此套接字的输入流。
- public int getPort()：返回此套接字连接到的远程端口。
- public int getLocalPort()：返回此套接字绑定到的本地端口。
- public SocketAddress getRemoteSocketAddress()：返回此套接字连接的端点的地址。
- public SocketAddress getLocalSocketAddress()：返回此套接字绑定的端点的地址。
- public OutputStream getOutputStream()throws IOException：返回此套接字的输出流。
- public int getPort()：返回此套接字连接到的远程端口。
- public boolean isConnected()：返回套接字的连接状态。

13.6.3　Socket 编程的一般流程

1．服务器端

Step1：创建一个 ServerSocket 对象。代码如下：

```
try{
    ServerSocket  server=new ServerSocket(9999);
}
    catch(IOException e1){ }
```

需要注意的是，服务器的端口号和客户端进程中指定的端口号应该一致，否则不能建立连接。

Step2：server 调用 accept()方法，以等待并侦听客户端的连接请求。如果客户端请求连接，则接受连接，返回通信套接字。代码如下：

```
try{
    Socket sc= server .accept();
}
catch(IOException e){}
```

Step3：sc 调用 getOutputStream()和 getInputStream()方法分别获取输出流和输入流，开始网络数据的发送和接收。代码如下：

```
{…
InputStream in=sc.getInputStream();
OutputStream out=sc.getOutputStream();
…
    发送和接收数据
…
}
```

Step4：最后关闭输入、输出流及套接字对象。

2．客户端

Step1：创建一个 Socket 对象。代码如下：

```
try{
    Socket socket=new Socket("127.0.0.1",9999);
}
catch(Exception e){}
```

Step2：socket 调用 getOutputStream()和 getInputStream()方法分别获取输出流和输入流，开始网络数据的发送和接收。代码如下：

```
try{
    Socket socket= server .accept();
}
catch(IOException e){}
```

Step3：socket 调用 getOutputStream()和 getInputStream()方法分别获取输出流和输入流，开始网络数据的发送和接收。代码如下：

```
{…
InputStream in=socket.getInputStream();
OutputStream out=socket.getOutputStream();
    …
    发送和接收数据
…
}
```

Step4:最后关闭输入、输出流及套接字对象。

【例题 13_4】一个简单的通信程序:客户端和服务器端程序互相发送消息。

```java
//服务器端程序
import java.io.*;
import java.net.*;
import java.util.Scanner;
public class Server{
    public static void main(String args[]){
      ServerSocket server=null;
      Socket socket=null;
      InputStream in=null;
      OutputStream out=null;
      DataInputStream  din=null;
      DataOutputStream dout=null;
      String str=null;
      Scanner scan=null;
      try{
       server=new ServerSocket(9999);     //创建ServerSocket对象
      }
      catch(IOException e1) {
         System.out.println(e1);
       }
      try{
       System.out.println("等待客户呼叫");
         socket=server.accept();           //侦听端口,看是否有来自于客户端的请求
         /*获取初级流**/
         in=socket.getInputStream();
         out=socket.getOutputStream();
         /*用DataInputStream和DataOutputStream包装初级流**/
         din=new DataInputStream(in);
         dout=new DataOutputStream(out);
         while(true){
           str=din.readUTF();              //读取来自客户端的消息
           System.out.println("来自客户端的消息: "+str);
          System.out.println("请输入发送给客户端的消息:");
           scan=new Scanner(System.in);
           str=scan.nextLine();
           dout.writeUTF(str);             //向客户端发送消息
         }
      }
      catch(Exception e) {
         System.out.println("客户已离开"+e);
      }
   }
}
//客户端程序
import java.io.*;
import java.net.*;
import java.util.Scanner;
public class Client{
```

```java
public static void main(String args[]){
    String s=null;
    Socket socket=null;
    InputStream in=null;
    OutputStream out=null;
    DataInputStream din=null;
    DataOutputStream dout=null;
    String str=null;
    Scanner scan=null;
    try{
        socket=new Socket("127.0.0.1",9999);       // 创建Socket对象
        /*获取初级流**/
        in=socket.getInputStream();
        out=socket.getOutputStream();
        /*用DataInputStream和DataOutputStream包装初级流**/
        din=new DataInputStream(in);
        dout=new DataOutputStream(out);
        while(true){
            System.out.println("请输入发送给服务器端的消息:");
            scan=new Scanner(System.in);
            str=scan.nextLine();
            dout.writeUTF(str);                    //向服务器端发送消息
            str=din.readUTF();                     //读取来自于服务器端的消息
            System.out.println("来自服务器端的消息: "+str);
        }
    }
    catch(Exception e){
        System.out.println("服务器已断开"+e);
    }
}
```

程序运行结果如图 13.4 和图 13.5 所示。

图 13.4　服务器端结果

图 13.5　客户端结果

说明：将服务器端和客户端程序分别保存、编译并执行。先执行服务器端程序，再执行客户端程序。

例题 13_4 的服务器端和客户端都是单线程程序，由于套接字连接中的输入流在读取信息时可能发生阻塞，所以客户端和服务器端都需要在一个单独的线程中读取信息；

【例题 13_5】编写 GUI 程序实现服务器和客户端的通信，要求将双方读取信息的代码以多线程方式组织。

//服务器端程序

```java
import java.awt.*;
import java.awt.event.*;
import java.io.*;
import java.net.*;
import javax.swing.*;
public class Server1 extends JFrame implements ActionListener, Runnable {
    JTextArea area;
    JLabel label;
    JTextField text;
    JButton send;
    JScrollPane scroll;
    JPanel panel;
    ServerSocket server=null;
    Socket socket = null;
    InputStream in=null;
    OutputStream out=null;
    DataInputStream din=null;
    DataOutputStream dout=null;
    public Server1() {
        area=new JTextArea("显示通信记录: \n");
        label=new JLabel("请输入消息: ");
        text=new JTextField(20);
        send=new JButton("发送");
        panel=new JPanel();
        panel.add(label);
        panel.add(text);
        panel.add(send);
        scroll=new JScrollPane(area);//使文本区带上滚动条
        this.setTitle("服务器端");
        this.add(scroll,BorderLayout.CENTER);
        this.add(panel,BorderLayout.SOUTH);
        send.addActionListener(this);
        this.setSize(400,200);
        this.setVisible(true);
        try {
            server = new ServerSocket(9999);//  服务器端开辟9993为服务端口
            socket=server.accept();//侦听端口,看是否有来自于客户端的请求
            new Thread(this).start();//创建服务线程并使之插入就绪队列排队
        } catch (Exception ex) {
        }
    }
    /*服务线程任务体**/
    public void run() {
        try {
            while(true) {
                in=socket.getInputStream();          // 得到初级输入流
                din=new DataInputStream(in);         // 包装初级输入流
                String str=din.readUTF();            //读取来自于客户端的信息
                area.append(str+"\n");               // 显示通信内容
            }
```

```java
            } catch (Exception ex) {
            }
        }
        /*send 的事件处理**/
        public void actionPerformed(ActionEvent e) {
            try {
                OutputStream out=socket.getOutputStream();      // 得到初级输出流
                dout=new DataOutputStream(out);                 // 包装初级输出流
                dout.writeUTF("来自服务器的消息: "+text.getText());//向客户端发送信息
                text.setText(null);
            } catch (Exception ex) {
            }
        }
        public static void main(String[] args) throws Exception {
            new Server1();
        }
}
//客户端程序
import java.awt.*;
import java.awt.event.*;
import java.io.*;
import java.net.*;
import javax.swing.*;
public class Client1 extends JFrame implements ActionListener,Runnable{
    JTextArea area;
    JLabel label;
    JTextField text;
    JButton send;
    JScrollPane scroll;
    JPanel panel;
    Socket socket = null;
    InputStream in=null;
    OutputStream out=null;
    DataInputStream din=null;
    DataOutputStream dout=null;
    public Client1(){
        area=new JTextArea("显示通信记录: \n");
        label=new JLabel("请输入消息: ");
        text=new JTextField(20);
        send=new JButton("发送");
        panel=new JPanel();
        panel.add(label);
        panel.add(text);
        panel.add(send);
        scroll=new JScrollPane(area);
        this.setTitle("客户端");
        this.add(scroll,BorderLayout.CENTER);
        this.add(panel,BorderLayout.SOUTH);
        send.addActionListener(this);
        this.setSize(400,200);
```

```java
            this.setVisible(true);
            try{
                socket=new Socket("127.0.0.1",9999);
                new Thread(this).start();
            }catch(Exception ex){}
        }
        public void run(){
            try{
                while(true){
                    in=socket.getInputStream();           // 得到初级输入流
                    din=new DataInputStream(in);          // 包装初级输入流
                    String str=din.readUTF();             // 读取来自于服务器的信息
                    area.append(str+"\n");                // 显示通信内容
                }
            }catch(Exception ex){}
        }
        public void actionPerformed(ActionEvent e){
            try{
                OutputStream out=socket.getOutputStream();// 得到初级输出流
                dout=new DataOutputStream(out);           // 包装初级输出流
                dout.writeUTF("来自客户端的消息: "+text.getText());//向服务器发送信息
                text.setText(null);
            }catch(Exception ex){}
        }
        public static void main(String[] args) throws Exception{
            new Client1();
        }
}
```

程序运行结果如图 13.6 和图 13.7 所示。

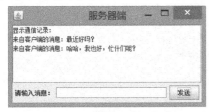

图 13.6　服务器端结果

图 13.7　客户端结果

分析：程序只是以 GUI 形式组织了服务器和客户端，同时，将服务器端和客户端通过 Socket 读取信息的代码放到了一个线程中进行，可以实现服务器端和客户端的"聊天"。但还不能实现服务器同时并发为多个客户端同时提供服务。

为了使得服务器端能够并发地为多个客户端同时提供服务，当服务器端收到一个客户的套接字后，就应该启动一个专门为该客户服务的线程。

【**例题 13_6**】编写一个聊天程序，要求可以有多个聊天客户端同时聊天，所有聊天信息在服务器端窗体和所有客户端窗体中显示。

```java
//服务器端程序
import java.awt.*;
import java.io.*;
```

```java
import javax.swing.*;
import java.net.ServerSocket;
import java.net.Socket;
import java.util.ArrayList;
class Server2 extends JFrame implements Runnable{
    JTextArea area;
    JScrollPane scroll;
    JPanel panel;
    ServerSocket server=null;
    Socket socket = null;
    ArrayList clients = new ArrayList();
    //用来保存为每个客户端服务线程的数据结构
    public Server2() throws Exception{
        area=new JTextArea("显示通信记录: \n");
        scroll=new JScrollPane(area);//使文本区带上滚动条
        this.setTitle("服务器端");
        this.add(scroll,BorderLayout.CENTER);
        this.setSize(400,200);
        this.setVisible(true);
        try {
            server = new ServerSocket(9997);// 服务器端开辟9993为服务端口
            new Thread(this).start();//创建服务线程并使之插入就绪队列排队
        }
        catch (Exception ex) {}
    }
    public void run(){
        try{
            while(true){
                socket = server.accept();//侦听端口，看是否有来自于客户端的请求
                ServerThread t = new ServerThread(socket);
                //当某个客户端请求连接服务器，则创建一个线程对象为之服务
                clients.add(t);//将新的服务线程添加到ArrayList对象clients中
                t.start(); //将服务线程插入就绪队列排队
            }
        }catch(Exception ex){}
    }
    //并发为多个聊天客户端服务的多线程类ServerThread,是一个内部类
    class ServerThread extends Thread{//为某个Socket负责接受信息
        Socket s = null;
        InputStream in=null;
        OutputStream out=null;
        DataInputStream din=null;
        DataOutputStream dout=null;
        public ServerThread(Socket s) throws Exception   {
            this.s = s;
        }
        public void run(){
            try{
                while(true){
                    in=socket.getInputStream();
```

```java
                    out=socket.getOutputStream();
                    // 得到初级输入、输出流
                    din=new DataInputStream(in);
                    dout=new DataOutputStream(out);
                    //包装初级输入、输出流
                    String str=din.readUTF();     //读来自于某个客户端的聊天信息
                    area.append(str+"\n");        // 显示聊天信息
                    sendMessage(str);             //将聊天信息str转发给所有客户端
                }
            }catch(Exception ex){}
        }
    }
    public void sendMessage(String msg){      //将信息发给所有客户端的方法
        for(int i=0;i<clients.size();i++){
            ServerThread t = (ServerThread)clients.get(i);//取出一个服务线程
            try {
                t.dout.writeUTF(msg);              //发送聊天信息
            } catch (IOException e) {
                e.printStackTrace();
            }
        }
    }
    public static void main(String[] args) throws Exception{
        new Server2();
    }
}
//客户端程序
import java.awt.*;
import java.awt.event.ActionEvent;
import java.awt.event.ActionListener;
import java.io.*;
import java.net.Socket;
import javax.swing.*;
public class Client2 extends JFrame implements ActionListener, Runnable {
    JTextArea area;
    JLabel label;
    JTextField text;
    JButton send;
    JScrollPane scroll;
    JPanel panel;
    Socket socket = null;
    InputStream in=null;
    OutputStream out=null;
    DataInputStream din=null;
    DataOutputStream dout=null;
    String nickName = null;
    public Client2() {
        area=new JTextArea("显示通信记录：\n");
        label=new JLabel("请输入消息: ");
        text=new JTextField(20);
        send=new JButton("发送");
```

```java
        panel=new JPanel();
        panel.add(label);
        panel.add(text);
        panel.add(send);
        scroll=new JScrollPane(area);
        this.setTitle("客户端");
        this.add(scroll,BorderLayout.CENTER);
        this.add(panel,BorderLayout.SOUTH);
        send.addActionListener(this);
        this.setSize(400,200);
        this.setVisible(true);
        nickName = JOptionPane.showInputDialog("输入昵称:");
        try{
            socket=new Socket("127.0.0.1",9997);    //创建Socket对象
            JOptionPane.showMessageDialog(this,"连接成功");
            //提示与服务器连接成功的对话框
            this.setTitle("客户端" + nickName);        //设置该客户端窗体标题栏标题
            new Thread(this).start();//创建通过Socket读取信息的线程对象并使之就绪
        }catch(Exception ex){}
    }
    public void run() {
        try{
            while(true){
                in=socket.getInputStream();           // 得到初级输入流
                din=new DataInputStream(in);          //包装初级输入流
                String str=din.readUTF();             //读取来自于服务器的信息
                area.append(str+"\n");
            }
        }catch(Exception ex){}
    }
    public void actionPerformed(ActionEvent e) {//发送按钮的事件处理代码
        try{
            out=socket.getOutputStream();             // 得到初级输出流
            dout=new DataOutputStream(out);           //包装初级输出流
            dout.writeUTF(nickName + "说:"+text.getText());//向服务器发送信息
            text.setText(null);                       //清空文本框
        }catch(Exception ex){}
    }
    public static void main(String[] args) throws Exception {
        new Client2();
    }
}
```

程序运行结果如图13.8（a）、（b）、（c）、（d）所示。

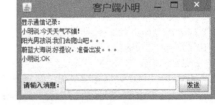

（a）服务器端窗体　　　　　　　　　　　　　　（b）小明聊天窗体

图13.8　多客户聊天

（c）阳光男孩聊天窗体

（d）蔚蓝大海聊天窗体

图 13.8　多客户聊天（续）

分析：服务器端声明内部类 class ServerThread extends Thread，当某个客户端请求连接服务器，则创建一个 ServerThread 线程对象为之服务，从而并发为多个聊天客户端并发提供服务。此外，本例是某个聊天客户端先将聊天信息发到服务器，在由服务器"群发"到各个聊天客户端。

13.7　基于 UDP 的网络编程

UDP 通信中通常先把应用层所传递来的数据封装为一个个独立的数据报文，称为数据报（Datagram），UDP 不能保证每个数据报是否能够完整地到达目的主机，不能保证数据报的到达时间和到达顺序，但由于不需要通信双方预先建立连接和进行可靠性保障等，其通信效率比基于 TCP 的 Socke 通信效率高。在对于即时性要最高，对可靠性要求相对较低的网络通信场合如聊天系统等就可以采用 UDP 协议进行通信，从而提高其网络管理数据接收和发送的效率。

Java 中提供的与 UDP 数据报相关的类有 DatagramPacket 类和 DatagramSocket 类，前者用于创建一个待发送或接收的 DatagramPacket 数据报对象，后者是创建一个用来发送或接收 DatagramPacket 数据报的 DatagramSocket 数据报套接字对象。它们封装了网络数据通信的底层细节，可以方便地实现基于 UDP 的网络编程。

13.7.1　DatagramPacket 类

要使用 UDP 数据报方式实现网络通信，必须先将数据打包（封装成数据报包）后才能进行传送和接收。java.net.DatagramPacket 类，就是用来创建数据包的。DatagramPacket 类的构造方法分为两类：一类是创建 DatagramPacket 对象用来接收数据报包，另一类是创建 DatagramPacket 对象用来发送数据报包。

用于创建接收数据报包的构造方法如下：

- public DatagramPacket(byte[] buf,int length)：创建 DatagramPacket 对象，buf 表示存放接收到数据报的缓冲区，length 表示要接收数据报的长度。
- public DatagramPacket(byte[] buf,int offset,int length)：创建 DatagramPacket 对象，buf 表示存放接收到数据报的缓冲区，length 表示要接收数据报的长度，在缓冲区中指定了偏移量 offset。

用于创建发送数据报包的构造方法如下：

- publicDatagramPacket(byte[] buf,int length,InetAddress address,int port)：创建 DatagramPacket 对象，用来将长度为 length 的包发送到指定主机上的指定端口号。

- public DatagramPacket(byte[] buf,int offset,int length,InetAddress address,int port)：创建 DatagramPacket 对象，用来将长度为 length 偏移量为 offset 的包发送到指定主机上的指定端口号。
- public DatagramPacket(byte[] buf,int length,SocketAddress address)throws SocketException：创建 DatagramPacket 对象，用来将长度为 length 的包发送到指定主机上的指定端口号。
- public DatagramPacket(byte[] buf,int offset,int length,SocketAddress address)throws SocketException：创建 DatagramPacket 对象，用来将长度为 length 偏移量为 offset 的包发送到指定主机上的指定端口号。

注意，以上各个构造方法中的参数 length 必须小于等于 buf.length。

DatagramPacket 类的常用方法如下：

- public InetAddress getAddress()：返回用 InetAddress 对象表示的对方的 IP 地址。
- public int getPort()：返回对方的端口号。
- public byte[] getData()：返回 DatagramPacket 对象中包含的数据。
- public int getLength()：返回将要发送或接收到的数据的长度。
- public int getOffset()：返回将要发送或接收到的数据在 byte[]中的偏移量。
- public int getPort()：返回对方的端口号。
- public void setData(byte[] buf,int offset,int length)：设置该 DatagramPacket 对象中包含的数据。

13.7.2　DatagramSocket 类

java.net.DatagramSocket 类的主要作用是对 DatagramPacket 对象进行接收和发送。其常用构造方法如下：

- public DatagramSocket()throws SocketException：创建绑定到本地主机上任何可用端口的数据报套接字对象。
- public DatagramSocket(int port)throws SocketException：创建绑定到本地主机上指定端口的数据报套接字对象。

DatagramSocket 类的常用方法有：

- public void receive(DatagramPacket p)throws IOException：从此套接字接收数据报包。
- public void send(DatagramPacket p)throws IOException：从此套接字发送数据报包。
- public void close()：关闭此数据报套接字。

13.7.3　基于 UDP 网络编程的一般流程

数据发送方：

Step1：用 DatagramPacket 类将待发送数据打包，即用创建发送数据报包的构造方法创建 DatagramPacket 类对象 dp。例如：
```
String str = "你好";
    byte[] data = str.getBytes();
InetAddress address = InetAddress.getByName("127.0.0.1");
    DatagramPacket dp = new DatagramPacket(data,data.length, address,9999);
```
Step2：创建一个 DatagramSocket()对象 ds，该对象负责发送数据包。例如：
```
DatagramSocket ds = new DatagramSocket();
```
Step3：发送数据如：

```
ds.send(dp);
```
数据接收方：

Step 1：用 DatagramSocket 的构造方法 DatagramSocket(int port) 创建一个对象 ds，其中的端口号必须和发送方数据包的端口号相同。例如，如果发送方发送的数据包的端口是 9999，则按如下方式创建 DatagramSocket 对象：

```
DatagramSocket ds = new DatagramSocket(9999);
```

Step 2：用 DatagramPack 类的构造方法如 DatagramPacket(byte[] buf,int length)创建一个接收数据的数据包对象 dp。例如：

```
byte[] data = new byte[1024];
DatagramPacket dp = new DatagramPacket(data,data.length);
```

Step 3：接收数据如：

```
ds.receive(dp);
```

Step4：将接收到的数据包拆封，取出数据。例如：

```
String content = new String(dp.getData(),0,dp.getLength());
```

【例题 13_7】编写一个基于 UDP 数据报的简单的聊天室程序，要求可以有多个聊天客户端同时聊天，所有聊天信息在服务器端窗体和所有客户端窗体中显示，同时，要求每个客户端都要通过登录窗体进行登录。

```java
//服务器端程序
import java.awt.BorderLayout;
import java.net.DatagramPacket;
import java.net.DatagramSocket;
import java.net.SocketAddress;
import java.util.ArrayList;
import javax.swing.JFrame;
import javax.swing.JTextArea;
public class Server3 extends JFrame implements Runnable{
    DatagramSocket socket=null;
    DatagramPacket packet=null;
    ArrayList<SocketAddress> clients=new ArrayList<SocketAddress>();
    //ArrayList对象clients保存客户端的IP地址及端口信息
    JTextArea area;
    public Server3() throws Exception{
        area=new JTextArea("服务器已启动……\n以下是各个聊天客户端的地址信息：\n");
        this.add(area, BorderLayout.CENTER);
        this.setTitle("服务器端");
        this.setDefaultCloseOperation(JFrame.EXIT_ON_CLOSE);
        this.setSize(400,300);
        this.setVisible(true);
        try {
            socket= new DatagramSocket(8899);//创建DatagramSocket对象
            new Thread(this).start();//创建并启动线程对象为多个聊天客户端服务
        } catch (Exception ex) {
            ex.printStackTrace();
        }
    }
    public void run(){
        try{
```

```java
            while(true){
                byte[] data = new byte[1024];
                packet= new DatagramPacket(data,data.length);//准备接收数据的包
                socket.receive(packet);        //接收数据
                 /*获取收到的数据包的宿主主机IP地址及端口信息并将其显示在文本区**/
                SocketAddress address = packet.getSocketAddress();
                area.append(address+"\n");
                /*若用户地址列表中不包含当前用户,则将其加入到地址列表**/
                if(!clients.contains(address)){
                    clients.add(address);
                }
                /*遍历地址列表,取出地址列表中每个客户发来的数据包中的内容并发送给所有客户端**/
                for(SocketAddress address1:clients){
                DatagramPacket datagram=new DatagramPacket(packet.getData(),
                 packet.getLength(),address1);
                    //构造发送数据的数据报包
                socket.send(datagram);//发送数据报包
                }
            }
        }
        catch(Exception ex){
            ex.printStackTrace();
        }
    }
    public static void main(String[] args) throws Exception{
        new Server3();
    }
}
//聊天登录窗体程序
import java.awt.BorderLayout;
import java.awt.Container;
import java.awt.event.ActionEvent;
import java.awt.event.ActionListener;
import javax.swing.*;
public class LoginFrame extends JFrame implements ActionListener {
    JLabel welcome=new JLabel("欢迎登录聊天系统");
    static JTextField name=new JTextField(10);
    static JPasswordField password=new JPasswordField (10);
    JButton confirm=new JButton("确定");
    Box vbox,hbox1,hbox2,hbox3,hbox4;
    JPanel panel;
    Container con;
    public LoginFrame(){
        panel=new JPanel();
        vbox=Box.createVerticalBox();
        welcome.setFont(new java.awt.Font("宋体",1,20));
        hbox1=Box.createHorizontalBox();
        hbox1.add(welcome);
        vbox.add(hbox1);
        vbox.add(Box.createVerticalStrut(20));
```

```java
        hbox2=Box.createHorizontalBox();
        hbox2.add(new JLabel("账号:"));
        hbox2.add(Box.createHorizontalStrut(30));
        hbox2.add(name);
        vbox.add(hbox2);
        vbox.add(Box.createVerticalStrut(20));
        hbox3=Box.createHorizontalBox();
        hbox3.add(new JLabel("密 码:"));
        hbox3.add(Box.createHorizontalStrut(30));
        hbox3.add(password);
        vbox.add(hbox3);
        vbox.add(Box.createVerticalStrut(20));
        hbox4=Box.createHorizontalBox();
        hbox4.add(confirm);
        hbox4.add(Box.createHorizontalStrut(30));
        hbox4.add(new JButton("取消"));
        vbox.add(hbox4);
        panel.add(vbox);
        con=this.getContentPane();
        con.add(panel,BorderLayout.CENTER);
        confirm.addActionListener(this);
        this.setTitle("聊天登录窗体");
        this.setBounds(0,0,360,240);
        this.setResizable(false);
        this.setVisible(true);
    }
    //"确定"按钮事件处理代码
    public void actionPerformed(ActionEvent e) {
        /*实用中还应连接数据库,判断是否合法账号和密码,此处简化了**/
        String n=name.getText().trim();
        String p=password.getText().trim();
        if(n.equals("")||p.equals(""))  {
            JOptionPane.showMessageDialog(this, "账号和密码不能为空!", "提示", JOptionPane.WARNING_MESSAGE);}
        else{
            new Client3();          //弹出聊天窗体
            this.dispose();         //关闭登录窗体
        }
    }
    public static void main(String args[]){
        new LoginFrame();
    }
}
//聊天客户端程序
import java.awt.*;
import java.awt.event.*;
import java.net.*;
import javax.swing.*;
public class Client3 extends JFrame implements ActionListener, Runnable {
    JTextArea area=new JTextArea("以下是聊天记录\n");
```

```java
            JTextField text=new JTextField(20);
            JButton send=new JButton("发送");
            JScrollPane scroll;
            JPanel panel=new JPanel();
            DatagramSocket socket = null;
            DatagramPacket packet=null;
            String userName=null;
            String msg=null;
            public Client3() {
                this.setTitle("客户端");
                scroll=new JScrollPane(area);
                this.add(scroll, BorderLayout.CENTER);
                panel.add(text);panel.add(send);
                this.add(panel, BorderLayout.SOUTH);
                text.addActionListener(this);
                send.addActionListener(this);
                //添加事件监听器
                this.setSize(400,300);
                this.setDefaultCloseOperation(JFrame.EXIT_ON_CLOSE);
                this.setVisible(true);
                userName=LoginFrame.name.getText();
                this.setTitle(userName);
                msg=userName+"上线了...";
                try{
                    socket=new DatagramSocket();
                    InetAddress address=InetAddress.getByName("127.0.0.1");
                    //*给服务器发送一个包含客户登录名的数据包**/
                    byte[] data = msg.getBytes();
                    DatagramPacket packet = new DatagramPacket(data,data.length,address,8899);
                    //构造发送数据报包
                    socket.send(packet);              //发送msg的内容
                    new Thread(this).start();         //创建线程对象并使之就绪
                }catch(Exception ex){}
            }
            public void run() {
                try {
                    while (true) {
                        /*获取来自服务器的数据包拆封后，将内容显示在文本区**/
                        byte[] data = new byte[1024];
                        packet = new DatagramPacket(data,data.length);//接收数据的包
                        socket.receive(packet);      //接收来自于服务器的数据包
                        String msg=new String(packet.getData(),0,packet.getLength());
                        //取出数据报包中的内容并将其转成字符串
                        area.append(msg+"\n");       // 文本区中显示
                    }
                } catch (Exception ex) {
                }
            }
            /*当在text中输入聊天内容并回车或单击send按钮都将聊天消息发出**/
            public void actionPerformed(ActionEvent e) {
```

```
        try {
            if(e.getSource()==text||e.getSource()==send){
            /*将聊天内容发给服务器**/
            String msg=userName+"说: "+text.getText();
            byte[] data = msg.getBytes();
            InetAddress address=InetAddress.getByName("127.0.0.1");
            DatagramPacket packet = new DatagramPacket(data,data.length,
            address,8899);
            //构造发送数据的数据报包
            socket.send(packet);//发送数据报包
            }
        }catch (Exception ex) {}
    }
}
```

说明：首先运行服务器端程序 Server3.java，结果如图 13.9（a）所示。接着，执行聊天登录程序 LoginFrame.java，结果如图 13.9（b）所示，当账号或密码为空时，弹出图 13.9（c）所示的对话框。

当在登录窗口中输入账号和密码并单击"确定"按钮后，结果如图 13.9（d）、（e）所示。

可以接着多次执行 LoginFrame.java，模拟有多个客户登录聊天系统。如"阳光男孩"和"蔚蓝大海"登录并参与聊天后，小明、阳光男孩、蔚蓝大海三人聊天的情况以及服务器端的实时变化情况如图 13.9（f）所示。

（a）服务器初始界面

（c）提示对话框

（d）客户登录

（e）开始聊天

（f）多客户聊天及服务器端界面

图 13.9　多客户聊天

13.8 基于组播的网络编程

网络数据传播按照接收者的数量，可分为"单播"（提供点对点的通信）、"广播"（发送者每次发送的数据可以被广播范围内的所有接收者接收）和"组播"（发送者每次发送的数据可以被组内的所有接收者接收）。

组播组内的所有主机共享同一个地址，这种地址称为组播地址，组播地址是范围在 224.0.0.0～239.255.255.255 之间的 IP 地址（D 类 IP 地址），主机可以在任何时候加入或离开组。

java.net.MulticastSocket 具有组播的功能，它是 DatagramSocket 的子类。MulticastSocket 类与 DatagramPacket 类配合使用，可以完成组播数据报的发送和接收。

MulticastSocket 类的常用构造方法：

- public MulticastSocket()throws IOException：创建组播套接字对象。
- public MulticastSocket(int port)throws IOException：创建组播套接字对象并将其绑定到特定端口。

MulticastSocket 类的常用方法：

- public void joinGroup(InetAddress mcastaddr)throws IOException：加入组播组。
- public void leaveGroup(InetAddress mcastaddr)throws IOException：离开组播组。
- public void setTimeToLive(int ttl)throws IOException：设置组播数据包的生存期，$0 \leqslant ttl \leqslant 255$。若 ttl 为 2，则表示最多经过 2 个路由器，否则数据报包被丢弃。

MulticastSocket 类从 DatagramSocket 类继承来的常用方法：

- public void send(DatagramPacket p)throws IOException：从此套接字发送组播数据报包。
- public void receive(DatagramPacket p)throws IOException：从此套接字接收组播数据报包。

【例题 13_8】编写程序实现：组播端（服务器端）不断向各个加入组播组中的客户端发送信息，客户端接收组播信息并显示在客户端窗体中。

```java
//组播服务器端程序
import java.awt.BorderLayout;
import java.awt.event.ActionEvent;
import java.awt.event.ActionListener;
import java.io.IOException;
import java.net.DatagramPacket;
import java.net.InetAddress;
import java.net.MulticastSocket;
import java.net.UnknownHostException;
import javax.swing.*;
public class MulticastServer extends JFrame implements ActionListener{
    JTextArea content;
    JScrollPane scroll;
    JTextField text;
    JButton send;
    JPanel panel;
    public MulticastServer(){
        this.setTitle("组播消息");
        this.setLayout(new BorderLayout());
        content=new JTextArea();
        scroll=new JScrollPane(content);
        text=new JTextField(40);
```

```java
        send=new JButton("组播");
        panel=new JPanel();
        panel.add(text);panel.add(send);
        this.add(scroll, BorderLayout.CENTER);
        this.add(panel, BorderLayout.SOUTH);
        text.addActionListener(this);
        send.addActionListener(this);
        this.setSize(550, 300);
        this.setVisible(true);
        this.setDefaultCloseOperation(JFrame.EXIT_ON_CLOSE);
    }
    /*事件处理代码**/
    public void actionPerformed(ActionEvent e) {
        String str=text.getText();
        int port=8999;//组播端口
        InetAddress group;
        try {
            group = InetAddress.getByName("239.255.0.7");//组播地址
            byte data[]=str.getBytes();//组播信息
            DatagramPacket datagramPacket =new DatagramPacket(data, data.length, group ,port);
            //构造组播信息数据报包
            MulticastSocket multicastSocket =new MulticastSocket();
            //创建组播 Socket 对象
            multicastSocket.send(datagramPacket);
            //发送组播信息
        }
        catch (UnknownHostException e1) {}
        catch (IOException e1) {}
        content.append("广播的内容为: "+"\n"+str+"\n");
        //将要组播的信息显示在文本区
        text.setText(null);//清空文本区
    }
    public static void main ( String [] args ){
        new MulticastServer();
    }
}
//组播客户端程序
import java.awt.BorderLayout;
import java.awt.event.ActionEvent;
import java.awt.event.ActionListener;
import java.io.IOException;
import java.net.DatagramPacket;
import java.net.InetAddress;
import java.net.MulticastSocket;
import java.net.UnknownHostException;
import javax.swing.*;
public class MulticastClient extends JFrame {
    JTextArea content;
    JScrollPane scroll;
```

```java
    public MulticastClient(){
      this.setTitle("组播消息接收");
      this.setLayout(new BorderLayout());
      content=new JTextArea();
      scroll=new JScrollPane(content);
      this.add(scroll, BorderLayout.CENTER);
      this.setSize(550, 300);
      this.setVisible(true);
      this.setDefaultCloseOperation(JFrame.EXIT_ON_CLOSE);
    }
    public void recieve() {                              //定义接收组播信息的方法
       InetAddress group;
       int port=8999;                                    //组播端口
    try {
       group=InetAddress.getByName("239.255.0.7");       //组播组
       MulticastSocket multicastSocket=new MulticastSocket(port);
       //创建组播 Socket 对象
       byte[] b = new byte[8192];
       multicastSocket.joinGroup(group);                 //加入该组播组
       content.append("收到的组播信息为: "+"\n");
       while(true){
          DatagramPacket dp =new DatagramPacket(b,b.length,group,port);
          //构造用于接收组播数据报的包
          multicastSocket.receive(dp);
          //接收组播信息
          content.append(new String(dp.getData(),0,dp.getLength())+"\n");
          //取出接收到的组播信息并显示在文本区
       }
    }
    catch (UnknownHostException e1) {}
    catch (IOException e1) {}
}
 public static void main ( String [] args ){
      new MulticastClient().recieve();
   }
}
```

程序运行结果如图 13.10（a）、（b）所示。

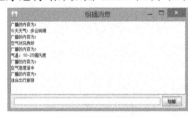
（a）组播服务器界面

（b）组播客户端界面

图 13.10　组播信息发送与接收

说明：当同时多次运行 MulticastClient.java 可以创建多个组播信息接收客户端，可以收到相同的组播信息。

小　　结

Java 网络编程的内涵很广，包括直接面向网络通信协议的网络编程、面向 Java EE 开发的网络编程、分布式网络编程以及 Applet 等。本章重点以 java.net 包中提供的几个网络编程类 InetAddress、URL、URLConnection、ServerSocket、Socket、DatagramSocket、DatagramPacket 及 MulticastSocket 类为主线，分别介绍了各个类的常用方法及其各自的实际应用。

习　题　13

上机实践题

1. 编写程序，运用 URL 类的相关方法，读取远程主机页面信息，要求：

（1）编写窗体，在其上布局一个文本框，用于输入 URL，一个名为"读文件"的按钮，单击该按钮，开始远程主机页面信息的读取。用一个标签对象显示读取文件是否成功。

（2）当读取文件成功，将读到的页面信息保存为.html 格式文件（保存在当前源程序文件目录下），若读取成功，用浏览器打开保存的.html 文件查看该页面。（用 JOptionPane 调用其方法 showInputDialog()弹出对话框，提示用户输入要保存的文件名）。

2. 编写程序实现：当在文本框中输入一个域名并回车后，在文本区显示该域名对应的所有域名/IP 地址。

3. 编写一个基于 TCP 的 Socket 的聊天程序，要求：

（1）服务器端可以为多个客户端同时提供连接服务，所有的客户聊天信息显示在服务器端窗口中。

（2）客户端窗口中包含两个文本区（一个用作显示所有客户的聊天信息，另一个用作输入聊天信息）以及一个"发送"按钮。

第 14 章　Java 泛型与集合类

【本章内容提要】
- 泛型；
- 集合类与接口。

14.1　泛　　型

14.1.1　泛型概述

JDK1.5 以后引入了"泛型"的概念。泛型实现了参数化类型的概念，使代码可以应用于多种类型。泛型这个术语的意思是"适用于许多种的类型"，目的是希望类或方法能够具备最广泛的表达能力。使用泛型编写的程序具有更好的安全性和可读性。

在 JDK1.5 以前，没有引入泛型的概念，通过对类型 Object 的引用来实现参数的"任意化"。"任意化"带来的缺点就是要做显示的强制类型转换，而这种转换是要求开发者对实际参数类型可以预知的情况下才可以进行。对于强制类型转换错误的情况，编译器可能不提示错误，但在运行的时候可能出现异常，而这是比较严重的安全隐患。泛型的引入很好地解决了这一问题。如下面例题。

【例题 14_1】用 Object 类实现参数"任意化"。

```java
class Test{
    private Object x;
    public Object getX() {
        return x;
    }
    public void setX(Object x) {
        this.x = x;
    }
}
public class Ch14_1{
    public static void main(String[] args) {
        Test test = new Test();
        test.setX(new Integer(100));
        System.out.println("test.getX="+(Integer)test.getX());
        test.setX(new Double(100.0));
        System.out.println("test.getX="+(Double)test.getX());
        System.out.println("test.getX="+(String)test.getX());
```

    }
}

程序运行结果如图 14.1 所示。

```
test.getX=100
test.getX=100.0
Exception in thread "main" java.lang.ClassCastException: java.lang.Double cannot be cast to java.lang.String
        at Ch14_1.main(Ch14_1.java:17)
```

图 14.1 Object 类实现参数 "任意化"

分析：程序中定义的 Test 类中 x 成员变量是 Object 类型，其 setter 方法的形参、getter 方法的返回值也都是 Object 类型。这样做的目的是由于 Object 是所有 Java 类的父类，因此可以通过上转型向 x 赋任意类对象并返回任意类对象，从而实现参数 "任意化"。但存在的问题是：必须预知当前 x 的类型，并在用 getX()方法取值时，强制类型转化为该类型，如 (Integer)test.getX()，(Double)test.getX()等。当不小心进行(String)test.getX()这样的强制类型转换时，由于当前的 x 是 Double 型，虽然编译正确，程序执行时就会发生 ClassCastException 类型的异常。编译时就应该发现的错误，不应该被推迟到执行时才发现，Java 泛型就很好地解决了这一问题，它要求编译时的严格类型检查。

14.1.2 泛型类

泛型类是指包含类型参数的类。在泛型类的内部，类型参数可以用作变量的类型或方法返回参数的类型。一个泛型类可以包含多个类型参数。

定义泛型类的语法格式：

```
class 类名<泛型类型参数 1，泛型类型参数 2，…,泛型类型参数 n>{
    [访问权限] 泛型类型参数 成员变量名；
    [访问权限] 泛型类型参数 成员方法名称(){};
    [访问权限] 返回值类型 成员方法名称(泛型类型参数 变量名){};
}
```

其中，< >为泛型标识，< >中为类型参数，多个类型参数之间用逗号分隔。[访问权限]表示可选。

【**例题 14_2**】泛型类测试—运用泛型类实现例题 14_1。

```java
class Test1<T>{
    private T x;
    public T getX() {
        return x;
    }
    public void setX(T x) {
        this.x = x;
    }
}
public class Ch14_2 {
    public static void main(String[] args) {
        Test1<Integer> test = new Test1<Integer>();
        test.setX(new Integer(100));
        System.out.println("test.getX="+test.getX());
        Test1<Double> test1 = new Test1<Double>();
        test1.setX(new Double(100.0));
```

```
            System.out.println("test1.getX="+test1.getX());
            Test1<String> test2 = new Test1<String>();
            test2.setX(new String("100"));
            System.out.println("test2.getX="+test2.getX());
        }
    }
```
程序运行结果如图 14.2 所示。

```
test.getX=100
test1.getX=100.0
test2.getX=100
```

图 14.2 泛型类实现参数 "任意化"

分析：本例是用泛型的定义解决了用 Object 类实现参数 "任意化" 中存在的类型转化的安全隐患。<>是泛型标识，T 是泛型形参占位符，即在给 Test 类创建对象，给 x 赋值时再确定具体类型，当在创建泛型类对象时，分别用具体的类型如 Integer、Double、String 使得 T 具体化，且确保在用 setX 给 x 赋值时也只能是当前某种具体的类型，取值时无须进行强制类型转化。

<T>中的字母 T 也可以写成 A，B 等任意字符，写成<T>是约定俗成的做法。

注意：对于数值类型，只能用其包装类作为泛型类型参数的实参，而不能用简单数据类型作为泛型类型参数的实参，如：Test<Integer> test = new Test<Integer>();不能写成 Test<int> test = new Test<int>();。

【例题 14_3】 泛型类测试。
```
class Test2<T>{
    private T x;
    public T getX() {
        return x;
    }
    public void setX(T x) {
        this.x = x;
    }
}
public class Ch14_3 {
    public static void main(String[] args) {
        Test2<Integer> test = new Test2<Integer>();
//已将泛型确定成了 Integer 类型,不能向其存入非 Integer 类型的内容
        //test.setX(new Double(100));//编译时就出错
        System.out.println("test.getX="+test.getX());
    }
}
```

分析：泛型的引入，使得强制在编译时进行类型检查，若给泛型类型参数传递的实参前后不一致时，在编译阶段就会报错。

14.1.3 泛型构造方法

泛型构造方法的一般格式：
[访问权限] 构造方法名([<泛型类型参数>参数名称])

【例题 14_4】 用泛型构造方法创建对象。
```
class Test3<T>{
    private T x;
```

```java
        public Test3(T x) {
            this.x = x;
        }
        public T getX() {
            return x;
        }
        public void setX(T x) {
            this.x = x;
        }
    }
    public class Ch14_4 {
        public static void main(String[] args) {
            Test3<Integer> test = new Test3<Integer>(new Integer(100));
            System.out.println("test.getX="+test.getX());
        }
    }
```

程序运行结果：test.getX=100

分析：泛型类的构造方法中可以用泛型类型参数指定形参，当用该构造方法创建泛型类对象时，给其传递实参如给 T x 传递 new Integer(100)。

当给一个泛型类型参数不传递实参时如将程序中 Test3<Integer> test = new Test3<Integer>(new Integer(100));改成 Test3 test = new Test3(new Integer(100));时，系统会提示安全警告（见图 14.3），虽不影响程序执行，但建议务必给泛型类的泛型类型参数指定实参。

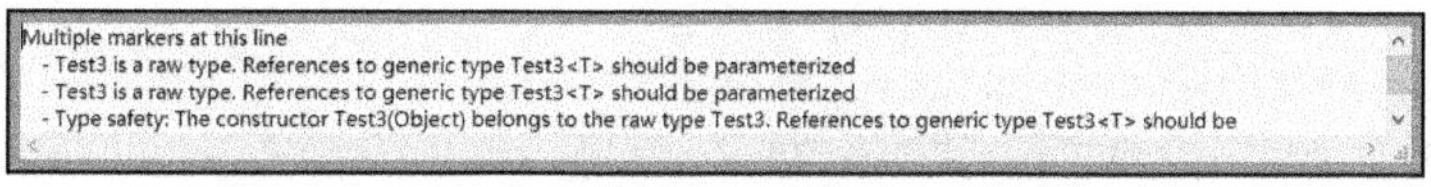

图 14.3　不指定泛型的具体类型时的警告信息

可以在泛型类中指定多个泛型类型参数。

【例题 14_5】具有多个类型参数的泛型类。

```java
class Test4<T1,T2>{
    private T1 x;
    private T2 y;
    public Test4(T1 x, T2 y) {
        this.x = x;
        this.y = y;
    }
    public T1 getX() {
        return x;
    }
    public void setX(T1 x) {
        this.x = x;
    }
    public T2 getY() {
        return y;
    }
    public void setY(T2 y) {
        this.y = y;
    }
```

```
}
public class Ch14_5 {
    public static void main(String[] args) {
        Test4<String,Integer> test = new Test4<String,Integer>(new String
("Hello"),new Integer(100));
        System.out.println("test.getX="+test.getX()+"\t"+"test.getY="+test.getY());
    }
}
```
程序运行结果：test.getX=Hello   test.getY=100

14.1.4 泛型方法

一个非泛型类中也可以定义泛型方法。

泛型方法的定义格式：

[访问权限] <泛型类型参数> 泛型参数 方法名称(泛型类型参数 参数名称){ }

【例题 14_6】 泛型方法测试。

```
class Test5{
    public <T> T getX(T t) {
        return t;
    }
}
public class Ch14_6 {
    public static void main(String[] args) {
        Test5 test = new Test5();
        Integer i=test.getX(new Integer(100));
        System.out.println("test.getX="+i);
    }
}
```
程序运行结果：test.getX=100

14.1.5 泛型通配符

在泛型中我们很多时候会用到泛型通配符。常用的泛型通配符主要有三种：无界通配符"？"、上限通配符 extends 和下限通配符 super。

1. 无界通配符"？"

泛型一个重要的特点是可以使用"？"作为通配符来匹配任何数据类型。泛型"？"通配符的格式：泛型类<?,?> test = new 泛型类<数据类型，数据类型>();

2. 上限通配符与下限通配符

有时我们会想限制可能出现的泛型类的类型范围上、下限。范围上限使用 extends 关键字声明，表示参数化的类型可能是所指定的类型或是此类型的子类；范围下限使用 super 进行声明，表示参数化的类型可能是所指定的类型，或是此类型的父类型，或为 Object 类。格式如下：

定义类：[访问权限] 类名称<泛型类型参数 extends 类>

声明对象：类名称<? extends 类名称> 对象名称

或：

定义类：[访问权限] 类名称<泛型类型参数 extends 类>

声明对象：类名称<? super 类名称> 对象名称

【例题 14_7】 设置泛型的实参类型上限为 Number 类。
```java
class Test6<T>{
    private T x;
    public Test6(T x) {
        this.x = x;
    }
    public T getX() {
        return x;
    }
    public void setX(T x) {
        this.x = x;
    }
}
public class Ch14_7 {
    public static void main(String[] args) {
        Test6<? extends Number> test = new Test6<Integer>(new Integer(100));
        //Integer 是 Number 类的子类
        //Test6<? extends Number> test1 = new Test6<String>(new String("100"));
         //非法，String 不是 Number 类的子类
        System.out.println("test.getX="+test.getX());
    }
}
```
程序运行结果：test.getX=100

说明：Byte、Double、Float、Integer、Long、Short、BigDecimal、BigInteger 等类都是 Number 类的子类，因此，当用<? extends Number>设置了泛型参数上限为 Number 时，只能以 Number 或其子类作实际参数化类型。String 不是 Number 的子类，不能作实际参数化类型。

可以将程序中 class Test6<T>改为 class Test6<T extends Number>，同时。将 Test6<? extends Number> test = new Test6<Integer>(new Integer(100));改为 Test6<Integer> test = new Test6<Integer>(new Integer(100));。

【例题 14_8】 设置泛型的实参类型下限。
```java
class Test7<T>{
    private T x;
    public T getX() {
        return x;
    }
    public void setX(T x) {
        this.x = x;
    }
}
public class Ch14_8 {
    public static void main(String[] args) {
        Test7<? super String> test =new Test7<String>();      //合法
        test.setX("abc");
        System.out.println("test.getX="+test.getX());
        Test7<? super Object> test1 =new Test7<Object>();     //合法
```

```
        test1.setX("abc");
        System.out.println("test1.getX="+test1.getX());
        // Test<? super Integer> test1 =new Test<Integer>();
        //非法，泛型实参只能是 String 类或其父类
    }
}
```
程序运行结果：
test.getX=abc
test1.getX=abc

14.1.6 泛型接口

JDK1.5 以后，Java 不仅支持泛型类，也支持泛型接口的定义和使用。格式如下：
[访问权限] interface 接口名<泛型类型参数 1, 泛型类型参数 2, …,泛型类型参数 n>{ }

【例题 14_9】泛型接口示例。

```
interface A<T1,T2>{
    public void setX(T1 x);
    public T1 getX();
    public void setY(T2 y);
    public T2 getY();
}
class Test8<T1,T2> implements A<T1,T2>{
    private T1 x;
    private T2 y;
    @Override
    public void setX(T1 x) {
        this.x=x;
    }
    @Override
    public T1 getX() {
        return x;
    }
    @Override
    public void setY(T2 y) {
        this.y=y;
    }
    @Override
    public T2 getY() {
        return y;
    }
}
public class Ch14_9{
    public static void main(String[] args) {
        Test8<String,Integer> test =new Test8<String,Integer>();
        test.setX(new String("abc"));
        test.setY(new Integer(100));
        System.out.println("test.getX="+test.getX()+"\t"+"test.getY="+test.getY());
    }
}
```
结果：test.getX=abc test.getY=100

14.1.7 子类泛型

Java 中允许子类泛型继承父类泛型。

【例题 14_10】 子类泛型测试。

```java
class Person<T1,T2> {
    private T1 x;
    private T2 y;
    public T1 getX() {
        return x;
    }
    public void setX(T1 x) {
        this.x = x;
    }
    public T2 getY() {
        return y;
    }
    public void setY(T2 y) {
        this.y = y;
    }
}
class Student<T1,T2,T3> extends Person<T1,T2>{
    private T3 z;
    public T3 getZ() {
        return z;
    }
    public void setZ(T3 z) {
        this.z = z;
    }
}
public class Ch14_10 {
    public static void main(String[] args) {
        Student<String,String,Integer> stu =new Student<String,String,Integer>();
        stu.setX(new String("2015001"));
        stu.setY(new String("张三"));
        stu.setZ(new Integer(18));
System.out.println("stu.getX="+stu.getX()+"\t"+"stu.getY="+stu.getY()+"\t"+"stu.getZ="+stu.getZ());
    }
}
```

程序运行结果：stu.getX=2015001   stu.getY=张三   stu.getZ=18

14.1.8 引入泛型的好处

1. 保障类型安全

泛型的主要目标是提高 Java 程序的类型安全。泛型允许编译器实施附加的类型约束，这样，类型错误就可以在编译时被捕获了，而不是在运行时当作 ClassCastException 展示出来。将类型检查从运行时提前到编译时有助于我们更容易地找到错误，并可提高程序的可靠性。

2. 消除强制类型转换

泛型的一个附带好处是，消除源代码中的许多强制类型转换。这使得代码更加可读，并且减少了出错机会。

14.2 集合类与接口

Java 集合类与接口是 Java 语言的重要组成部分，它包含了系统而完整的集合层次体系，封装了大量的数据结构的实现，这些集合类和接口都包含在 java.util 包中。Java 集合的核心接口为 Collection、List（列表）、Set（集合）和 Map（映射）。

14.2.1 Collection 接口

Collection 是 Java 集合继承树中最顶层的接口，几乎所有的 Java 集合框架成员都继承或实现了 Collection 接口，或者与其有密切关系。

Collection 提供了关于集合的通用操作。Set 接口和 List 接口都继承了 Collection 接口，而 Map 接口没有继承 Collection 接口。因此，Set 对象和 List 对象都可以调用 Collection 接口的方法，而 Map 对象则不可以。

Collection 接口的定义原型为：public interface Collection<E> extends Iterable<E>{}

Collection 接口采用了泛型定义，这样可以保证集合操作的安全性，避免发生 ClassCastException 异常。

Collection 接口的直接子接口是 Set 和 List 接口，Collection 接口包含了这两个直接子接口的通用方法：

- boolean add(E e)：向容器中添加一个元素。
- boolean addAll(Collection<? extends E> c)：将指定 collection 中的所有元素都添加到此 collection 中。
- boolean contains(Object o)：如果此 collection 包含指定的元素，则返回 true。
- boolean containsAll(Collection<?> c)：如果此 collection 包含指定 collection 中的所有元素，则返回 true。
- boolean isEmpty()：如果此 collection 不包含元素，则返回 true。
- Iterator<E> iterator()：返回一个 Iterator<E>，用来遍历集合框架中所有元素。
- int size()：返回此 collection 中的元素数。
- Object[] toArray()：返回包含此 collection 中所有元素的数组。
- <T> T[] toArray(T[] a)：返回包含此 collection 中所有元素的数组。

14.2.2 Iterator 接口

Iterator 迭代接口是专门用来进行迭代输出的接口，使用的比较普遍。其定义是：
```
public interface Iterator<E>{}
```
Iterator 接口定义了对 Collection 类型对象中所含元素的遍历的处理功能,可以通过 Collection 接口中定义的 iterator()方法获得一个对应的 Iterator(实现类)对象。

Iterator 接口常用方法：

- boolean hasNext()：如果集合框架中没有被遍历完毕，则返回 true。
- E next()：返回迭代的下一个元素。
- void remove()：从迭代器指向的 collection 中移除迭代器返回的最后一个元素。必须先调用一次 next()方法后，才能调用一次 remove()方法。

14.2.3 List 接口

List 接口的主要特征把加入集合的对象以线性方式存储，即按照对象加入集合的顺序存放，并且允许存放重复的对象。该接口的实现类主要有 ArrayList 和 LinkedList。

List 接口的定义：public interface List<E> extends Collections<E>{}。该接口除了继承 Collection 接口的方法外，还扩展了很多新的方法：

- boolean add(E e)：向列表的尾部添加指定的元素。
- void add(int index,E element)：在列表的指定位置插入指定元素。
- E get(int index)：返回列表中指定位置的元素。
- int indexOf(Object o)：返回此列表中第一次出现的指定元素的位置；如果此列表不包含该元素，则返回 -1。
- int lastIndexOf(Object o)：返回此列表中最后出现的指定元素的位置；如果列表不包含此元素，则返回 -1。
- List<E> subList(int fromIndex,int toIndex)：返回集合中指定的 fromIndex（包括）和 toIndex（不包括）之间的子集合。
- E set(int index,E element)：用指定元素替换列表中指定位置的元素。
- boolean isEmpty()：如果列表不包含元素，则返回 true。

14.2.4 ArrayList 类

在实际应用中常常会遇到需要动态操纵数组，比如在运行时增加和删除数组元素，而且有时在编译时又不想确定数组大小希望它可以动态伸缩，在 java 中解决这一问题的方法是使用 java.util 包中的 ArrayList 类。ArrayList 是 List 接口的一个可变长数组实现。

【例题 14_11】ArrayList 常用方法测试。

```java
import java.util.ArrayList;
import java.util.Iterator;
public class Ch14_11 {
    public static void main( String[] args) {
    ArrayList<Integer> al = new ArrayList<Integer>();
    for(int i=0;i<10;i++){
    al.add(new Integer(i));                //给 al 添加 Integer 对象元素
    }
    System.out.println("al 中的元素为: ");
    /**遍历输出 al 中的各元素值*/
    for(int i=0;i<al.size();i++){
       System.out.printf("%4d",al.get(i));
    }
     al.remove(7);                          //移除 al 中 7 下标的元素
     al.set(7,new Integer(100));            //在 al 的 7 下标处插入元素 100
```

```
        Iterator<Integer> iter=al.iterator();     //得到迭代器对象
        System.out.println("\nal 中的元素为: ");
        /**用迭代器方式遍历输出 al 中的各元素值*/
        while(iter.hasNext()) {
            Integer integer =(Integer)iter.next();
            System.out.printf("%4d",integer);
        }
    }
}
```

程序运行结果如图 14.4 所示。

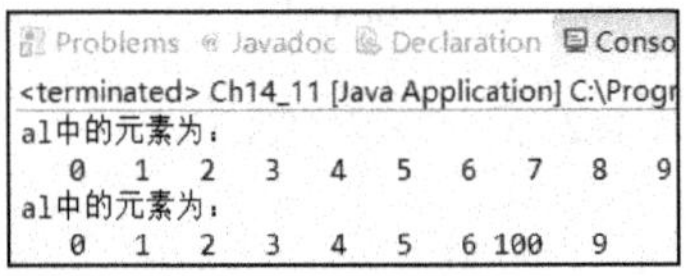

图 14.4 ArrayList 遍历结果

【例题 14_12】List 及 ArrayList 常用方法测试。

```java
import java.util.ArrayList;
import java.util.Iterator;
import java.util.List;
public class Ch14_12 {
    public static void main(String[] args) {
        ArrayList<String> list = new ArrayList<String>();
        System.out.println("集合是否为空"+list.isEmpty());
        list.add("红楼梦");
        list.add("西游记");
        list.add("三国演义");
        System.out.println("集合是否为空"+list.isEmpty());
        /**正序、逆序遍历输出 list 中各元素的值*/
        System.out.println("正序输出 list 元素: ");
        for(int i=0;i<list.size();i++){
            System.out.print(list.get(i)+"\t");
        }
        System.out.println("\n 逆序输出 list 元素: ");
        for(int i=list.size()-1;i>=0;i--){
            System.out.print(list.get(i)+"\t");
        }
        /**判断 list 中是否包含"水浒传",若不包含则添加*/
        if(!list.contains("水浒传"))
            list.add("水浒传");
        /**截取 list 并 用迭代器方式输出*/
        List<String> sublist=list.subList(0, 2);
        Iterator<String> iter=sublist.iterator();
        System.out.println("\nsublist 遍历结果: ");
        while(iter.hasNext())
            System.out.print(iter.next()+"\t");
        /**将 list 转为数组并输出*/
        String str[]=list.toArray(new String[]{});
        System.out.println("\n 数组遍历结果: ");
        for(String s:str)
            System.out.print(s+"\t");
    }
}
```

程序运行结果如图 14.5 所示。

图 14.5 对 ArrayList 的各种操作结果

14.2.5　LinkedList 类

利用 ArrayList 实现了数据结构的顺序存储结构，利用 List 的另一个常用实现类是 LinkedList，它实现了数据结构的链式存储结构。除了 List 提供的方法之外，LinkedList 还提供了其他方法来支持堆栈、队列和双向队列的操作方法：

- public void addFirst(E e)：将指定元素插入此列表的开头。
- public void addLast(E e)：将指定元素添加到此列表的结尾。
- public boolean offer(E e)：将指定元素添加到此列表的末尾（最后一个元素）。
- public boolean offerFirst(E e)：在此列表的开头插入指定的元素。
- public boolean offerLast(E e)：在此列表末尾插入指定的元素。
- public E peek()：获取但不移除此列表的头（第一个元素）。
- public E peekFirst()：获取但不移除此列表的第一个元素；如果此列表为空，则返回 null。
- public E peekLast()：获取但不移除此列表的最后一个元素；如果此列表为空，则返回 null。
- public E poll()：获取并移除此列表的头（第一个元素）
- public E pollFirst()：获取并移除此列表的第一个元素；如果此列表为空，则返回 null。
- public E pollLast()：获取并移除此列表的最后一个元素；如果此列表为空，则返回 null。
- public E pop()：从此列表所表示的堆栈处弹出一个元素。
- public void push(E e)：将元素推入此列表所表示的堆栈。
- public E remove()：获取并移除此列表的头（第一个元素）。
- public E remove(int index)：移除此列表中指定位置处的元素。
- public E removeLast()：移除并返回此列表的最后一个元素。

【例题 14_13】运用 LinkedList 模拟队列的先进先出操作。

```
import java.util.LinkedList;
public class Ch14_13 {
    public static void main(String[] args) {
        LinkedList<Integer> queue=new LinkedList<Integer>();
        for(int i=1;i<=10;i++){
            queue.add(new Integer(i));//向队列中添加元素
        }
        System.out.println("队列的元素个数："+queue.size());//输出队列的元素个数
        System.out.print("输出队列元素（先进先出方式）：");
        for(int j=1;j<=10;j++){
            System.out.print(queue.poll()+" ");//以先进先出方式输出队列元素
        }
        System.out.println("\n 队列的元素个数："+queue.size());//输出队列的元素个数
    }
}
```

程序运行结果如图 14.6 所示。

图 14.6　LinkedList 实现队列结果

【例题 14_14】 运用 LinkedList 模拟堆栈的先进后出操作。

```java
import java.util.LinkedList;
public class Ch14_14 {
    public static void main(String[] args) {
        LinkedList<Integer> stack=new LinkedList<Integer>();
        for(int i=1;i<=10;i++){
            stack.push(new Integer(i));              //向栈中添加元素
        }
        System.out.println("栈中的元素个数: "+stack.size());//输出栈中的元素个数
        System.out.print("输出栈中元素（先进后出方式）: ");
        while(!stack.isEmpty()){                     //栈不为空
            System.out.print(stack.pop()+" ");       //弹出栈顶元素
        }
        System.out.println("\n 栈中的元素个数: "+stack.size());//输出栈中的元素个数
    }
}
```

程序运行结果如图 14.7 所示。

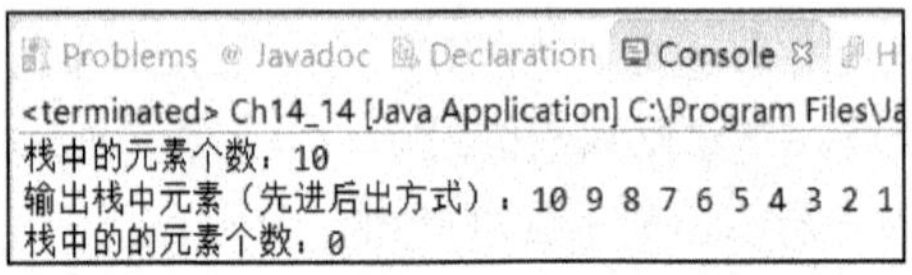

图 14.7 LinkedList 实现堆栈结果

除了运用 LinkedList 可以实现堆栈的功能外，Java 本身提供了 Stack 类，专门用于实现堆栈操作。

Stack 类的定义：public class Stack<E> extends Vector<E>。Stack 类表示后进先出（LIFO）的堆栈。

Stack 类的常用方法：

- public boolean empty()：判断堆栈是否为空。
- public E peek()：查看堆栈顶部的对象，但不从堆栈中移除它。
- public E pop()：移除堆栈顶部的对象，并作为此函数的值返回该对象。
- public E push(E item)：把 item 压入堆栈顶部。
- public int search(Object o)：在栈中查找，返回对象到堆栈顶部的位置，以 1 为基数；返回值 -1 表示此对象不在堆栈中。

【例题 14_15】 Stack 类测试。

```java
import java.util.Stack;
public class Ch14_15 {
    public static void main(String[] args) {
        Stack<String> stack=new Stack<String>();
        stack.push(new String("A"));
        stack.push(new String("B"));
        stack.push(new String("C"));                    //向栈中添加元素
        System.out.println("栈顶元素为: "+stack.peek());   //输出栈顶元素
        System.out.println("在栈中查找元素 B: " +stack.search("B"));
        System.out.print("输出栈中元素（先进后出方式）: ");
        while(!stack.isEmpty()){
```

```
            System.out.print(stack.pop()+" ");              //弹出栈顶元素
        }
        System.out.println("\n在栈中查找元素C: " +stack.search("C"));
    }
}
```
程序运行结果如图 14.8 所示。

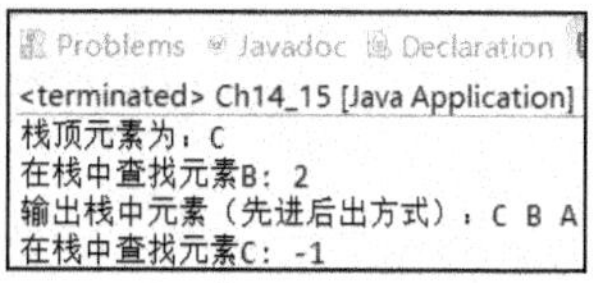

14.2.6 Collections 类

图 14.8 对 Stack 类的操作结果

有时需要对链表中的结点进行排序，以便查找一个数据是否与链表中某结点数据相等。或将链表中结点数据进行重新随机排列、移位或置逆等操作，Collections 类提供了进行这些操作的方法。

- public static <T> boolean addAll(Collection<? super T> c,T... elements)：将所有指定元素添加到指定 collection 中。
- public static <T extends Comparable<? super T>> void sort(List<T> list)：该方法可以将 list 中的元素升序排列。
- public static <T> int binarySearch(List<? extends T> list,T key,Comparator<? super T> c)：使用折半法查找 list 是否含有和参数 key 相等的元素，如果 key 链表中某个元素相等，方法返回和 key 相等的元素在链表中的索引位置（链表的索引位置从 0 考试），否则返回-1。
- public static void shuffle(List<?> list)：将 list 中的数据重新随机排列。
- public static void rotate(List<?> list,int distance)：根据指定的距离轮换指定列表中的元素。
- public static void reverse(List<?> list)：反转指定列表中元素的顺序。
- public static <T> boolean replaceAll(List<T> list,T oldVal,T newVal)：使用另一个值替换列表中出现的所有某一指定值。

【例题 14_16】Collections 常用方法测试。
```
import java.util.ArrayList;
import java.util.Collections;
import java.util.Iterator;
public class Ch14_16 {
    public static void main(String[] args) {
        ArrayList<Integer> al = new ArrayList<Integer>();
        Collections.addAll(al,45,23,67,1,78,90,10,111);
        //给 al 添加 Integer 对象元素
        System.out.println("初始时 al 中的元素为:");
         Iterator<Integer> iter=al.iterator();
         while(iter.hasNext()) {
            System.out.print(iter.next()+" ");
        }//遍历链表并输出各元素值
        Collections.sort(al);          //对 al 中的数据按升序排序
        System.out.println("\n 排序后 al 中的元素为:");
        iter=al.iterator();
         while(iter.hasNext()) {
            System.out.print(iter.next()+" ");
         }
        int index=Collections.binarySearch(al, 78,null);//二分查找元素 78 的下标位置
```

```
            System.out.println("\n 数据 78 在链表中的位置是: "+index);
            Collections.reverse(al);       ////对 al 中的数据置逆
            System.out.println("对 al 中的数据进行置逆后的结果是:");
            iter=al.iterator();
            while(iter.hasNext()) {
                System.out.print(iter.next()+"  ");
            }
            Collections.rotate(al, -2);  //让 al 中的数据左移 2 位
            System.out.println("\n 让 al 中的数据左移 2 位后的结果是:");
            iter=al.iterator();
            while(iter.hasNext()) {
                System.out.print(iter.next()+"  ");
             }
            Collections.shuffle(al);    //对 al 中的数据进行随机排列
            System.out.println("\n 对 al 中的数据进行随机排列后的结果是:");
            iter=al.iterator();
            while(iter.hasNext()) {
                System.out.print(iter.next()+"  ");
            }//执行结果不唯一


    }
}
```
程序运行结果如图 14.9 所示。　　　　　　　　　　　　　　　图 14.9　Collections 测试结果

【例题 14_17】将类 Person 对象作为 ArrayList 结点，对各个对象按年龄排序并进行二分查找。
```
package myproject.collections;
import java.util.ArrayList;
import java.util.Collections;
import java.util.Iterator;
class Person implements Comparable{  //实现 Comparable 接口
    private String name;
    private int age;
    public Person(String name, int age) {
        this.name = name;
        this.age = age;
    }
    public String getName() {
        return name;
    }
    @Override
    public int compareTo(Object o) {//实现 Comparable 接口的 compareTo(Object o)方法
        Person person=(Person)o;
        if(this.age>person.age)
            return 1;
        else if(this.age<person.age)
            return -1;
        else
            return this.name.compareTo(person.name);
    }
    @Override
    public boolean equals(Object obj) {//重写 Object 类的 equals 方法
```

```java
            Person person=(Person)obj;
            if(this.age==person.age&&(this.name).equals(person.name))
                return true;
            else
                return false;
        }
        @Override
        public String toString() {//重写 Object 类的 toString()方法
            return name+" "+age;
        }
    }
    public class Ch14_17 {
        public static void main(String[] args) {
            ArrayList<Person> al = new ArrayList<Person>();
            al.add(new Person("张三",21));
            al.add(new Person("李四",18));
            al.add(new Person("王五",19));
            al.add(new Person("赵六",20));
            Iterator<Person> iter=al.iterator();
            System.out.println("初始时 al 中的元素为:");
            while(iter.hasNext())
              System.out.println(iter.next());
            Collections.sort(al);//排序
            iter=al.iterator();
            System.out.println("排序后 al 中的元素为:");
            while(iter.hasNext())
              System.out.println(iter.next());
            Person p=new Person("赵六",20);//创建 Person 对象
            int index=Collections.binarySearch(al,p, null);
            //二分查找与对象 p 相等的结点并返回其下标值
            if(index>=0){  //若找到
                System.out.println(p.getName()+"在链表 al 中的下标位置为: "+index);
            }
        }
    }
```

程序运行结果如图 14.10 所示。

图 14.10 类对象的排序与查找

14.2.7 Set 接口

Set 接口继承 Collection 接口, 而且它不允许集合中存在重复项, 每个具体的 Set 实现类依赖添加的对象的 equals()方法来检查独一性。Set 接口没有引入新方法, 所以 Set 就是一个 Collection, 只不过比 Collection 接口要求更加严格, 即不许增加重复元素。

Set 接口是 Collection 的子接口。其定义是
```
public interface Set<E> extends Collection<E>{}
```
Set 接口主要有 HashSet 和 TreeSet 两个实现类。

14.2.8 HashSet 类

该类实现了 Set 接口, 该类按照哈希算法来存取集合中的对象, 该类表示的集合是无序的, 当向容器中加入一个对象时, HashSet 会调用对象的 hashCode()方法来获取哈希码, 然后根据这

个哈希码进一步计算出对象的存放位置。

【例题 14_18】 向 HashSet 中增加元素并输出。

```java
import java.util.HashSet;
import java.util.Set;
public class Ch14_18 {
    public static void main(String[] args) {
        Set<String> set= new HashSet<String>();
        set.add("A");
        set.add("A");
        set.add("B");
        set.add("C");
        set.add("C");
        set.add("D");
        System.out.println("set 中的元素为: "+set);
    }
}
```

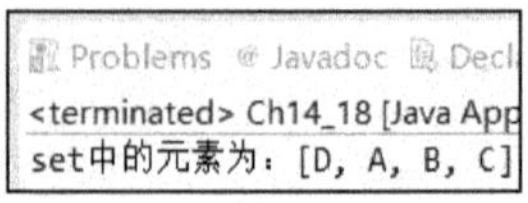

程序运行结果如图 14.11 所示。

图 14.11　HashSet 输出结果

分析：程序中虽然多次向 set 中增加了元素"A"、"C"，但对于重复元素只增加一次，表明不可以向 HashSet 增加重复元素；输出结果并不是按照元素的增加顺序依次输出的，表明 HashSet 中的元素是无序排列的。

14.2.9　TreeSet 类

TreeSet 类实现了 SortedSet 接口，能够对容器中的对象进行排序。当向 TreeSet 中加入一个对象后，会继续保持对象间的排序的次序。

在 JDK 类库中，Integer、Double 和 String 等系统类已经实现了 java.lang.Comparable 接口，并重写了 Comparable 接口的 compareTo(Object o)方法，使得 x.compareTo(y)时，如果返回值为 0，则表示 x 和 y 相等；如果返回值大于 0，则表示 x 大于 y；如果返回值小于 0，则表示 x 小于 y。TreeSet 调用对象的 compareTo()方法比较容器中对象的大小，然后进行升序排列。

向 TreeSet 集合加入没有实现 Comparable 接口并重写其 compareTo(Object o)方法的类对象时，要求该类必须先实现 Comparable 接口并重写其 compareTo(Object o)方法，以在 compareTo(Object o)方法中指定比较规则，否则将出现编译错误。

【例题 14_19】 TreeSet 测试。

```java
import java.util.Set;
import java.util.TreeSet;
public class Ch14_19 {
    public static void main(String[] args) {
        Set<String> set= new TreeSet<String>();
        set.add("A");
        set.add("A");
        set.add("B");
        set.add("C");
        set.add("C");
        set.add("D");
```

```
        System.out.println("set 中的元素为: "+set);
    }
}
```

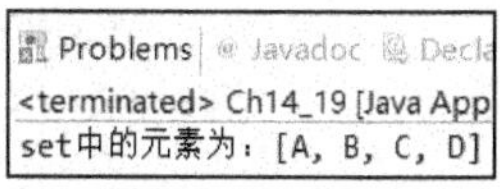

程序运行结果如图 14.12 所示。

图 14.12 TreeSet 输出结果

分析：结果表明 TreeSet 中不能加入重复元素，同时将 String 类型的元素按 Unicode 字符编码升序进行了排序。

【例题 14_20】对自定义类 Person 对象按年龄升序排序。

```java
package myproject.treeset;
import java.util.Iterator;
import java.util.Set;
import java.util.TreeSet;
class Person implements Comparable{   //实现 Comparable 接口
    private String name;
    private int age;
    public Person(String name, int age) {
        this.name = name;
        this.age = age;
    }
    @Override
    public int compareTo(Object o) {//实现 Comparable 接口的 compareTo( )方法
        Person person=(Person)o;
        if(this.age>person.age)
            return 1;
        else if(this.age<person.age)
            return -1;
        else
            return this.name.compareTo(person.name);//若年龄相等，则按姓名排序
    }
    @Override
    public int hashCode() {   //重写 Object 类的 hashCode()方法
        return name==null?0:name.hashCode();
    }
    @Override
    public boolean equals(Object obj) {   //重写 Object 类的 equals()方法
        if(!(obj instanceof Person))
            return false;
        Person person=(Person)obj;
        if(this.name.equals(person.name)&&this.age==person.age)//姓名、年龄都相等
            return true;
        else
            return false;
    }
    @Override
    public String toString() {            //重写 Object 类的 toString()方法
        return "姓名: "+name+"\t 年龄: "+age;
    }
}
public class Ch14_20 {
```

```java
    public static void main(String[] args) {
        Set<Person> set= new TreeSet<Person>();   //创建TreeSet对象
        set.add(new Person("张三",19));
        set.add(new Person("李四",18));
        set.add(new Person("王五",20));
        set.add(new Person("赵六",19));
        //在TreeSet中添加对象
        Iterator<Person> iter=set.iterator();
        while(iter.hasNext()){                    //用迭代器遍历输出
          System.out.println(iter.next());
        }
    }
}
```

程序运行结果如图 14.13 所示。

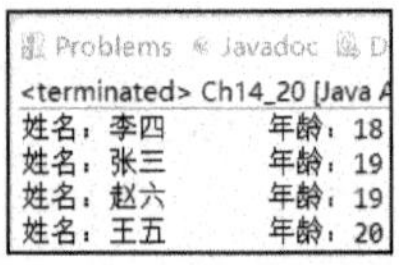

图 14.13　Person 类对象排序结果

分析：对于自定义类 Person，必须通过实现 Comparable 接口并给其 compareTo()方法给出具体的方法实现并指明排序规则，以达到预期的排序目的。重写 Object 类的 equals ( )方法以及 hashCode()方法的目的是为了剔除 set 中重复的元素。

14.2.10　EnumSet 类

EnumSet 类是实现过 Set 接口的类，因此，也不允许其中的元素重复。可以用 EnumSet 调用其一系列静态方法得到类 EnumSet 对象,不能通过 new 给类 EnumSet 创建对象。通过 EnumSet 类调用其 allOf()方法，可以将枚举成员放到 EnumSet 集合中。

【例题 14_21】EnumSet 类方法测试。

```java
package myproject.enumset;
import java.util.EnumSet;
enum WeekDay {
    MONDAY,TUESDAY,WEDNESDAY,THURSDAY,FRIDAY,SATURDAY,SUNDAY;
//定义枚举成员
}
public class Ch14_21 {
    public static void main(String[] args) {
        EnumSet<WeekDay> enumset=EnumSet.allOf(WeekDay.class);
        //将所有枚举成员放到 EnumSet 对象 enumset 中
        System.out.println("enumset 中的元素为: ");
        for(WeekDay w:enumset){
          System.out.print(w+" ");
        }
    }
}
```

程序运行结果如图 14.14 所示。

图 14.14　enumset 输出结果

14.2.11　Map 接口

Map（映射）是用于存储{关键字，值}对的对象。关键字必须是唯一的。给定一个关键字，就可以得到它的值。

Map 接口定义：public interface Map<K,V>{}

Map 接口采用泛型，其中泛型类型参数 K 表示键对象(Key)，泛型类型参数 V 表示值对象（Value），当向 Map 集合加入元素时，必须指定 K 和 V。

Map 接口的主要方法：

- void clear()：清空 Map。
- boolean containsKey(Object key)：判断指定的 Key 是否存在。
- boolean containsValue(Object value)：判断指定的 value 是否存在。
- Set<Map.Entry<K,V>> entrySet()：将 Map 对象变为 Set 集合。
- V get(Object key)：通过 key 取得 value，若此映射不包含该键的映射关系，则返回 null。
- boolean isEmpty()：判断集合是否为空。
- Set<K> keySet()：取得所有的 key。
- V put(K key, V value)：向集合中加入元素。
- void putAll(Map<? extends K,? extends V> m)：将一个 Map 集合中的内容加入到另一个 Map。
- V remove(Object key)：根据 key 删除 value。
- int size()：返回此映射中的键-值映射关系数。
- Collection<V> values()：取出全部的 value。

14.2.12　HashMap 类

HashMap 是 Map 的实现类，通过计算键对象的哈希码(hasCode())来保存保存键对象，该集合没有进行排序，并且 key 值不能重复(通过 key 对象的 equals()方法进行比较没有重复)，若 key 值有重复，则最后一次的对象会覆盖原来的对象。

【例题 14_22】给 HashMap 中添加结点，每个结点包括一个学生的（学号，姓名）信息并输出结点的值。

```java
import java.util.HashMap;
import java.util.Set;
public class Ch14_22{
    public static void main(String[] args) {
        HashMap<String,String> hm=new HashMap<String,String>();
        hm.put("20150001","张三");
        hm.put("20150002","李四");
        hm.put("20150003","王五");//给 hm 添加结点，每个结点都是(key,value)形式
        System.out.println("键值为 20150003 的学生姓名为: "+hm.get("20150003"));
        //通过 key 获得 value 值
        hm.put("20150003", "赵六");//通过 key 修改 value 值
        System.out.println("修改后键值为 20150003 的学生姓名为: "+hm.get("20150003"));
        /**遍历输出所有的 HashMap 信息*/
        Set<String> keySet = hm.keySet();
        for(String key:keySet){
            System.out.println("key:"+key +"  value:"+hm.get(key));}
    }
}
```

程序运行结果如图 14.15 所示。

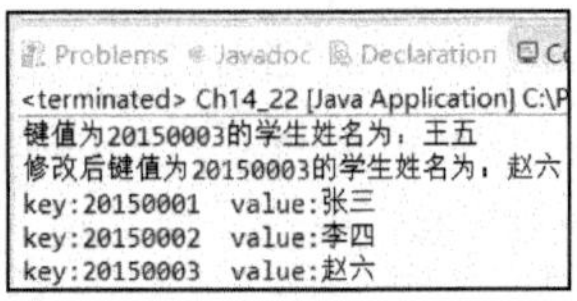

在对 Map 集合进行操作时，经常需要分别取得 key 和 value 的信息，其中 key 和 value 也分别是集合，可以利用迭代器 Iterator 进行获取。

【例题 14_23】 输出 HashMap 中全部的 key 或 value。

图 14.15　HashMap 测试结果

```java
import java.util.Collection;
import java.util.HashMap;
import java.util.Iterator;
import java.util.Set;
public class Ch14_23 {
    public static void main(String[] args) {
        HashMap<String,String> hm=new HashMap<String,String>();
        hm.put("20150001","张三");
        hm.put("20150002","李四");
        hm.put("20150003","王五");
        Set<String> keys=hm.keySet();//得到全部 key
        Iterator<String> iter=keys.iterator();
        System.out.println("hm 中所有的 key:");
        while(iter.hasNext())
            System.out.print(iter.next()+" ");
        Collection<String> values=hm.values();//得到全部 value
        iter=values.iterator();
        System.out.println("\nhm 中所有的 value:");
        while(iter.hasNext())
            System.out.print(iter.next()+"\t");
    }
}
```

程序运行结果如图 14.16 所示。

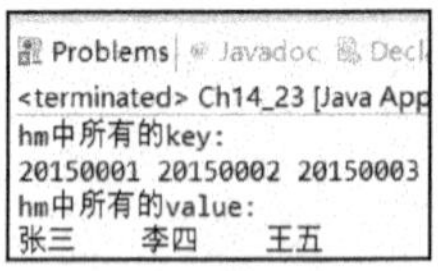

图 14.16　HashMap 中的 key 与 value

14.2.13　SortedMap 接口与 TreeMap 类

使用 SortedMap 接口，可以确保键处于排序状态。SortedMap 接口的常用方法：

- Comparator<? super K> comparator()：返回当前 Map 使用的比较器 Compartor。
- K firstKey()：返回 Map 中的第一个键。
- K lastKey()：返回 Map 中的最后一个键。
- SortedMap<K,V> subMap(K fromKey,K toKey)：生成此 Map 的子集，由键大于或等于 fromKey，小于 toKey 的所有键值对组成。
- SortedMap<K,V> tailMap(K fromKey)：生成此 Map 的子集，由键大于或等于 fromKey 的所有键值对组成。
- SortedMap<K,V> headMap(K toKey)：生成此 Map 的子集，由键小于 toKey 的所有键值对组成。

TreeMap 类是接口 SortedMap 的唯一实现。该映射根据创建映射时提供的 Comparator 进行排序。TreeMap 是唯一带有 subMap() 方法的 Map，它可以返回一个子树。

【例题 14_24】 TreeMap 常用方法测试。
```java
import java.util.Iterator;
import java.util.Set;
```

```
import java.util.TreeMap;
public class Ch14_24 {
    public static void main(String[] args) {
    TreeMap<String,String> tm=new TreeMap<String,String>();
    tm.put("20150018","张三");
    tm.put("20150001","李四");
    tm.put("20150045","王五");
    tm.put("20150023","赵六");
    System .out.println("tm中第一个元素的key:"+tm.firstKey()+"tm中第一个元素的key对应的Value:"+tm.get(tm.firstKey()));
    System .out.println("tm中最后一个元素的key:"+tm.lastKey()+"tm中最后一个元素的key对应的Value:"+tm.get(tm.lastKey()));
    System.out.println("tm中的键-值对为: ");
    System.out.println(tm);//以{键1=值1，…，键n=值n}形式输出tm
    Set<String> set=tm.keySet();//得到所有的key
    Iterator <String> iter=set.iterator();//得到迭代器
    System.out.println("以键—值形式输出tm");
    while(iter.hasNext()){
        String key=iter.next();//得到一个键值
        String value=tm.get(key);//得到对应的值
        System.out.print(key+"—"+value+"  ");
    }
    System.out.println("\n以{键1=值1,……键n=值n}形式输出某范围的键=值对: ");
    System.out.println(tm.subMap("20150018", "20150045"));//输出Map的子集
    }
}
```

程序运行结果如图 14.17 所示。

图 14.17 TreeMap 测试结果

14.2.14 EnumMap 类

EnumMap 是一种特殊的 Map，它要求所有的键(Key)都必须来自同一个枚举类型。该类与前面提到的 EnumSet 类都是集合框架对枚举类型的操作类。要使用 EnumMap，需要用其构造方法如 public EnumMap(Class<K> keyType)创建 EnumMap 对象。

【例题 14_25】EnumMap 测试。
```
package myproject.enummap;
import java.util.EnumMap;
import java.util.Map;
enum WeekDay {
    MONDAY,TUESDAY,WEDNESDAY,THURSDAY,FRIDAY,SATURDAY,SUNDAY;//定义枚举成员
}
public class Ch14_25 {
```

```java
    public static void main(String[] args) {
        Map<WeekDay,String> map=new EnumMap<WeekDay,String>(WeekDay.class);
        //创建EnumMap对象
        map.put(WeekDay.MONDAY, "星期一");
        map.put(WeekDay.TUESDAY, "星期二");
        map.put(WeekDay.WEDNESDAY, "星期三");
        map.put(WeekDay.THURSDAY, "星期四");
        map.put(WeekDay.FRIDAY, "星期五");
        map.put(WeekDay.SATURDAY, "星期六");
        map.put(WeekDay.SUNDAY, "星期日");
        /**两种不同方法取得map中{键，值}中的值*/
        for(WeekDay w:WeekDay.values())
            System.out.print(map.get(w)+" ");
        System.out.println();
        for(String s:map.values())
            System.out.print(s+" ");
    }
}
```

程序运行结果如图 14.18 所示。

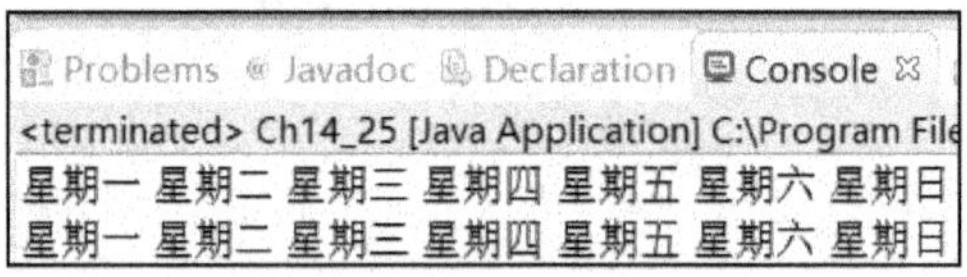

图 14.18　EnumMap 测试结果

说明：map 通过调用 put()方法将 enum 的成员作为键，并给出该键对应的值来给 EnumMap 添加元素。通过调用 get()方法取得某个键值对应的值。也可以通过遍历 map.values()得到 EnumMap 中个元素的值。

小　　结

JDK1.5 以后，Java 通过引入泛型的概念，很好地解决了此前"类型安全"的问题。本章较系统地阐明了 Java 泛型类、泛型构造方法、泛型方法、泛型接口、子类泛型各自的含义及典型用法。同时，对 Java 中以泛型方式定义的接口 Collection、Iterator、List、Map，以及与之相关的集合类 ArrayList、LinkedList、HashSet、TreeSet、EnumSet、HashMap、TreeMap、EnumMap 等的典型应用方法进行了系统阐述。

习　题　14

上机实践题

1. 自定义一个泛型类 Generic<T>，在其中定义含参构造方法以及一个 getter()方法。自定义一个学生类 Student，在其中定义成员变量 name、sex、age，并以这 3 个成员变量定义一个含参构造方法。编写主类，在 main()方法中创建 3 个 Generic<T>对象，分别将 T 确定为一个 String 对象、一个 Integer 对象以及一个 Student 对象并输出这些对象信息（用这 3 个 Generic<T>对象调用其

getter()方法)。

2. 用至少两种方法编写程序实现：当堆栈为空时，将 0、1、2、3、4 压入堆栈；当堆栈不为空时，从堆栈中弹出这些元素并输出。

3. 可以利用 List 接口的 Collections.sort(List a)对 a 进行排序，利用 add(int index, Long element)方法可将 element 插入到 index 指定的位置。编程实现：先对一个数据序列排序，当输入一个数，要求按序将它插入该数据序列中。

第 15 章　Java Applet 编程简介

【本章内容提要】
- Java Applet 简介；
- Apple 程序的编写方法；
- Applet 类的主要方法；
- <apple>标记的属性及 Applet 参数传递；
- Applet 的组件与事件处理。

15.1　Java Applet 简介

Java Applet 被称为小应用程序，它是嵌入在网页中被浏览器内置的 Java 虚拟机加载执行的一种 Java 程序。

Java Applet 可以实现图形绘制、字体和颜色控制、实现动画和声音的播放、人机交互等功能。 随着网络编程技术的日益丰富多样，现在 Java Applet 的应用已远不如以前频繁。但它作为 Java SE 程序的一种重要类型，还是有必要了解一下。

Java Applet 源程序被编译后，会产生.class 字节码文件，将.class 的文件嵌在.html 的网页中并随着网页文件被下载到客户端，由支持 Java（具有内置 JVM）的浏览器解释执行。

15.2　Apple 程序的编写方法

一个 Java Applet 程序中必须有一个类是系统类 java.applet.Applet 类或 javax.swing.JApplet 的子类，该类被称为该 Applet 的主类，同时，该主类要被修饰为 public 的。

java.applet.Applet 类的继承关系如图 15.1 所示，Applet 类是一个 AWT 类，设计它时没有考虑与 Swing 组件一起工作。要 Java Applet 中使用 Swing 组件，需要通过扩展 javax.swing.JApplet 来创建 Java Applet，javax.swing.JApplet 是 javax.applet.Applet 的一个子类。JApplet 继承了 Applet 类的所有方法，而且支持放置 Swing 组件。javax.swing.JApplet 类的继承关系如图 15.2 所示。

【例题 15_1】在 Applet 中绘制文本。
```
import java.applet.Applet;
import java.awt.Color;
import java.awt.Graphics;
import java.awt.Font;
public class Ch15_1 extends Applet{        //继承 Applet 类
```

```
    public void init(){                      //重写Applet的init()方法
        this.setBackground(Color.PINK);      //设置背景色
    }
    public void paint(Graphics g){           //重写Container类的paint方法
        Color c1=new Color(255,0,0);
        Color c2=new Color(10,0,245,100);
        Color c3=Color.GREEN;
        Font font1=new Font("宋体",Font.BOLD+Font.ITALIC,20);
        String s="好好学习";
        g.setColor(c1);
        g.setFont(font1);
        g.drawString(s,40,40);
        g.setColor(c2);
        g.drawString(s,40,80);
        g.setColor(c3);
        g.drawString(s,40,120);
    }
}
```

程序运行结果如图 15.3 所示。

图 15.1　Applet 类的继承关系　　图 15.2　JApplet 类的继承关系　　图 15.3　Applet 上绘制文字

若在 Eclipse 环境下运行 Ch15_1.java，只需在 Ch15_1.java 上右击，选择"Run As"→"Java Applet"命令，或用 Run 菜单中的 Run 命令，或单击工具栏中的 Run 按钮，执行程序得到图 15.3 所示的结果。

若在记事本等编辑环境下编辑了该源程序，则按下面步骤测试该程序：

- 先将源程序以 Ch15_1.java 进行保存（E:\\下）；
- 在命令提示符下用 javac 命令编译 Ch15_1.java；
- 编辑一个 .html 格式文件并以 Ch15_1.html 为名保存在与 Ch15_1.java 相同的目录下。
  Ch15_1..html 的源码如下：

```
<html>
    <applet code= Ch15_1.class  width=800  height=600>
</applet>
</html>
```

- 在命令提示符下用 appletviewer Ch15_1.html 并回车（如图 15.4），也会得到如图 15.3 所示的结果。

图 15.4　编译和浏览 Applet

也可以通过浏览器浏览 Ch15_1.html，但由于有些浏览器设置禁用 Applet，需要对浏览器进行相应设置后方可浏览。

15.3　Applet 类的主要方法

- public void init()：由浏览器或 Appletviewer 调用，通知此 Applet 它已经被加载到系统中。它经常在第一次调用 start 方法前被调用。如果 Applet 的子类要执行初始化，则应该重写此方法。主要作用是为 Applet 的正常运行做一些初始化工作。通常在该方法中完成从网页向 Applet 传递参数、创建程序所需的其他对象等工作。该方法只被执行一次。
- public void start()由浏览器或 appletviewer 调用，通知此 applet 它应该开始执行。它在 init 方法调用后以及在 Web 页中每次重新访问 applet 时被调用，start()方法可以被多次执行。如果 Applet 子类在包含它的 Web 页被访问时有想要执行的操作，则它应该重写此方法。
- public void stop()由浏览器或 appletviewer 调用，通知此 applet 它应该终止执行。当包含此 Applet 的 Web 页已经被其他页替换时，在 applet 被销毁前调用此方法。主要作用是停止一些耗用系统资源的工作以免影响系统的运行速度，如 Applet 中包含播放动画、声音等的代码，则需要实现该方法完成当用户离开包含该 Applet 的页面时，停止播放功能。
- destroy()方法：浏览器关闭的时候才调用该方法，它在浏览器关闭的时候自动执行。
- paint(Graphics g)方法：该方法是 Applet 继承自 Container 类的方法。作用是在 Applet 容器上显示某些信息，如文字、色彩、背景或图像等。在 Applet 生命周期内可以被多次调用。比如改变浏览器窗口的大小，Applet 被其他页面遮挡，然后又重新放到最前面时，主类创建的对象都会自动调用 paint()方法。Graphics g 是绘图类的对象，相当于一只"画笔"，可以用其完成在 Applet 上绘制文本、图形、图像等。

【例题 15_2】Applet 生命周期测试。

```
import java.applet.Applet;
import java.awt.Graphics;
public class Ch15_2 extends Applet {
    public void init() {
        System.out.println("init()方法被调用");        //控制台输出
    }
    public void start() {
        System.out.println("start()方法被调用");       //控制台输出
    }
    public void stop() {
        System.out.println("stop()方法被调用");        //控制台输出
    }
    public void destroy() {
        System.out.println("destroy()方法被调用");     //控制台输出
    }
    public void paint(Graphics g) {
        System.out.println("paint()方法被调用");       //控制台输出
        g.drawString("Applet 生命周期测试", 40, 60);    //在 Applet 绘制输出
    }
}
```

程序运行结果如图 15.5（a）、（b）、（c）所示。

（a）程序首次被执行的输出

（b）Applet 输出

（c）关闭 Applet 后的输出

图 15.5　例题 15_2 运行结果

【例题 15_3】在 Applet 中绘制图形。
```
import java.applet.*;
import java.awt.*;
public class Ch15_3 extends Applet{
  public void paint(Graphics g) {
     g.setColor(Color.GREEN);
     g.drawRect(10,20,80,40);//绘制矩形框
     g.setColor(Color.RED);
     g.fillOval(60,80,50,50);//绘制实心圆
     g.setColor(Color.BLUE);
     g.drawOval(100,120,100,60);//绘制空心椭圆
  }
}
```
程序运行结果如图 15.6 所示。

图 15.6　例题 15_3 运行结果

若需要在 Applet 上显示图像，则需用到 Applet 类的以下方法：
- public URL getCodeBase()：获得包含此 applet 的目录的 URL 对象。
- public Image getImage(URL url,String name)以 url 指定的路径和 name 指定的图像名返回能被绘制到屏幕上的 Image 对象。

【例题 15_4】在 Applet 中显示（绘制）图像。
```
import java.applet.Applet;
import java.awt.*;
public class  Ch15_4 extends Applet{
  Image image;
  public void init(){
      image=this.getImage(getCodeBase(),"a.jpg");
      //image=this.getImage(getCodeBase(),"b.gif"); //装载图像
      System.out.println(this.getCodeBase());
  }
  public void paint(Graphics g){
      g.drawImage(image,0,0,this);
//在 Applet 的(0,0)坐标开始显示图像，由 Applet 对象本身 this 监视图像变化
  }
}
```
程序运行结果如图 15.7 和图 15.8 所示。

图 15.7　显示 a.jpg

图 15.8　显示 b.gif

说明：在 Ch15_4 同一目录下存放一个 a.jpg 图像文件以及一个 b.gif 文件，分别测试 image=this.getImage(getCodeBase(),"a.jpg"); 以及 image=this.getImage(getCodeBase(),"b.gif"); 的执行结果。

15.4　<applet>标记的属性及 Applet 参数传递

在 HTML 文件中嵌入了 Applet 要使用标记：<applet>...</applet>，其中 3 个参数 code、height 和 width 是必有的，其他都是可选的。在<applet>...</applet>中，可以包含<param>标记，通过该标记可以向网页动态地传递参数。

```
<applet
    [archive= archiveList]
    code= Applet 源文件名.class
    width= pixels
    height= pixels
    [codebase= codebaseURL ]
    [alt= alternateText ]
    [name= appletInstanceName ]
    [align= alignment ]
    [vspace = pixels ] [hspace= pixels ]
>
    [<param name= appletAttribute1 value= value >]
    [<param name= appletAttribute2 value= value >]
    ...
</applet>
```

<applet> 标记的重要属性含义如下：

- code = Applet 源文件名.class：给定了含有已编译好的 Applet 字节码文件。这个文件与要装入的 HTML 文件的 URL 有关，它不能含有路径名。一般情况下，Applet 子类的类文件与 HTML 文件放在同一个目录中，所以无须路径。如果类文件和 HTML 文件不在一个目录下，需要用到<codebase>选项，也就是说，要改变 Applet 的 URL，可使用<codebase>选项。
- width = pixels height = pixels：给出了 Applet 显示区域的初始宽度和高度（以像素为单位）。
- codebase = codebaseURL：这一可选的选项指定了 Applet 的 URL——包含有 Applet 代码的目录。如果这一选项未指定，则认为 Applet 的类文件与 HTML 文件在同一个目录中。
- <param name = applet Attribute value = value>：可以将 value 的值作为 HTML 参数传递给 Applet 去处理，Applet 用 getParameter()方法来存取参数。

【例题 15_5】给 Applet 传递参数测试。

（1）编写以下 Applet 程序，以 ParamTrans.java 命名并保存在 E:\\目录下。
```
import java.awt.Color;
import java.awt.Font;
import java.awt.Graphics;
import java.applet.Applet;
public class ParamTrans extends Applet{
  String s=null;
  public void init(){
     s = getParameter("str");
   }
  public void paint(Graphics g){
     g.setColor(Color.RED);
     g.setFont(new Font("宋体",Font.BOLD,40));
     g.drawString(s,10,50);
   }
}
```
（2）编写网页文件，以 ParamTrans.html 命名并保存在 E:\\目录下。
```
<html>
    <applet code=ParamTrans.class width=700 height=200>
       <param name="str" value="我是网页传来的字符串！">
    </applet>
</html>
```
（3）按如图 15.9 的步骤，先用 javac 命令编译 ParamTrans.java，在用 appletviewer 命令浏览 ParamTrans.html（如图 15.10）。

图 15.9 编译、浏览 Applet

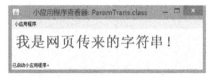
图 15.10 Applet 显示结果

15.5 Applet 的组件与事件处理

在 Applet 中，也可以添加组件并对事件源做相应的事件处理。下面通过几个实例来了解 Applet 的组件与事件处理方法。

【例题 15_6】在 Applet 中添加 3 个文本框，分别用于输入运算数 1、运算数 2 及求和结果，添加一个"求和"按钮。完成当单击"求和"按钮时，将运算数 1、运算数 2 的求和结果显示在结果文本框中。
```
import java.awt.FlowLayout;
import java.awt.event.ActionEvent;
import java.awt.event.ActionListener;
import javax.swing.*;
public class Ch15_6  extends JApplet implements ActionListener{//继承 JApplet
   JLabel l1,l2,l3;
   JTextField t1,t2,result;
```

```java
        JButton add;
        int a,b,sum=0;
        public void init() {
            this.setLayout(new FlowLayout());      //将组件布局方式设置为FlowLayout
            /*创建组件对象*/
            l1=new JLabel("运算数1: ");
            l2=new JLabel("运算数2: ");
            l3=new JLabel("结果: ");
            t1=new JTextField(10);
            t2=new JTextField(10);
            result=new JTextField(10);
            add= new JButton("求和");
            /*将组件对象添加到JApplet*/
            this.getContentPane().add(l1);
            this.getContentPane().add(t1);
            this.getContentPane().add(l2);
            this.getContentPane().add(t2);
            this.getContentPane().add(l3);
            this.getContentPane().add(result);
            this.getContentPane().add(add);
            add.addActionListener(this);           //注册事件监听器
        }
        public void actionPerformed(ActionEvent e) {   //事件处理代码
            a=Integer.parseInt(t1.getText().trim());
            b=Integer.parseInt(t2.getText().trim());
            sum=a+b;
            result.setText(sum+"");
        }
    }
```

程序运行结果如图 15.11 所示。

说明：由于本例是通过继承 JApplet 类来实现的，可以在其中添加 javax.swing 组件。由于 JApplet 的默认布局方式为 BorderLayout，本例是将布局管理方式改为 FlowLayout 布局方式。

图 15.11　例题 15_6 运行结果

【例题 15_7】 在 Applet 中从拖动鼠标的起点到终点绘制一条线段。

```java
import java.applet.Applet;
import java.awt.Color;
import java.awt.Graphics;
import java.awt.event.MouseEvent;
import java.awt.event.MouseListener;
import java.awt.event.MouseMotionListener;
public class Ch15_7 extends Applet implements MouseListener,MouseMotionListener{
//继承Applet类,实现MouseListener,MouseMotionListener接口
    int x1,y1,x2,y2;
    public void init() {
        this.addMouseListener(this);               //注册鼠标事件监听器
        this.addMouseMotionListener(this);         //注册鼠标移动事件监听器
    }
    public void mouseClicked(MouseEvent e) {}
```

```java
    public void mousePressed(MouseEvent e) {   //鼠标按下时
       x1=e.getX();
       y1=e.getY();
       //获得起点坐标x1,y1
    }
    public void mouseReleased(MouseEvent e) {}
    public void mouseEntered(MouseEvent e) {}
    public void mouseExited(MouseEvent e) {}
    public void mouseDragged(MouseEvent e) {   //鼠标拖动时
       x2=e.getX();
       y2=e.getY();
       //获得终点坐标x2,y2
       this.repaint();
    }
    public void mouseMoved(MouseEvent e) {}
    public void paint(Graphics g) {
       g.setColor(Color.GREEN);
       g.drawLine(x1,y1,x2, y2);          //绘制线段
    }
}
```
程序运行结果如图 15.12 所示。

图 15.12 例题 15_7 运行结果

【例题 15_8】在 Applet 上以 "HH:mm:ss"（时、分、秒）格式实时显示当前时间。

```java
import java.awt.BorderLayout;
import java.awt.Color;
import java.awt.Font;
import java.text.SimpleDateFormat;
import java.util.Date;
import javax.swing.JApplet;
import javax.swing.JLabel;
public class Ch15_8 extends JApplet implements Runnable{
    JLabel label;
    public void init() {
       label=new JLabel("",JLabel.CENTER);
       label.setForeground(Color.RED);
       label.setFont(new Font("宋体",Font.BOLD,60));
       this.getContentPane().add(label,BorderLayout.CENTER);
       this.setSize(300, 120);
       new Thread(this).start();//创建线程对象并使之就绪
    }
    public void run() {//实现Runnable接口的run()方法
       while (true){
         try {
            SimpleDateFormat sdf=new SimpleDateFormat("HH:mm:ss");
            //日期、时间格式化
            label.setText(sdf.format(new Date()));
            //以指定格式显示系统当前日期、时间
            Thread.sleep(1000); //休眠1000 ms
         }
         catch(InterruptedException e){
```

```
            }
          }
        }
      }
```

程序运行结果如图 15.13 所示。

图 15.13　例题 15_8 运行结果

说明：由于需要实时获取系统当前时间，可以考虑用多线程编程方式，让程序每隔 1 s 就实时获取系统时间一次。

小　　结

本章简单介绍了 Java Applet 程序的编写方法；Applet 类的主要方法及应用；Applet 参数传递方法及其用法；Applet 的组件与事件处理应用实例等。

习　题　15

上机实践题

1. 编程在 Applet 上绘制字符串"拖动鼠标改变文字显示位置"，并当在 Applet 上拖动鼠标时，实现字符串位置的改变。

2. 编程在 Applet 上添加 Choice（java.awt 中的下拉列表组件类）对象，在该下拉列表中添加待选项"绘制线段""绘制圆形""绘制矩形"，实现当在下拉列表选择不同的选项时，在 Applet 中通过拖动鼠标绘制相应的图形。

参 考 文 献

[1] 张跃平，耿祥义. Java 2 实用教程[M]. 4 版. 北京:清华大学出版社，2012.
[2] 郑阿奇，姜乃松，殷红先，等 Java 实用教程[M]. 2 版. 北京:电子工业出版社，2009.
[3] 郭克华. Java 语言程序设计[M].北京:清华大学出版社，2012.
[4] 李兴华. JAVA 开发实战经典[M].北京:清华大学出版社，2009.
[5] 郭广军,刘安丰,阳西述. Java 程序设计教程 [M]. 2 版. 武汉: 武汉大学出版社，2015.
[6] 张孝祥. Java 基础与案例开发详解[M].北京:清华大学出版社，2009.
[7] 高宏静. Java 从入门到精通[M].北京:化学工业出版社，2009.
[8] 叶核亚. Java 程序设计实用教程[M]. 4 版. 北京:电子工业出版社，2013.
[9] ECKEL B. Java 编程思想[M]. 4 版. 陈昊鹏，译. 北京:机械工业出版社，2007.
[10] HORSTMANN CS，CORNELL G. Java 核心技术卷 1 基础知识[M]. 9 版. 周立新，陈波，叶乃文，等，译. 北京:机械工业出版社，2013.
[11] 郭克华. Java EE 程序设计与应用开发[M].北京:清华大学出版社，2011.